INDUSTRIAL ENERGY MANAGEMENT: PRINCIPLES AND APPLICATIONS

THE KLUWER INTERNATIONAL SERIES IN ENGINEERING AND COMPUTER SCIENCE

POWER ELECTRONICS AND POWER SYSTEMS

Consulting Editor

Thomas A. Lipo
University of Wisconsin-Madison

Other books in the series:

SPOT PRICING OF ELECTRICITY, F. C. Schweppe, M. C. Caramanis, R. D. Tabors, R. E. Bohn
ISBN 0-89838-260-2

RELIABILITY ASSESSMENT OF LARGE ELECTRIC POWER SYSTEMS, R. Billington, R. N. Allan
ISBN 0-89838-266-1

MODERN POWER SYSTEMS CONTROL AND OPERATION, A. S. Debs
ISBN: 0-89838-265-3

ELECTROMAGNETIC MODELLING OF POWER ELECTRONIC CONVERTERS, J. A. Ferreira
ISBN: 0-7923-9034-2

ENERGY FUNCTION ANALYSIS FOR POWER SYSTEM STABILITY, M. A. Pai
ISBN: 0-7923-9035-0

Library of Congress Cataloging-in-Publication Data

Petrecca, Giovanni.
Industrial energy management : principles and applications / Giovanni Petrecca.
p. cm. -- (Kluwer international series in engineering and computer science ; 207. Power electronics and power systems.)
Includes bibliographical references and index.
ISBN 978-0-7923-9305-4 ISBN 978-1-4615-3160-9 (eBook)
DOI 10.1007/978-1-4615-3160-9
1. Industry--Energy conservation. I. Title. II. Series: Kluwer international series in engineering and computer science ; 207.
III. Series: Kluwer international series in engineering and computer science. Power electronics & power systems.
TJ163.3.P47 1992 92-36761
670.42.--dc20 CIP

Printed on acid-free paper.

INDUSTRIAL ENERGY MANAGEMENT: PRINCIPLES AND APPLICATIONS

by

Giovanni Petrecca
Faculty of Engineering
University of Pavia, Italy

SPRINGER SCIENCE+BUSINESS MEDIA, LLC

CONTENTS

LIST OF FIGURES

LIST OF TABLES

LIST OF EXAMPLES

LIST OF SYMBOLS

Letter symbols include symbols for physical quantities and symbols for units in which these quantities are measured.

For symbol units see Tables 2.4, 2.4 and 2.5.

Quantity symbols are listed below.

Efforts have been made to attribute a single letter symbol to each physical quantity. Because of the number of topics introduced, some symbols have different meanings according to the context and to the common practice; in these cases, the list of symbols indicates the chapter to which these symbols apply.

A, S	area, section
A_n	transformer rated power
AE	Available energy (isoentropic expansion)
ASF	Actual Steam Flow
ASR	Actual Steam Rate
c	specific heat
c_d	dispersion coefficient (Chapter 13)
C	capacitor (Chapter 7)
C	heat capacity flow-rate (Chapter 15)
C_p	ratio between shaft power and power available in the wind (Chapter 4)
C_p	present monetary value (Chapter 20)
COD	liquid Chemical Oxygen Demand
COP	coefficient of performance (Chapter 12)
CU	coefficient of utilization (Chapter 14)
d	length (thickness, diameter)
E	energy
E,P	heat exchanger effectiveness (Chapter 13)
E_p	electric active energy
E_q	electric reactive energy
f	electric frequency
f, f*	inflation rate (Chapter 20)
FT	correction factor (Chapter 15)
FWF	Future Worth Factor

g	gravitational acceleration
g_c	conversion factor
h	enthalpy per unit mass
h	heat transfer coefficient (Chapter 8)
H	head (Chapter 10)
H	enthalpy
i	coefficient for life cycle costing
I	investment (Chapter 20)
I	electric current
I_p	present investment (Chapter 20)
IRR	Internal Rate of Return
ln	natural logarithm
k	thermal conductivity
K_c	unburned combustible losses coefficient (Chapter 6)
K_m	Karmaan coefficient (Chapter 2)
K_s	Hassenstein coefficient (Chapter 6)
K_v	velocity coefficient (Chapter 2)
l, L	length
m	mass flow-rate
M	mass
n	number of moles (Chapter 2, Table 2.5)
n	life of investiment (Chapter 20)
n	number of phase conductors (Chapter 7)
n_v	rate of ventilation (Chapter 13)
NTU	Number of Transfer Units (Chapter 15)
p	pressure
ppm	gas concentration (1 ppm = 1 SmL/Sm^3 = 0.0001%)
P	Power (mechanical, electric active power)
P, E	heat exchanger effectiveness (Chapter 15)
PAF	Present annuity factor
P_{cc}	transformer load losses (Chapter 5)
P_{cn}	transformer load losses at rated power (Chapter 5)
P_o	transformer no-load losses (Chapter 5)
PL	electric line losses (Chapter 7)
PW	Present Worth

PWF	Present Worth Factor
q	volume flow-rate
q_s	volume flow-rate in standard conditions
Q	power as heat transfer rate
Q	electric reactive power (Chapter 7)
Q	annual revenues (Chapter 20)
r	radius
r	interest or discount rate (Chapter 20)
r	P_{cn}/P_o (Chapter 5)
r_p	compression ratio (Chapter 11)
R	electric resistance (Chapter 7)
R	radius of gyration (Chapter 4)
R	parameter for heat exchanger calculation (Chapter 15)
R	universal gas constant (Chapter 2, Table 2.5)
RH, Φ	relative humidity
ROR	Investors Rate Of Return
$R_{th} = 1/U$	overall thermal resistance (Chapter 8)
S, A	section, area
t, T	temperature (capital letter is generally used for absolute temperatures)
TOE	Ton Oil Equivalent
TSR	Theoretical Steam flow Rate
u	internal energy
$U=1/R_{th}$	overall heat transfer coefficient
v	specific volume
V	voltage (Chapter 7)
V	linear velocity
V	volume (Chapter 2, Table 2.5; in Chapter 6)
VB	volume of the building
x	load factor (Chapter 5)
x	steam quality index (Chapters 6, 9)
X	electric reactance
Z	compressibility factor

α	instrument flow coefficient (Chapter 2)
α	ratio between gross and net heating value for fuel (Chapter 6)
δ	electric current density
ε	thermal emissivity (Chapter 8)
ε	expansibility factor (Chapter 2)
η	efficiency
ρ	mass density
ρ	electrical resistivity of conducting material (Chapter 7)
σ	Stefan-Boltzmann constant
Φ, RH	relative humidity
Ω	rotational speed
ω	humidity ratio or specific humidity
$\cos \varphi$	electric power factor

Institutions and Associations

ABMA	American Boiler Manufacturers Association
ASHRAE	American Society of Heating, Refrigerating and Air Conditioning Engineers
ASME	American Society of Mechanical Engineers
IEC	International Electrotechnical Commission
IEEE	The Institute of Electrical and Electronics Engineers
ISO	International Organization of Standardization
TEMA	Tubular Exchanger Manufacturers Association

PREFACE

After many years of researching, teaching and consulting I am firmly convinced that all problems, in order to be solved, must first be reduced to their essentials.

That does not mean ignoring or skating over some aspects, but on the contrary it means going deeply into the core of the problem to understand it completely, and then focusing on the few elements really necessary for its solution.

This is quite a difficult approach because it requires a clear comprehension of what is really important and of what is not. Engineering universities generally train students to achieve that, but during the professional career a great effort must always be made to apply these principles to practical problems.

I became engaged in the research for a unifying approach to energy management after experiences in teaching electrical machines and drives with the unified theory, which in 1976 was rather innovative for the student curriculum, but was afterwards demonstrated to be an excellent approach to an intimate and all-embracing comprehension of the related phenomena.

Professional experiences in the field of industrial energy management have stimulated me to organize a course on this subject in the Engineering Faculty of the University of Pavia in Italy, and to write a text providing a global view of energy management in industry and in the tertiary sector.

The stream of energy, in different forms, which is transformed step by step inside any production unit, has been assumed as the unifying factor for understanding all the topics, which at first glance seem to be comprehensible only in terms of independent theories.

In conclusion, this book is not a new specialized textbook or handbook and so it does not compete with those already existing which remain indispensable tools for solving technical problems in detail. Basically, it aims to help the reader by giving him the fundamental elements to simplify and tackle problems within a global vision of the production units by exploiting the correlations always present among many topics apparently distant from one another.

Having completed this work, I should like to thank university and professional colleagues who reviewed parts of the book and made suggestions to facilitate the comprehension of many topics. In addition, thanks are due to CSE Bilanci Energetici Srl, Pavia, Italy, from whose data bank most of the examples have been drawn.

On a more personal note, I should like to thank Peter Farries, who assisted me for two years in the task of writing the book in English.

Giovanni Petrecca

Pavia, August 1992

INDUSTRIAL ENERGY MANAGEMENT:
PRINCIPLES AND APPLICATIONS

CHAPTER 1

INTRODUCTION

1.1 GENERAL PRINCIPLES OF ENERGY MANAGEMENT

In spite of a great diversity among energy end-user technologies, due to technical, economic and environmental factors, energy management at all levels must be based on a few general principles of proven validity.

Basically, energy management means ensuring that users get all the energy necessary, when and where it is needed, and of the quality requested, supplied at the lowest cost. Of course, this aim must be achieved while duly safeguarding both production and environmental needs.

To be effective, energy management programs should include four main sections: (i) analyses of historical data; (ii) energy audits and accounting; (iii) engineering analyses and investment proposals based on feasibility studies; (iv) personnel training and information. An energy management program can be organized in several ways, by employing either internal or external consultants, according to the company's size and to the incidence of energy costs on the company's budget.

To obtain the best results, the main sections listed above must be carefully implemented and correlated with one another.

Energy management began to be considered one of the main functions of industrial management only in the 1970s. Faced with the rising price of energy and reports about the approaching exhaustion of world energy resources, both national governments and industrial companies had to cope with this situation with no further delay.

Inadequate knowledge, however, of energy-management techniques and the lack of strong traditions in this kind of capital investment militated against a widespread introduction of energy-saving strategies.

Large plants with high energy consumption tackled the problem by retrofitting process plants and facilities. Other industrial sectors, less sensitive to energy, resorted to investments with the shortest possible payback such as heat recovery and reduction of losses, but they put off process modifications, which often involve a more drastic change of production strategies.

That decade of energy shortage, the first that has seriously affected the industrial era, and the following oscillations in energy prices have made people aware that the energy problem does and will always exist. Therefore, energy-saving technologies have markedly affected both component and plant designs and users have become accustomed to making decisions on energy with regard to more than mere payback considerations.

Now energy managers are sought for large plants and buildings and play an important role in developing industrial strategies, so a knowledge of technical and economic disciplines is generally requested. Mechanical and thermal sciences have traditionally been the base in this educational field; but the spread of computer control and power electronic systems suggests that electrical and thermal approaches need to be interrelated and that technicians must be prepared to work within this broad context.

Students should have the opportunity at the university to fit themselves for these jobs; people already working in industry should try to correlate practical knowledge with basic theoretical principles in order to derive the maximum benefit from previous experience. Managers, who have not time to study technical problems in detail, need a guide to the essentials in order to make well-founded decisions.

The aim of this book is to give an overall view of the energy management approach by following the stream of energy from factory boundaries to end users. All the topics will be examined from the point of view of plant users rather than from that of designers and only the basic concepts necessary to clarify the operation of the plants are outlined.

1.2 ENERGY TRANSFORMATIONS IN AN INDUSTRIAL SITE

The overall approach to energy management is summarized in Fig. 1.1 which shows the energy streams from factory boundaries to end users and the related transformations:

— energy is transported to the factory site as purchased fuel or utilities (electric network, gas pipeline, water pipe, etc.) and it is transformed mostly inside the factory (electrical substation with transformers, boiler plants) before reaching the energy users.

Further transformations in facilities plants should be made in order to obtain different forms of derived energy suitable for the end users.

Obviously, the efficiency of all the transformation plants must be checked and kept as high as possible;

— energy streams, in different forms (electric energy at secondary voltage, steam, hot water, chilled water, hot oil, compressed air, etc.), are distributed around the factory to process and facility end users. Distribution systems are responsible for losses, which should be reduced by means of correct planning and thermal insulation;

— the energy end users around the factory perform different operations which lead to end products or services. These make up the output of the industrial site, together with waste (which may or may not contain energy in some form) and wasted energy. Waste and wasted energy include water, solid and liquid materials (combustible or not), and gases.

Energy can be saved at the end-user stage in many ways: by introducing more efficient equipment or process systems, by improving controls, by improving monitoring or metering of energy streams, by recovering heat and waste. Waste recovery is closely related to waste quality and to the pollution produced.

Energy flow which crosses the site boundaries can be schematized as in Fig. 1.1.

INPUT ENERGY =

INPUT ENERGY =	+	**individual end-user consumption**
	–	**waste and energy recovery such as heat recovery, etc.**
	+	**difference between the energy contained in raw material entering the site and that in the products and services leaving the site**
	+	**energy losses (waste, heat, distribution line losses, transformation losses, etc.)**

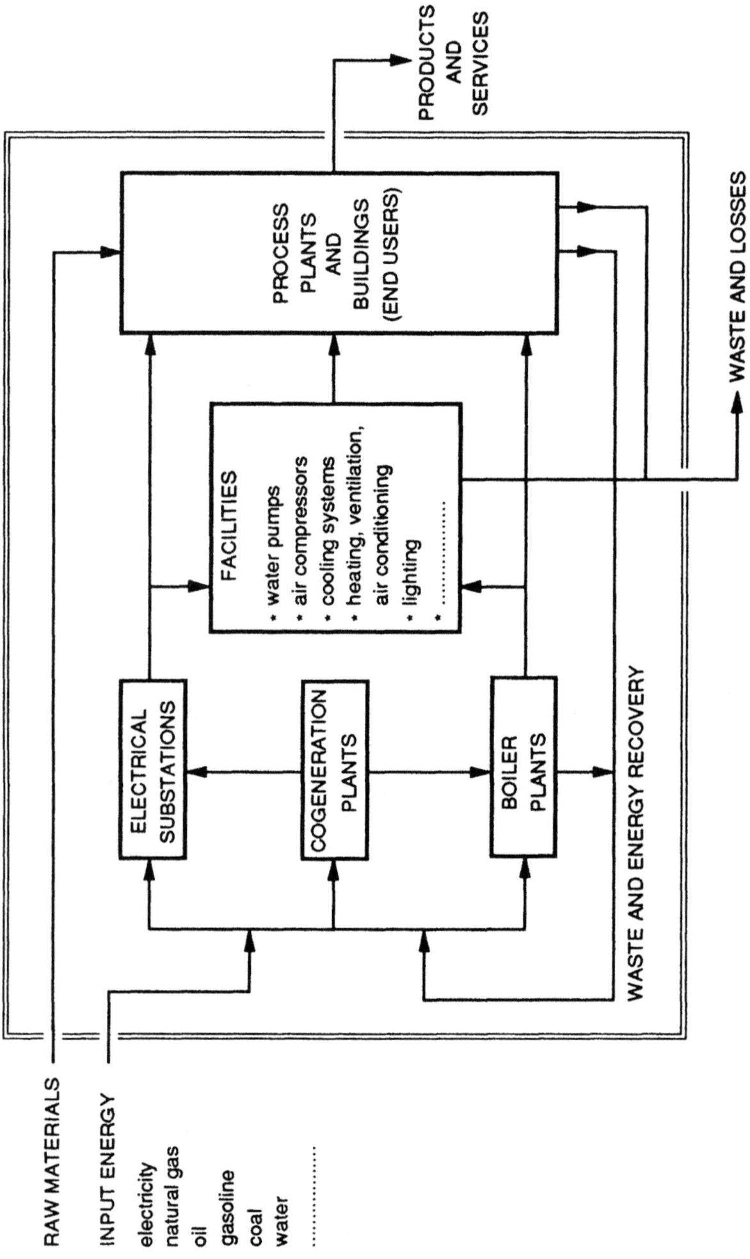

Fig. 1.1 Energy flow through the site boundary

This energy flow-chart allows an overall approach to site energy management and provides the guidelines for an energy survey as well as an educational program.

1.3 THE PLAN OF THE BOOK

The book has been conceived both as a textbook for university courses in engineering and as a work of reference for professionals in energy management. Readers are assumed to have a basic knowledge of thermodynamics, heat and mass transfer, electric systems and power electronics, as well as computer programming.

The book comprises 21 chapters which can be grouped as follows:

— General principles of energy transformation and energy management, and energy sources (Chapters 1, 2, 3, 4).
  This part aims to give readers a general overview of energy management, from the energy sources to the end users in the factory.
— Transformation plants at factory borders (Chapters 5, 6). Electrical substations and boiler plants are examined and suggestions are made on how to improve efficiency.
— Internal electric networks and thermal fluid distribution systems from facilities to end users (Chapters 7, 8).
— Cogeneration plants (Chapter 9).
— Facilities plants such as pumps and fans, air compressors, cooling systems, HVAC systems, lighting systems (Chapters 10, 11, 12, 13, 14).
— Heat recovery from process and facilities. Heat exchangers (Chapter 15).
— Industrial waste management and correlation with energy management (Chapter 16).
— Computer systems to control facilities and process by improving energy efficiency. Energy accounting for control and planning (Chapters 17,18).
— The role of education in energy management (Chapter 19).
— Economic analysis of energy-saving investments (Chapter 20).
— Conclusion (Chapter 21).

Throughout the book boxes highlight the fundamental concepts and some important figures for practical applications.

Examples have been given only for basic cases and mainly for facilities, but they can easily be extended to more complicated situations, including process systems which at first glance seem quite different from one another, if these are reduced to their essentials. Technical evaluations are shown at the end of each chapter; data are generally organized in tables to facilitate elaboration by a standard spreadsheet. Economic evaluations of each example are shown in Table 20.3.

The SI system has been used; for ease of understanding, other units commonly used have been added.

Efforts have been made to attribute a single letter symbol to each physical quantity. Because of the number of topics introduced, some symbols have different meanings according to the context and to the common practice; in these cases, the list of symbols indicates the chapters to which these symbols apply.

References are at the end of the book in chronological order.

CHAPTER 2

POWER AND ENERGY MEASUREMENT UNITS AND TECHNIQUES

2.1 THE SI SYSTEM AND CONVERSION FACTORS

The SI system (International System of Units) is a set of definitions and rules which provides an organic and consistent approach for obtaining the units for each quantity.

The SI system is based on seven basic and two supplementary units. The basic units are the following: the kilogram (mass), the meter (length), the second (time), the ampere (electric current), the kelvin (temperature), the candela (luminous intensity), the mole (molecular substance). The supplementary units are the radian (plane angle) and the steradian (solid angle). The choice, which has been made by the General Conference on Weights and Measures, is always susceptible of changes and integrations.

The SI basic and supplementary units are reported in Table 2.1. Starting from these nine units any other SI units can be derived by using the following expression:

$$\text{derived unit} = \text{m}^{\alpha_1} \cdot \text{kg}^{\alpha_2} \cdot \text{s}^{\alpha_3} \cdot \text{A}^{\alpha_4} \cdot \text{K}^{\alpha_5} \cdot \text{cd}^{\alpha_6} \cdot \text{mol}^{\alpha_7} \cdot \text{rad}^{\alpha_8} \cdot \text{sterad}^{\alpha_9}$$

where the exponents $\alpha_1, \alpha_2, \alpha_3, \alpha_4, \alpha_5, \alpha_6, \alpha_7, \alpha_8, \alpha_9$ are positive or negative whole numbers. Each exponent is equal to zero when the derived unit does not depend on the corresponding basic or supplementary unit.

Table 2.2 shows prefixes which must be used to derive multiples or submultiples of any SI unit. Table 2.3 shows a few derived units currently used.

It is worth noticing that E.E.C. (European Economic Community) countries must use the SI units reported in Table 2.1 (they are obliged to by E.E.C. recommendations and related national laws). This system has been adopted also by the International Organization of Standardization (ISO).

Table 2.4 lists units commonly used which, however, do not belong to the SI system. Technicians should refrain from using these units. Conversion factors from these units to SI units are shown; conversion between two non-SI units can be made by using the ratio between the conversion factors of the single units. This is valid also for temperature, if temperature changes are considered (as often occurs in technical relationships).

Table 2.5 lists some parameters frequently used for technical calculations.

2.2 PRIMARY ENERGY MEASUREMENT UNITS

The primary energy content of combustibles (solid, liquid or gaseous) is expressed by net calorific values such as kJ/kg or kJ/m^3 (SI units). There are other units in common use.

The quantity of combustibles is generally referred to TOE (Ton Oil Equivalent, Net Heating Value 41860 kJ/kg) by using net calorific values.

Primary energy (hydro, geothermal, nuclear or other renewable energy sources) from which electric energy is produced, is converted into TOE on the basis of the specific consumption (kJ/kWh) of conventional fuel-fed power plants. Typical values range between 9628 and 10465 kJ/kWh (corresponding to 9126-9919 Btu/kWh; 2300-2500 kcal/kWh) if both power plant and distribution line losses are taken into account.

Tables 2.6, 2.7 report values from international statistics and conversion factors:
— Table 2.6 Net and gross calorific values of solid, liquid and gaseous fuels;
— Table 2.7 Conventional densities of liquid fuels.

Some definitions, which will be discussed in detail in later chapters, are summarized below in order to allow a clearer understanding of Chapter 2:
— calorific value (heating value). This is a measure of the heat any given fuel can release during the combustion process.
  Combustion of fuels consisting of carbon, hydrogen and sulfur requires oxygen, which is normally supplied from atmospheric air. It starts at different ignition temperatures depending on the fuel. Typical values range between

573.15 and 973.15 K (300-700°C, 572-1292°F).
Once ignition temperature has been reached, combustion continues until all the fuel or oxygen has been consumed.
Hydrogen, combined with oxygen, produces water (roughly 9 kg of water for 1 kg of hydrogen) which is discharged as liquid water as well as water vapor into the atmosphere together with combustion gaseous waste at the same temperature;

— gross calorific value (gross heating value). This is the number of heat units measured as being liberated when unit mass of fuel is burned in oxygen saturated with water vapor in a bomb in standard conditions, the residual materials being gaseous oxygen, carbon dioxide, sulfur dioxide and nitrogen, ash and liquid water (the water produced during the combustion is assumed to be discharged as liquid water). The standard conditions are defined by ISO. The international reference temperature for combustion is 25°C (77°F), but in some countries different temperatures are used;
— net calorific value (net heating value). This is the number of heat units measured as before, the residual materials being gaseous oxygen, carbon dioxide, sulfur dioxide and nitrogen, ash and water vapor (the water produced during the combustion is assumed to be discharged as water vapor). The latent heat is then completely wasted.

2.3 END-USER ENERGY MEASUREMENT UNITS

Energy which enters a factory goes through many transformations before reaching end users (this energy is commonly called final energy). Units referring to transformed energy rather than to purchased energy (fuels and electric energy) are conveniently used at this stage. Table 2.8 lists typical ranges of values for the commonest ratios between end-user and purchased energy which depend on the efficiency of the transformations as well as on the level of energy exploitation:

— kg of steam/kg of fuel;
— kg of hot water/kg of fuel;
— cubic meter of compressed air/kWh consumed by mechanical compressor;
— kJ available for the process/kWh consumed by mechanical compressor;
— others.

Table 2.1 SI base and supplementary units

QUANTITY	UNIT	SYMBOL
Length	meter	m
Mass	kilogram	kg
Time	second	s
Electric current	ampere	A
Thermodynamic temperature	kelvin	K
Luminous intensity	candela	cd
Molecular substance	mole	mol
Plane angle	radian	rad
Solid angle	steradian	sr

Table 2.2 Prefixes commonly used

FACTOR	PREFIX NAME	SYMBOL
Multiple		
10^{18}	exa	E
10^{15}	peta	P
10^{12}	tera	T
10^{9}	giga	G
10^{6}	mega	M
10^{3}	kilo	k
10^{2}	hecto	h
10^{1}	deka	da
Submultiple		
10^{-1}	deci	d
10^{-2}	centi	c
10^{-3}	milli	m
10^{-6}	micro	μ
10^{-9}	nano	n
10^{-12}	pico	p
10^{-15}	femto	f
10^{-18}	atto	a

Table 2.3 Units derived from SI

QUANTITY	UNIT	SYMBOL
Space and Time		
Area	square meter	m^2
Volume	cubic meter	m^3
Velocity	meter per second	m/s
Acceleration	meter per second squared	m/s^2
Angular velocity	radian per second	rad/s
Angular acceleration	radian per second squared	rad/s^2
Frequency	hertz	Hz=cycle/s
Rotational speed	radiant per second	rad/s
Mechanics		
Density	kilogram per cubic meter	kg/m^3
Momentum	kilogram meter per second	$kg \cdot m/s$
Moment of inertia	kilogram meter squared	$kg \cdot m^2$
Force	newton	$N=kg \cdot m/s^2$
Torque, moment of force	newton meter	$N \cdot m$
Energy, work, heat quantity	joule	$J=N \cdot m$
Power	watt	W=J/s
Pressure, stress	pascal	$Pa=N/m^2$
Electricity and magnetism		
Electric charge	coulomb	$C=A \cdot s$
Electric potential, voltage	volt	V=W/A
Electric field strength	volt per meter	V/m
Capacitance	farad	$F=C/V=A \cdot s/V$
Current density	ampere per square meter	A/m^2
Magnetic field strength	ampere per meter	A/m
Magnetic flux	weber	$Wb=V \cdot s$
Magnetic flux density	tesla	$T=Wb/m^2$
Inductance	henry	$H=V \cdot s/A$
Permeability	henry per meter	H/m
Resistance	ohm	$\Omega=V/A$
Conductance	siemens	S=A/V
Magnetomotive force	ampere	A
Light		
Luminous flux	lumen	$lm=cd \cdot sr$
Luminance	lux	$lx=lm/m^2$
Viscosity		
Kinematic viscosity	square meter per second	m^2/s
Dynamic viscosity	pascal second	$Pa \cdot s$

Table 2.4 SI Units and conversion factors

To convert from	Symbol	to	Symbol	Multiply by
LENGTH				
foot	ft	meter	m	0.3048
inch	in	meter	m	0.0254
mile	mi	meter	m	1609
AREA				
square foot	ft^2	square meter	m^2	0.0929
square inch	in^2	square meter	m^2	0.000645
VOLUME				
cubic foot	ft^3	cubic meter	m^3	0.02832
cubic inch	in^3	cubic meter	m^3	0.00001639
USA liq gallon	gal	cubic meter	m^3	0.0037854
liter	L	cubic meter	m^3	0.001
MASS				
pound	lb	kilogram	kg	0.45359
ton(short)	ton	metric ton, tonne	$t=10^3 \cdot kg$	0.9072
ton(long)	ton	metric ton, tonne	$t=10^3 \cdot kg$	1.016
barrel(oil)	barrel	metric ton, tonne	$t=10^3 \cdot kg$	0.137
FORCE				
pound-force	lbf	newton	N	4.448
kilogram-force	kgf	newton	N	9.807
PRESSURE				
pound-force/square foot	lbf/ft^2	pascal	Pa	47.8788
pound-force/square inch	lbf/in^2	pascal	Pa	6895
kilogram-force/square meter	kgf/m^2	pascal	Pa	9.807
bar	bar	pascal	Pa	100000
atmosphere	atm	pascal	Pa	101325
mm H_2O	mm H_2O	pascal	Pa	9.7739
inch H_2O	in H_2O	pascal	Pa	248.7
SPEED, VELOCITY				
foot/second	ft/s	meter per second	m/s	0.3048
foot/min	ft/min	meter per second	m/s	0.00508
mile/hour	mi/h	meter per second	m/s	0.4469
kilometer/hour	km/h	meter per second	m/s	0.2777
ACCELERATION				
foot/second^2	ft/s^2	meter per second squared	m/s^2	0.3048

Table 2.4 SI Units and conversion factors (continued)

To convert from	Symbol	to	Symbol	Multiply by
ENERGY, WORK				
British thermal unit	Btu	Joule	J	1055
foot pound-force	ft · lbf	Joule	J	1.356
calorie	cal	Joule	J	4.1868
watthour	Wh	Joule	J	3600
POWER				
Btu/hour	Btu/h	Watt	W	0.2931
Btu/second	Btu/s	Watt	W	1055
horsepower	hp	Watt	W	745.7
calorie/hour	cal/h	Watt	W	1.1628
REFRIGERANT CAPACITY				
tons	tons	Watt	W	3520
frigorie/hour	frig/h	Watt	W	1.1628
TORQUE				
pound-force foot	lbf · ft	newton meter	N · m	1.356
kilogram-force meter	kgf · m	newton meter	N · m	9.807
DENSITY				
pound/cubic foot	lb/ft^3	kilogram per cubic meter	kg/m^3	16.018
VOLUME FLOW-RATE				
cubic foot/minute	ft^3/min	cubic meter per second	m^3/s	0.00047
SPECIFIC ENERGY				
Btu/pound	Btu/lb	Joule/kilogram	J/kg	2326
calorie/kilogram	cal/kg	Joule/kilogram	J/kg	4.186
SPECIFIC HEAT				
Btu/pound · °F	Btu/lb · °F	Joule/kilogram · K	J/kg · K	4186
calorie/kilogram · °C	cal/kg · °C	Joule/kilogram · K	J/kg · K	4.186
LIGHT				
footcandle	fc	lux	lx	10.764
TEMPERATURE				
Celsius °C change	°C	Kelvin change	K	1
Fahrenheit °F change	°F	Kelvin change	K	5/9

Note that conversion between two non SI units can be made by using the ratio between the conversion factors of the single unit.

Examples

To convert from	to	Multiply by
Celsius change	Fahrenheit change	1/(5/9)
Btu/h	cal/h	0.2931/1.1628

Table 2.5 Parameters frequently used

DESCRIPTION	Other systems		SI system
Specific heat	kcal/kg · °C	Btu/lb · °F	kJ/kg · K
water	1	1	4.18
superheated steam (1)	0.5	0.5	2.09
air	0.24	0.24	1
iron	0.114	0.114	0.477
copper	0.092	0.092	0.385
mineral oil	0.486	0.486	2.034
Density		lb/ft^3 (2)	kg/m^3 (2)
water		62.5	1000
air (standard conditions)		0.08	1.29
mineral oil		57.75	925
iron		490	7850
copper		557.5	8930
natural gas		0.047	0.750

Notes

(1) average value in industrial boiler. In air-water mixture (see Chapter 13) the specific heat of superheated steam equals 1.8 kJ/kg · K (steam pressure < 0.1 MPa)

(2) for gas, density is referred to standard conditions: 273.15 K (0°C; 32°F) and 0.1 MPa (1.013 bar; 14.69 psi).

For an ideal gas, but widely accepted for most real gases, basic relationships (where V is the standard volume) are:

$$V_1 = V\,T_1/273.15 \qquad p_1V_1 = nRT_1 = \text{constant} \cdot T_1$$

The density of air at T_1(K) and at standard pressure is:

$$\text{air density}(T_1) = \frac{1.293}{(T_1/273.15)}$$

Notice that in some countries and in some applications the standard conditions can be referred to 288.15 K (15.6°C; 60°F) and 0.1 MPa (1.013 bar; 14.69 psi). The user should ascertain the reference conditions for each application.

Table 2.6 Net and gross calorific values of solid, liquid and gaseous fuels

FUELS	AVERAGE VALUES			
	NET	GROSS	NET	NET
Solid fuels	*kJ/kg*		*kcal/kg*	*Btu/lb*
Vegetal fuels	10465	16700	2500	4499
Pitch lignite	18000	36000	4300	7738
Standard coal	29302	33500	7000	12597
Charcoal	31395	34750	7500	13497
Cokery coke	29302	33000	7000	12597
Gas coke	26790	32650	6400	11517
Petroleum coke	34744	37250	8300	14936
Liquid fuels	*kJ/kg*		*kcal/kg*	*Btu/lb*
Crude oil	41860	44400	10000	17995
Petroleum condensates	44372	47000	10600	19075
Light petroleum distillates	43534	46150	10400	18715
Gasoline	43953	46600	10500	18895
Jet fuel	43534	46150	10400	18715
Refined kerosine	43116	45700	10300	18535
Gasoil	42697	45250	10200	18355
Fuel oil	41023	43500	9800	17636
Liquified petroleum gas	46046	49700	11000	19795
Gaseous fuels	*kJ/m³* (*)		*kcal/m³* (*)	*Btu/ft³* (*)
Natural gas	34325	38450	8200	921
Cokery gas	17791	19900	4250	478
Blast furnace gas	3767	4200	900	101
(*) see note Table 2.5				

Table 2.7 Conventional densities of fuels

FUELS	kg/m³ (*)	lb/ft³ (*)
Gasoline	734	45.8
Gasoil	825	51.5
Oil	925	57.7
LPG	565	35.3
Natural gas	0.75	0.047
Standard coal	800	43.9
Vegetal fuels	400	24.3
(*) see note Table 2.5		

Table 2.8 Ratios between end-users energy and purchased energy

TYPES OF TRANSFORMATION		TRANSFORMATION COEFFICIENT	UNIT	NOTE
To convert from	to			
oil	steam	12-14	kgsteam/kgoil	boiler
natural gas	steam	9-11	kgsteam/Sm^3	boiler
electricity	cold fluid (HVAC)	12000-16000	kJ/kWh	compressor
electricity	cold fluid (below ice point)	3000-10000	kJ/kWh	compressor
electricity	compressed air (0.8 MPa)	7-9	Sm^3/kWh	compressor
electricity	heat	3600	kJ/kWh	resistor
electricity	heat	10000-15000	kJ/kWh	heat pump
oil	electricity	4	kWh/kg	utility plant
oil	electricity	8	kWh/kg	cogeneration plant
electricity	water storage	10	t · m/kWh	pump
electricity	lighting	50-100	lm/W	lamps

2.4 AN OUTLINE OF THE MAIN MEASUREMENT TECHNIQUES

Energy measurement is important because on the one hand it is a means of improving energy management and control of plant operations and on the other it determines the incidence of energy cost on the total cost of production.

Energy in industrial processes can be found in many different forms:

— solid, liquid and gaseous fuels;

— electric energy;

— fluids by which thermal energy is distributed throughout a site, such as steam, hot water, hot oil, heated air, etc.;

— fluids by which motive power is distributed throughout a site, such as compressed air, compressed oil, etc..

All these energies can be measured by means of two categories of system:

— meters that measure energy flow over time by performing the integral function inside the instrument;

— transducers of instantaneous values such as flow and power. Energy is then calculated by means of independent systems able to perform an integral function. Notice that it is also possible to calculate energy if physical parameters such as temperature, pressure, etc. are measured and models of the phenomena are known.

Liquid, Gas and Vapor Measurements

Flow meters for combustibles or fluids can be grouped in different types: volume meter; velocity meter; head meter; pitot tube; vortex; rotameter; sonic meter; magnetic meter; others.

i) Volume meter. Volume meters generally operate on the principle of giving an indication proportional to the quantity that has flowed into fixed-volume containers over time. With alternate filling and emptying of the container, an indication of the number of times the container is filled and emptied can easily be transferred to a counter with a calibrated dial in order to determine the quantity of the flow.
   Gasoline and gas meters were widely used in the past but now they are outdated; other types of volume meters are still in use.
   The error ranges from 0.5% to 2.5% depending on the meter model; a good degree of accuracy can be obtained from full capacity down to zero (working range 100:1).
   Notice that volume meters can be installed anywhere along the pipe, but particular attention must be paid to floating solid matter. Filters must be installed and they should be cleaned on a routine basis.

ii) Velocity meter. The quantity or the volume of the fluid is derived by means of a velocity measure which is transferred to a counter generally calibrated in rates of total flow.
   The operating principle is as follows: the meter has an orifice with a shunt circuit around it. Fluid flows through it and compels a turbine or a helix to rotate (see Fig. 2.1). The rotation speed, which is proportional to the fluid

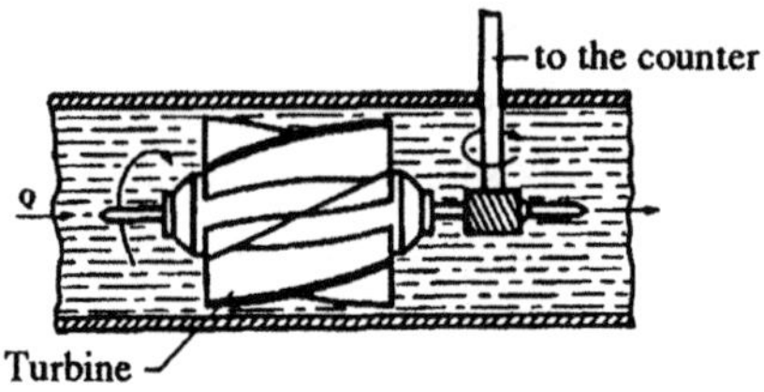

Fig. 2.1 Velocity meter with a turbine rotor

velocity, is transmitted to a counter calibrated in rates of flow through the orifice (volume flow-rate = velocity · area).

The meter can be installed anywhere along the pipe; in bypass installation, the counter is calibrated by taking into account the diameters of the orifice and the nozzle of the main pipe.

Overloads up to 150-200 per cent of rated capacity are handled temporarily without loss of accuracy.

Both volume and velocity meters are able to work with liquid and gas, but they are not interchangeable (for instance, if gas flows through a meter originally designed for a liquid, the rotor velocity will be too high and mechanical damage may occur).

Notice that these meters always measure volume flow-rate at actual temperature and pressure. In the case of gas, the meter indication must be corrected in order to obtain volume flow-rate in standard conditions by taking into account pressure and temperature as follows:

$$q_s = q \cdot \frac{p}{p_s} \cdot \frac{T_s}{T} \cdot \frac{Z_s}{Z}$$

where

q_s = standard condition volume flow-rate (Sm^3/s)
q = meter indication volume flow-rate (m^3/s)
p_s = standard atmospheric pressure (0.1013 MPa)
p = absolute actual pressure (MPa)
T_s = standard temperature (273.15 K)
T = actual gas temperature (K)
Z_s, Z = compressibility factors.

iii) Head meter. The operating principle is based on the measurement of a differential pressure (pressure losses) between two points of the pipe; this is produced by introducing an orifice or a nozzle. In this way the fluid speed increases and the pressure decreases.
The differential pressure between the two points is proportional to the square of the mass flow-rate, thus:

$$m = \alpha \cdot \varepsilon \cdot S \cdot \sqrt{\rho \cdot (p_1 - p_2)}$$

where

m = mass flow-rate (kg/s)
α = instrument flow coefficient
ε = expansibility factor (for gas)
S = section of the orifice or the nozzle (m^2)
ρ = mass density (kg/m^3)
p_1, p_2 = upstream and downstream pressures (Pa).

Head meters differ with the type of device producing a differential pressure:

— thin plate and sharp edge orifices (see Fig. 2.2a,b). An orifice with a diameter d is inserted by means of two shafts anywhere along the pipe with a diameter D. Values of the diameter ratio d/D ranging from 0.05 to 0.7 and straight pipes upstream and downstream are required. The length of the straight pipes must be at least $20 \cdot D$ upstream and $5 \cdot D$ downstream;
— Venturi tube (see Fig. 2.2c). This meter utilises a Venturi nozzle which permits flow measurement with low differential pressure values. Then shorter straight pipes upstream and downstream are required, generally less than $15 \cdot D$.

Notice that the error of head meters is generally less than 2%, but it increases as the meter works in the lowest part of the scale because of the squared relationship between flow and differential pressure. Care must be taken not to oversize the full-scale capacity on the basis of the real flow-rate; a good degree of accuracy is obtained from full capacity down to 25% (working range 4:1).

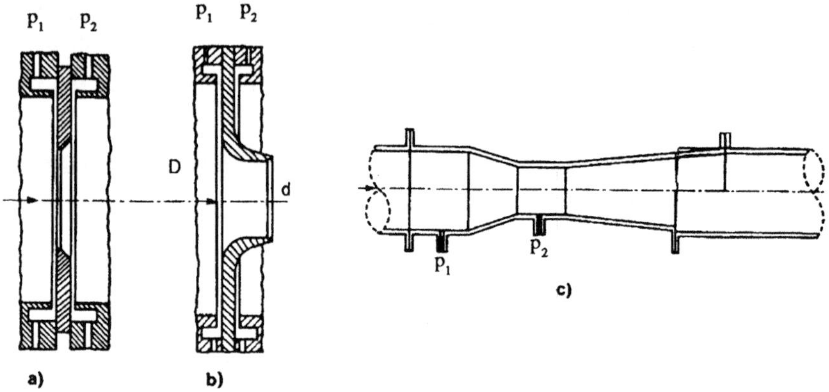

Fig. 2.2 Different types of head meters. a) thin plate orifice; b) sharp edge orifice; c) Venturi tube. Upstream (p_1) and downstream (p_2) pressure measurement points are also shown.

iv) Pitot tube (see Fig. 2.3). This operates on the principle that the total pressure p_t, detected at a small orifice facing into the flow, is the sum of the static and the dynamic pressures. The static pressure p_s can be detected by measurement at 90° to the direction of the flow. The dynamic pressure is then obtained by subtraction. Notice that straight pipes are required upstream and downstream in order to avoid turbulence near the point of measurement.

Fluids without solid matter are required. Flow speed must be relatively high because low speed involves a differential pressure too slight to measure easily. A squared relationship exists between flow and differential pressure, as in head meters.

The basic operating formula is as follows:

$$m = \alpha \cdot \varepsilon \cdot K_v \cdot S \cdot \sqrt{2 \cdot \rho \cdot (p_t - p_s)}$$

where

m = mass flow rate (kg/s)
α = instrument flow coefficient (for gas)
ε = expansibility factor (for gas)
S = section of the orifice or the nozzle (m^2)

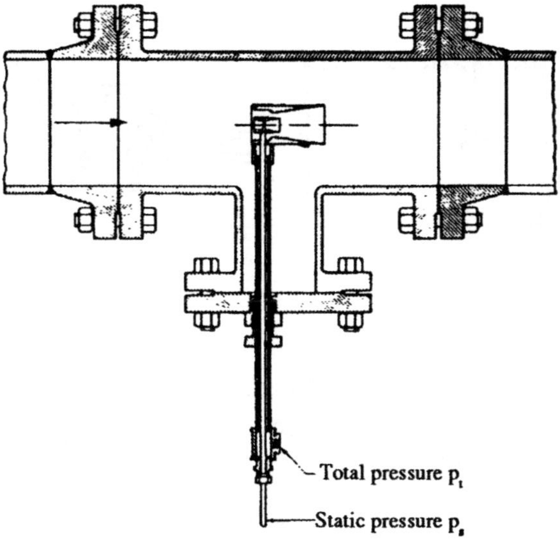

Fig. 2.3 Pitot tube

K_v = velocity coefficient which depends on the Reynolds number
ρ = mass density (kg/m^3)
p_t, p_s = total and static pressures (Pa).

In case of gas, head meter and Pitot tube indication must be corrected to take into account variations of mass density, expansibility factor and instrument flow coefficient from the design values because of different values of temperature and pressure. Correction factors are generally elaborated inside the measurement system to give the real value of the mass flow-rate.

v) Vortex (see Fig. 2.4). The operating principle is based on Karmaan's law which concerns vortices produced by a fluid. A vortex generator, which is an appropriately shaped solid, is placed inside the pipe and the fluid vortex frequency is detected. The working equation is as follows:

$$m = S \cdot V \cdot \rho = S \cdot (K_m \cdot f) \cdot \rho$$

where

S = section of the pipe (m^2)

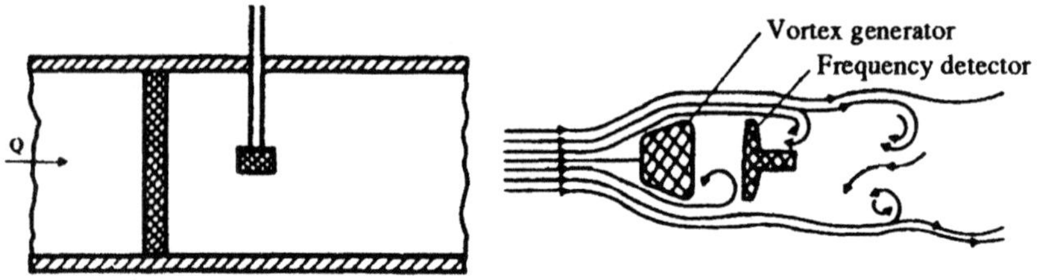

Fig. 2.4 Vortex meter

V = fluid velocity (m/s) which is equal to the vortex frequency f(Hz) multiplied by the vortex meter constant (K_m = Karmaan constant)
ρ = mass density (kg/m^3).

Vortex meters are able to measure volume flow-rates (mass flow-rates are derived by introducing the mass density) of gas, liquid and vapor. A good degree of accuracy is obtained from full capacity down to 10% (working range 1:10). In the case of gas, a correction factor as with velocity meters is required. It is generally elaborated inside the measurement system to give the value of the standard volume flow-rate and the real value of the mass flow-rate.

vi) Rotameter (see Fig. 2.5). This operates on the principle that a flowing fluid exerts a force proportional to the flow. A rotameter is a vertical conic tube in which a ball or any other calibrated device made of plastic or glass moves freely. The calibrated device is subject to two opposite forces: the force of gravity which works downwards and the fluid force which works upwards. The device moves vertically and indicates the rate of the fluid flow.
Notice that no straight pipes upstream or downstream are required; the only condition is that fluid must flow upwards from below.
The rotameter is suitable for measuring low rates of flow and must be differently calibrated for each kind of fluid.

vii) Sonic meter, magnetic meter, others.
The operating principle of a sonic meter is based on the fluid property of propagating a sound at a speed which depends on fluid volume, pressure and

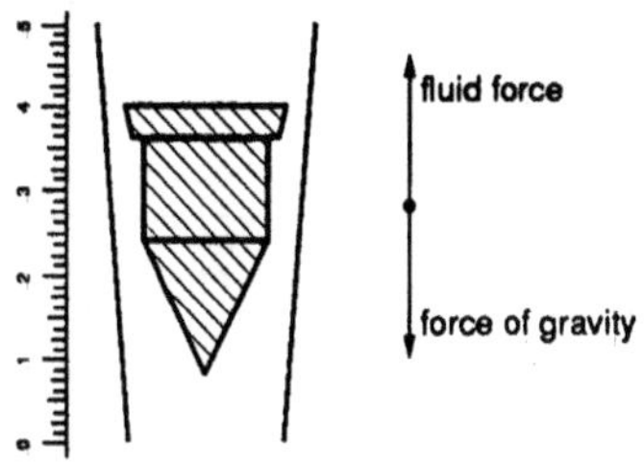

Fig. 2.5 Operating principles of a rotameter

temperature. Electric pulses are generated and transformed into sonic ones through the fluid; the sonic pulses are then detected and the fluid speed is calculated. The flow is calculated as fluid speed multiplied by area, as already shown for the velocity meters.

A magnetic meter works on the basis of Faraday's law which says that at the ends of a conductor (the fluid in this case) moving inside a magnetic field there is an electromotive force linearly depending on both the conductor speed and the magnetic field itself. This meter, which does not require the pipes to be cut, can be used only with liquid with a conductivity of not less than 500 μS/mm.

Other systems based on different principles are also available but it is not necessary to describe all of them in detail. They can readily be found in specialized books.

Table 2.9 shows basic operating parameters able to qualify flow meters and their applications. Note that the commonest devices are head meters, with orifice or Venturi tube, and Pitot tube meters, depending on the fluid to be measured.

Electric Energy Flow Measurements

Equipment for measuring energy flow and power demand in the distribution system consists of conventional voltmeters, ammeters and kilowatthour meters. Potential transformers (PT) or current transformers (CT) are generally installed to reduce the level of voltages and currents for technical as well as for security reasons. Current is generally stepped down to 5 A or less, line voltage to 120 V or 100 V or less.

Table 2.9 A review of the main flow meters

Type of device	Pipe diameter	Type of fluid to measure			Max. pressure		Max. temperature		Reynolds number	Operating range	Typical accuracy	Pressure losses
	mm	Liquid	Gas	Vapor	MPa	bar	°C	°F			%	kPa
Volume meter	5-500	x	x	—	20	200	100	212	any value	50:1	02.-0.5	50-100
Velocity meter	10-600	x	x	—	10	100	200	392	>10000	15:1	0.2-0.5	5-10
Head meters: orifice and Venturi tube	>50	x	x	x	50	500	500	932	>2500	4:1	1-2	2050
Pitot	>100	x	x	x	50	500	500	932	any value	4:1	2-5	1-2
Vortex	25-200	x	x	x	10	100	200	392	>10000	15:1	0.75-1.5	10-20
Rotameter	5-150	x	x	x	4	40	200	392	>10000	10:1	2-3	5-10
Sonic meter	any value	x	x	—	20	200	200	392	any value	10:1	2-3	0
Magnetic meter	any value	x	—	—	20	200	200	392	any value	20:1	0.5	0

Specific equipment, which performs combinations of the above-mentioned functions, is also available to provide kilovarhour, kilovoltamperehour, kilowatt, kilovoltampere, power factor.

Typical connection schemes of the three basic meters are shown in Fig. 2.6, where basic relationships between current, voltage, and power are also illustrated.

Notice that:

— voltmeter is a high-resistance coil device and it must always be connected in parallel;

— ammeter is a low-resistance coil device and it must always be connected in series with the current. With portable meters, clamp-around current transformers are generally used;

— wattmeter, which measures real power, reactive or apparent power, is a combination of potential coils (parallel connected) and current coils (series connected);

— when current transformers or potential transformers are used, the CT and the PT ratios must be taken into account. The meter readings multiplied by the meter constant and the CT or PT ratios (or both in the case of wattmeter) give the values of current, voltage and power.

  Typical CT ratios are 100:5, 1000:5, 5000:5. These values show how many amperes flow in the primary conductor or line conductor when a 5 A current flows in the secondary winding. For security reasons the secondary winding must never work with no-load, and consequently it must always be connected to a low-resistance ammeter or shorted by means of a jumper wire, screw, or switch on a CT shorting terminal strip.

Complex metering systems can also be introduced by using current and voltage transducers whose signals are elaborated by means of computers and transformed into electric power and energy. The introduction of computerized systems allows both the measurement of energy flow and the control of electric systems based on different optimization criteria.

These systems can also be integrated conveniently with flow and other physical parameter measurements in order to obtain complete process control. This approach, quite common in energy-intensive industry, is spreading in many manufacturing sectors (see Chapter 17).

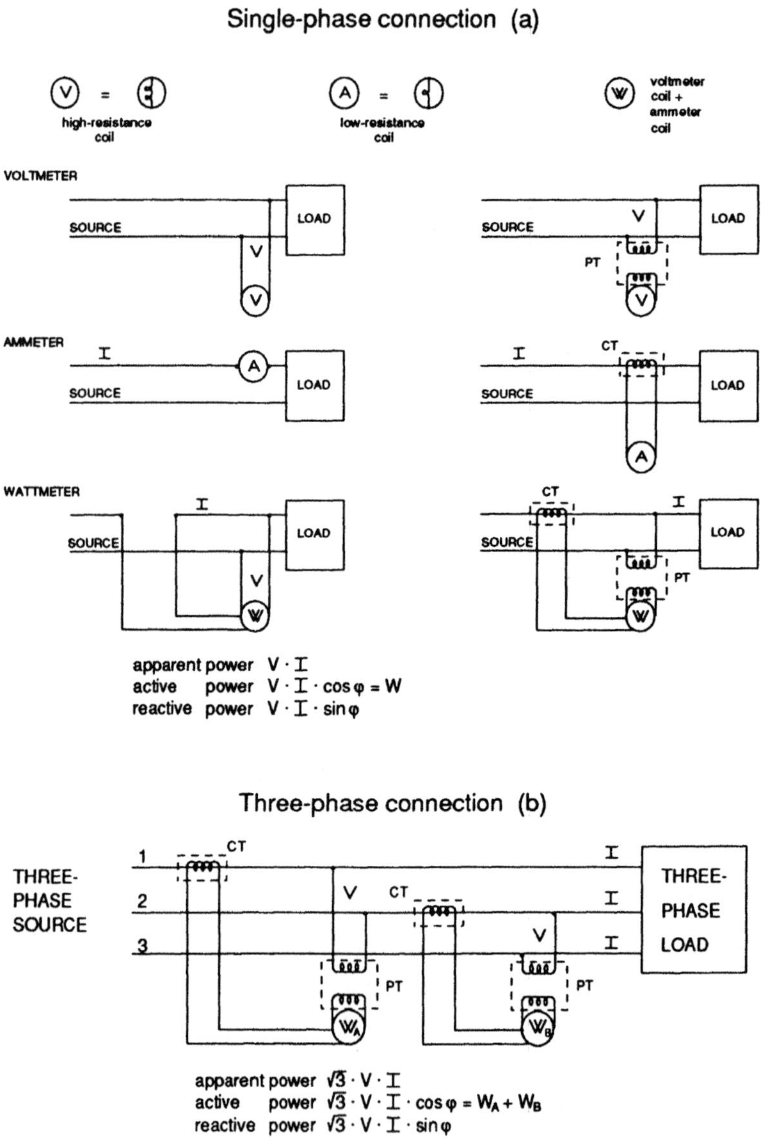

Fig. 2.6 Typical connection schemes: voltmeter, ammeter and wattmeter with single-phase line (a) and with three-phase line (b)

Temperature, Pressure and Other Measurements

Devices for measuring temperatures can be grouped according to the principle of operation and the temperature range. The commonest devices are: liquid-in-glass thermometers, bi-metallic thermometers, resistance temperature devices, thermocouples, infrared thermometers, optical pyrometers. A correct choice must be based on temperature ranges, environmental conditions and uses of the measurements for reporting or process control.

Fig. 2.7 shows relationships among the absolute, Celsius, Rankine and Fahrenheit temperatures.

Devices for pressure measurement can be listed as Bourdon gage, diaphragm gage and manometer. They measure the difference between the absolute pressure in a system and the absolute pressure of the atmosphere outside the measuring device. This difference is called differential or gage pressure. If the pressure of the system is lower than the atmospheric pressure, the term vacuum pressure is used instead of gage pressure.

The Bourdon gage consists of a curved tube closed at one end with the other end connected to the pressure to be measured. When the pressure inside the tube is greater than the pressure outside, the tube tends to straighten and the amount of change in length or curvature can be translated into a gage reading.

The diaphragm gage is based on the detection of the diaphragm movement if the pressures against its two sides are different.

The manometer is a tube with liquid in it; one end is open to the air and the second end is exposed to a different pressure, so that the end with the higher pressure has a lower liquid level and the pressure difference is detected.

Fig. 2.8 shows relationships among the absolute, atmospheric, gage and vacuum pressures.

Other physical parameter measurements can be made to meet specific needs. A careful study of specialized books and manufacturers' technical specifications is suggested in order to make the right choice for each application.

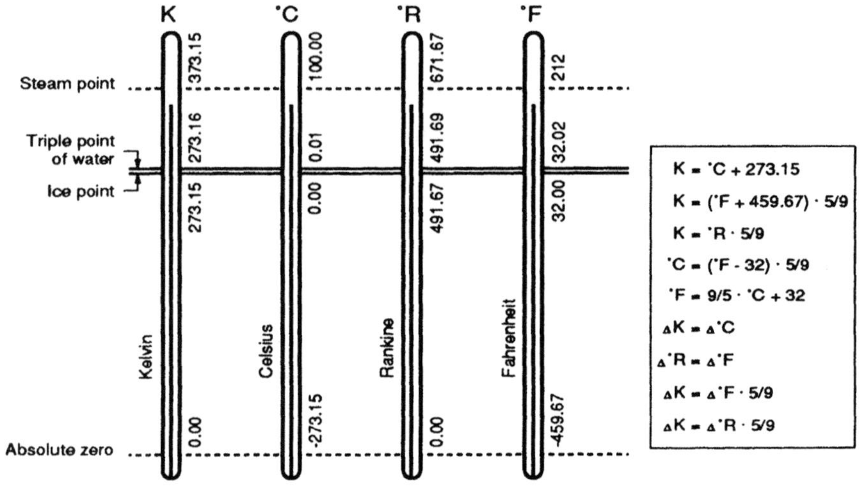

Fig. 2.7 Relationships among the absolute, Celsius, Rankine and Fahrenheit temperatures

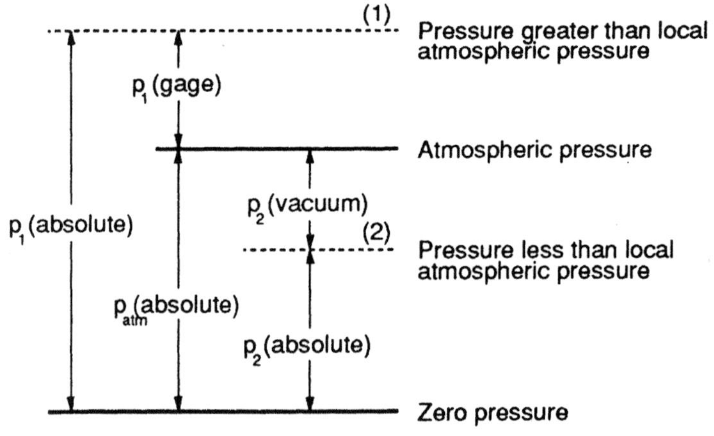

p_1 (gage) = p_1 (absolute) - p_{atm} (absolute)

p_2 (vacuum) = p_{atm} (absolute) - p_2 (absolute)

Fig. 2.8 Relationships among the absolute, atmospheric, gage and vacuum pressures

CHAPTER 3

WORLD ENERGY RESOURCES AND INDUSTRIAL ENERGY BALANCE

3.1 WORLD ENERGY RESOURCES AND CONSUMPTION

Trends in world energy production during the first decades after 2000 are expected not to diverge greatly from the last period of the twentieth century because resources of solid and liquid combustibles and natural gas are still able to cover the needs of final users and other primary energy exploitation is slowly increasing all over the world.

Consumption of all forms of primary energy should expand all over the world and by 2000, according to international projections, it should reach 10-12 thousand million TOE. This figure can be assumed as an average reference and it is the basis of the considerations made below; any more detailed reference would be related to a specific period and could become outdated within a short time.

Table 3.1 reports a projection for 2000 of world production of primary energy, distinguishing the contributions of different forms of energy from industrialized countries, developing countries and centrally-planned economies (see note in Table 3.1). A more detailed energy balance by major geographical regions is reported in Table 3.2.

Gross world consumption of primary energy classified in the same way as production (as in Table 3.1) is shown in Table 3.3. Notice that: the industrialized countries would produce only 70-80% of their energy needs; the developing countries would continue to account for nearly 50-60% of world oil production of which about half would be for export; the centrally-planned economies (this definition refers to the situation existing before 1990) are expected to remain self-sufficient and would even be able to export part of their production. A more detailed energy balance by major geographical regions is reported in Table 3.4.

Table 3.1 Annual world production of primary energy in million TOE

Projection for 2000	SOLID FUEL	OIL	NATURAL GAS	NUCLEAR	HYDRO & OTHER	TOTAL
Industrialized countries	1578	671	651	589	231	3720
Developing countries	158	1921	431	38	707	3255
Centrally-planned economies (1)	1716	825	1003	147	156	3847
TOTAL WORLD	3452	3417	2085	774	1094	10822
(1) This definition refers to the situation existing before 1990						

Table 3.2 Annual world production of primary energy in million TOE by major geographical areas

Projection for 2000	SOLID FUEL	OIL	NATURAL GAS	NUCLEAR	HYDRO & OTHER	TOTAL
EEC (2)	172	108	108	215	21	624
USA	939	415	387	180	60	1981
Japan	10	2	0	90	55	157
Other industrialized countries	457	146	156	104	95	958
OAPEC (3)	0	1000	151	0	14	1165
OLADE (4)	21	330	139	11	130	631
Other developing countries	137	591	141	27	563	1459
USSR	420	650	928	105	30	2133
China	800	150	15	2	100	1067
Other countries with centrally-planned economies (1)	496	25	60	40	26	647
TOTAL WORLD	3452	3417	2085	774	1094	10822

(1) This definition refers to the situation existing before 1990

(2) European Economic Community in 1986 (Belgium, Denmark, France, Germany, Greece, Ireland, Italy, Luxembourg, Netherlands, United Kingdom)

(3) Organization of Arab Petroleum Exporting Countries

(4) Latin-American Energy Organization

Table 3.3 Annual world gross primary energy consumption in million TOE

Projection for year 2000	SOLID FUEL	OIL	NATURAL GAS	NUCLEAR	HYDRO & OTHER	TOTAL
Industrialized countries	1530	1677	847	589	231	4874
Developing countries	238	919	329	38	709	2233
Centrally-planned economies (1)	1685	840	887	147	156	3715
TOTAL WORLD	3453	3436	2063	774	1096	10822

(1) This definition refers to the situation existing before 1990

Table 3.4 Annual world gross primary energy consumption in million TOE by major geographical areas

Projection for year 2000	SOLID FUEL	OIL	NATURAL GAS	NUCLEAR	HYDRO & OTHER	TOTAL
EEC (2)	264	439	196	215	21	1135
USA	838	680	455	180	60	2213
Japan	109	233	57	90	55	544
Other industrialized countries	319	325	139	104	95	982
OAPEC (3)	5	104	122	0	14	245
OLADE (4)	25	321	138	11	130	625
Other developing countries	208	494	69	27	565	1363
USSR	420	605	715	105	30	1875
China	775	90	15	2	100	982
Other countries with centrally-planned economies (1)	490	145	157	40	26	858
TOTAL WORLD	3453	3436	2063	774	1096	10822

(1) This definition refers to the situation existing before 1990

(2) European Economic Community in 1986 (Belgium, Denmark, France, Germany, Greece, Ireland, Italy, Luxembourg, Netherlands, United Kingdom)

(3) Organization of Arab Petroleum Exporting Countries

(4) Latin-American Energy Organization

3.2 ENERGY END USERS

Final energy consumption can be analyzed by taking into account the energy demand in each sector: industry; transport; residential, tertiary and agriculture; non-energy uses. Final consumption in 2000 will amount to roughly 70% of the gross world consumption because of losses mainly in electric power production plants and in distribution and in other transformations inside energy industries, that is 7-8.4 thousand million TOE. For industrialized countries the final consumption could be shared among sectors roughly as follows: 31% industry; 22% transport; 31% residential, tertiary and agriculture; 6.5% non-energy uses; 9.5% energy industry.

The electric energy, in TOE, is roughly one third of the total primary energy consumption.

These balances are shown in Figure 3.1.

Industry

Industrial energy consumption is concentrated mainly in only a few sectors. Iron and steel, non-ferrous metals, non-metallic minerals, and chemicals account for 40% of the energy consumed by industry with a tendency toward lower values. Four other sectors, i.e. paper and pulp, textiles, food processing and glass are responsible for another 40%. The remaining energy is shared among other manufacturing sectors such as engineering, plastic, electronic, etc..

Transport

The form of transport mainly responsible for energy consumption is that effected by road. Average values for an industrialized country are as follows: 80-85% for road transport, 10-15% for air, 3-5% for rail, 1-3% for water transport.

Differences will be found depending on local conditions.

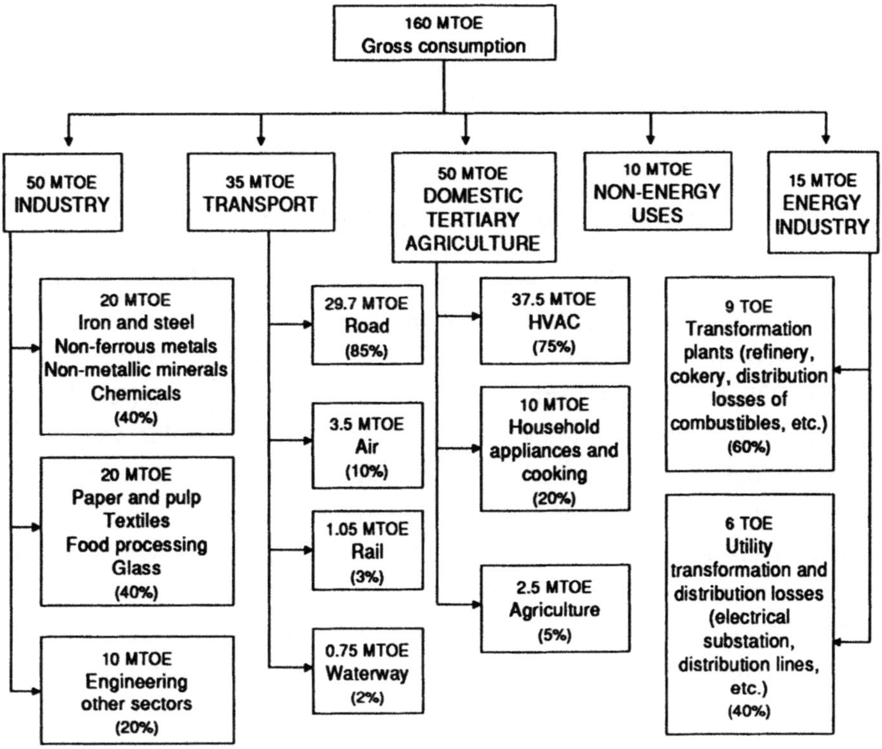

Fig. 3.1 Gross and final consumption in a typical industrialized country with a gross consumption of 160 MTOE

Residential, Tertiary and Agriculture

Residential energy consumption in industrialized countries occurs mainly in heating, ventilation and air conditioning, which account for roughly 75-80% if hot water production is included.

Household appliances are responsible for 15-20% of the total consumption and cooking for a further 5%. Agriculture can be assimilated to this sector with 5% of energy consumption.

Non-energy Uses

Non-energy uses include oils and lubricants, bitumens and oil products, components for motor vehicles, chemical industries, etc..

3.3 AN OUTLINE OF INDUSTRIAL ENERGY CONSUMPTION

The outline of industrial energy consumption ranges in Table 3.5 is based on data from international reports and specific energy audits.

Final consumptions (thermal TOE, MWh) are transformed into primary energy, i.e. TOE, by using transformation factors reported in Chapter 2 (see also note in Table 3.5) and referred to the quantity of production (t). In this way specific consumptions can be listed as TOE/t and different sectors can be compared. Notice that these values must be assumed as average values, but they are able to qualify industrial sectors from the point of view of energy. The cost of the total primary energy as TOE for final users changes according to the use (directly in furnaces and dryers, to produce steam, to produce electricity, etc.), to the quantity and to the geographical area. It is generally equal to the barrel price multiplied by a number ranging from 8 to 16, the lower if mainly combustibles are used inside the factory and the higher if electricity is the main energy used.

The parameter TOE/t has been chosen in order to avoid any relationship between energy consumption and other reference units such as number of employees or company incomes which are often considered but which may be outdated in a short period, independently of energy influence.

Table 3.5 Specific energy consumption of manufacturing industries

Industrial sector	End-user energy		Primary energy
	Thermal TOE/t	Electricity MWh/t	Total energy TOE/t
Brewing	0.05	0.10	0.069
Brick	0.075	0.05	0.087
Cement	0.08	0.11	0.105
Chocolate	0.20	2.00	0.660
Dairy industry	0.15	0.5	0.265
Engineering	0.30	2.75	0.933
Flour products (pasta, etc.)	0.04	0.16	0.077
Foundry	0.3	0.9	0.507
Ham, sausage	0.1	0.35	0.181
Hot pressing	0.25	0.75	0.423
Ice cream/cake	0.1	0.75	0.273
Milk processing	0.02	0.10	0.043
Non-ferrous metal	0.07	0.30	0.134
Paper and pulp	0.15	0.43	0.249
Plastic	0.05	0.55	0.177
Rubber	0.10	5.00	1.250
Textiles (dyeing)	0.75	0.75	0.923
Textiles (spinning)	0.45	7.5	2.175
Wood	0.02	0.06	0.037
	(a)	(b)	(c) $= (a) + (b) \cdot 0.23$

Notes
— To convert from end-user electric energy to primary energy multiply the MWh/t value by 0.23 (see Chapter 4.2 for this assumption).
— Consumption range is ± 25% around average values shown in the table

CHAPTER 4

UTILITY PLANTS AND RENEWABLE SOURCES

4.1 FACTORY INPUT ENERGY

Energy is transported to the factory site as purchased fuel (oil, gasoil, LPG, natural gas) or as electricity purchased from utilities. Water, to which attention must be paid when an energy management program is implemented, can be either purchased from utility or pumped from wells.

In addition, energy can be used from so-called renewable sources, i.e. waste recovery and permanent natural sources such as sun, wind and water. The importance of these sources, in comparison with utility supplies, is generally very slight, with the exception of those used in hydraulic plants which can cover a significant amount of the factory demand.

4.2 UTILITY PLANTS AND RENEWABLE SOURCES FOR INPUT ENERGY TO THE FACTORY

Utility plants, mainly to produce electric energy and heat, may be classified as fossil fuel or nuclear plants. In some local conditions, other sources such as, above all, hydro and geothermal primary energy, can be used by utility plants.

Utility plants generally use steam-condensing turbines, gas turbines and combined cycles. Heat recovery for district heating is a common way to improve overall efficiency; of course the local regulations, the climatic conditions and the plant location determine the attractiveness of this possibility.

As a general rule, utility plants have an average efficiency of less than 40%; if electric distribution losses are considered, an average value of 35-37% can be assumed for the overall efficiency from the primary energy to the end users. That means that 0.25-0.23 kg of equivalent oil are required to deliver 1 kWh to the end user.

Industrial cogeneration plants (see Chapter 9), that is, plants which realize either the coincident generation of necessary heat and mechanical or electric power, or the production of power by recovering low-level heat from process, may reach overall efficiency ranging from 55% to 85%, depending on the type of cogeneration plant. This is a considerable primary energy saving, generally accompanied by a significant energy cost saving, more or less important depending on the local regulations and tariffs.

The main renewable sources of energy are solar, wind and hydraulic energy and energy from waste. Most of these sources are inadequate for industrial users because they yield only low specific power in standard conditions while the current capital and operating costs, and those projected for the next decades, are extremely high. Nevertheless, in spite of these limitations, renewable sources should be considered in the light of local conditions and efforts made to exploit them.

4.3 RENEWABLE SOURCES: SOLAR ENERGY

The solar energy which reaches the surface of the earth can be used to produce hot water or to produce electric energy by means of photovoltaic cells. Typical values of the rate of the solar energy that reaches the earth's surface are 300-900 W/m^2 depending on the latitude, on the time of day and on the atmospheric conditions.

Hot water from solar energy

Water can be heated by solar collectors which collect a fraction of the incident sunlight ranging between 90% and 20% depending on the number and type of glazings and on operating parameters such as rate of insulation, ambient air temperature and input cooling fluid temperature.

With average values of 300-900 W/m² of solar energy rate, a standard collector may yield an actual specific power ranging from 200 to 400 W/m².

For a collector installed to produce hot water at 50-60°C (122-140 °F) with a difference of 30°C (48 °F) between the average internal collector temperature and ambient temperature, the value of the heat available can be determined approximately from diagrams as in Fig. 4.1 if the solar insulation rate in W/m² is introduced.

Heat losses from a collector are generally defined by a coefficient (W/m² · °C) which varies from 7 with a single glazed flat plate to 2.5 in the case of a double one with selective surfaces (which have a selective coating such as black chrome able to reduce the emission of energy, particularly at high temperatures). The solar energy, when converted into hot water or hot air, must be stored to match the end-user demand throughout the day.

Insulated water-storage tanks are always installed; the average capacity is at least 100 liters per square meter of collector. If intermediate fluid is used, heat exchangers with high effectiveness must be installed.

With hot-air systems, thermal storage with rock-bed can be used, if necessary. Generally, these systems are difficult to justify from the economic and technical points of view for industrial applications or for use in factories.

Photovoltaic systems

Electric energy can be produced from solar energy by using photovoltaic silicium cells. As a general indication, the voltage across one cell is 0.5 V and it is independent of the solar energy rate to which the electric current fed to the load is linearly related. In practice, cells are mounted on a module and the modules are arranged in panels or arrays in series, in parallel or in combined series/parallel.

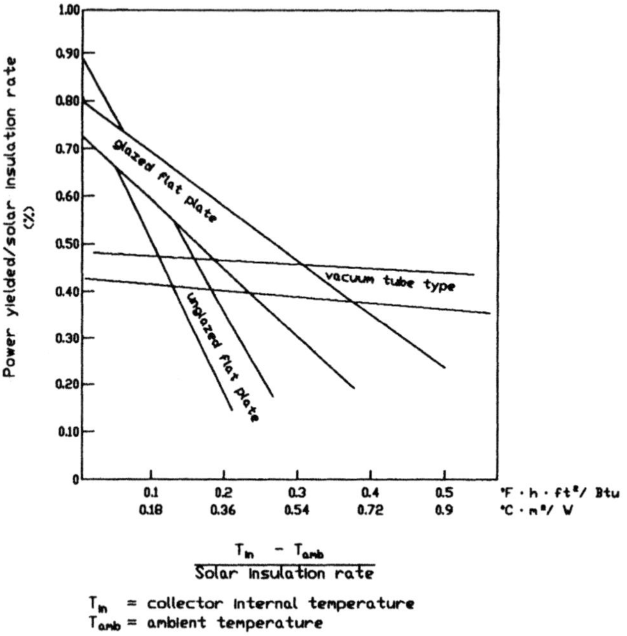

Fig. 4.1 Various types of collector and their efficiency

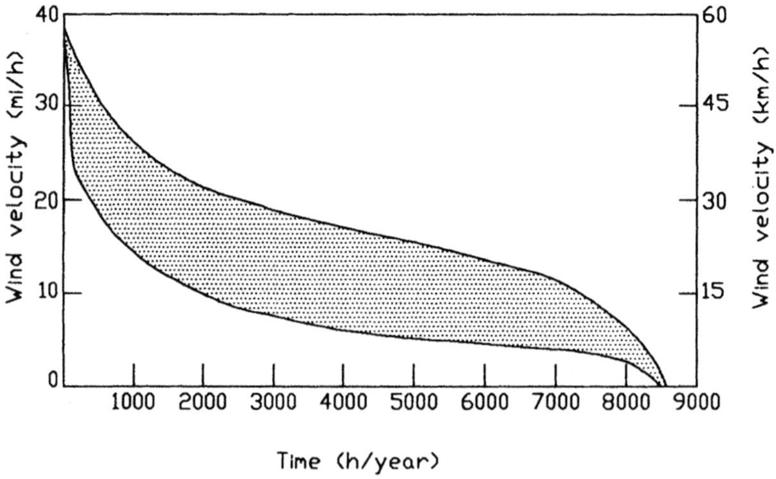

Fig. 4.2 Range of typical variations for annual average wind velocity – duration curves

The efficiency, that is the ratio between the electric power generated and the solar energy rate, is not more than roughly 15% for standard applications. That means that 1 m^2 of cells will produce not more than 75 W with an average solar energy rate of 500 W/m^2.

The electric energy produced by a series of cells must be stored in an accumulator storage system.

The low value of the power yielded and the high cost of installation make this system unattractive for industrial applications.

4.4 RENEWABLE SOURCES: WIND ENERGY

Wind energy has traditionally been exploited for mills at sites where wind velocities are high and quite steady all the year round.

A basic relationship which relates the wind velocity to the power available and to the device surface normal to the wind velocity is:

$$P = \frac{1}{2} \cdot A \cdot \rho \cdot V^3$$

where

P = power yielded by the wind source (W)

A = area normal to the wind velocity (m^2)

ρ = density of air (1.29 kg/Sm^3 or 0.08 lb/Sft^3 in standard conditions)

V = velocity of air stream (m/s)

For practical applications, the ratio P/A is generally introduced; then

$$(P/A) = \frac{1}{2} \cdot \rho \cdot V^3 = 0.645 \cdot V^3 \quad \left(\frac{W}{m^2}\right)$$

$$(P/A) = \frac{1}{2} \cdot \rho \cdot V^3 = 0.059 \cdot V^3 \quad \left(\frac{W}{ft^2}\right)$$

The theory of wind devices shows that not more than 59% of the power available in the wind can be converted into mechanical power. Practical values are lower, due to aerodynamic efficiency generally being not more than about 75%.
The power coefficient C_p, defined as the ratio between the power actually converted to mechanical power (P_m) and the power available in the wind (P), has an average value of less than 75% · 59% = 44%.

Thus:

$$\left(\frac{P_m}{A}\right) = C_p \cdot \left(\frac{P}{A}\right)$$

Note that the power actually converted into mechanical power by a specific device varies with the cube of the wind velocity. Mechanical power may be used to drive equipment such as mills or to produce electric energy by means of direct current (d.c.) or alternating current (a.c.) machines acting as generators. The energy can be stored as d.c. energy in batteries and then converted to a.c. energy by means of a d.c./a.c. static converter.

Wind power converted to mechanical power with a wind velocity of 7-8 m/s is roughly 100-150 W/m^2 (square meter of area normal to the wind velocity).
Wind power installations, with a wind velocity of 7-8 m/s, reach up to about 2000 kW of mechanical power per unit 70-80 m in diameter with blades shaped to increase the surface normal to the wind velocity.

In order to quantify the energy generated by wind devices, speed duration curves for the specific site, that is wind velocity versus hours per year, must be known. Fig. 4.2 shows a range of values which can be found in various sites where wind power installations can conveniently be used.

Notice that wind devices generally work in a range of wind speed between cut-in speed (bottom speed) and cut-off speed (upper speed). The wind device works at variable speed; more or less sophisticated controls and power electronic conversion systems are used to improve the exploitation of the available energy. The blade pitch can also be varied.

Of course, the choice among different systems must be based on the capital cost, energy cost saving and operating costs (see Chapter 20).

4.5 RENEWABLE SOURCES: HYDRAULIC ENERGY

Hydro-electric power stations are designed to convert the gravitational energy of water into mechanical and then into electric energy. Both large and small power stations can be classified as follows: (i) run-of-river power station when no significant regulating reservoir exists; (ii) pondage power station which is run-of-river with small reservoirs able to delay the production of electricity for a short period (hours, days); (iii) seasonal power station with a reservoir to regulate the water supply to the turbines; (iv) pumped-storage power station when a reservoir is filled exclusively or partially by pumps.

Large power plants, greater than a few MW, belong exclusively to utilities; small and mini-power stations down to few hundreds of kW usually belong to single factories for their own use.

Three main types of turbine are employed in hydro-electric power stations.

Propeller turbines, such as the Kaplan turbine, are employed for low heads of not more than 40-50 m (130-160 ft) and high flow rates ranging between 2 and 40 m^3/s (70-1400 ft^3/s). They realize high rotor velocity for relatively low water through-flow velocities and may attain a good efficiency over a wide range of loads by blade pitch variation.

Francis turbines are used for higher heads (20 to 300 m, 66 ft to 1000 ft) and medium flow rates (0.2-20 m^3/s, 7-700 ft^3/s); these turbines are radial inflow units where water enters the rotor through a set of variable-angle inlet guide vanes and flows radially inward and axially downward, with a pressure drop within the turbine wheel itself.

Pelton turbines, which are impulse units, are used for heads from 100 m (330 ft) up to 300 m (1000 ft) or higher, with a very low flow rate, generally lower than 1 m^3/s (35 ft^3/s). All the static head is converted into velocity and all this energy is absorbed in the wheel so that water leaves at a very low velocity.

Since electric-power generation at constant frequency requires constant speed, problems arise in the partial-load operating mode when shaft torque is reduced because of a reduction in electric-power demand.

Kaplan and Francis turbines can accept partial flow without significant loss of efficiency by varying either the rotor blade or the inlet guide vane angles. Pelton turbines can be regulated by an adjustable nozzle.

The basic relationship giving the power yielded at the turbine shaft by hydraulic energy is as follows:

$$P = q \cdot H \cdot \rho \cdot \eta \cdot g/g_c$$

where

P = turbine power shaft output (W, $\frac{ft \cdot lbf}{s}$)

q = volume flow-rate ($\frac{m^3}{s}$, $\frac{ft^3}{s}$)

H = total head = Hstatic-Hlosses (m, ft)

ρ = density of water ($\frac{kg}{m^3}$, $\frac{lb}{ft^3}$)

η = turbine efficiency

g = gravitational acceleration at the location which may differ from the standard value equal to $9.81\ \frac{m}{s^2} = 32.17\ \frac{ft}{s^2}$

g_c = conversion factor = $\frac{1\ kg \cdot (m/s^2)}{N} = 32.17 \frac{lb \cdot (ft/s^2)}{lbf}$

On the assumption $\eta = 1$, the basic relationship shows that with 10 m of head the power related to a flow rate of 10 m^3/s equals 981 kW. Thus, the energy related to 10 m^3 of water with 10 m of head is $9.81 \cdot 10^5$ J or 0.2725 kWh.

Fig. 4.3 and Fig. 4.4 show the ranges of application of the various types of hydraulic turbine units and the effects of load on the efficiency of typical turbines, which is generally higher than 90%.

The same relationship can be used in pumping mode by introducing the total head H = Hstatic + Hlosses (see Chapter 10.2).

In choosing small and mini plants one must consider that:

— the useful available inflow may fluctuate during the year or the day;

— usually, such plants are economically viable only if fully automated and free from maintenance;

— the lower the installed capacity, the higher the capital and operating costs per kW installed. Notice that the only operating costs are those for maintenance, which are generally low in terms of kWh.

In conclusion, hydraulic sources are the most easily used renewable sources and their exploitation is suggested whenever they are available. Capital investment is generally high, depending on the site and on the power of the plant, but it has to be considered as a long-term investment which presents no significant risk.

4.6 RENEWABLE SOURCES: ENERGY FROM WASTE

Urban and industrial waste can conveniently be used to produce energy in different ways. Fig. 4.5 shows how waste can be treated for use inside or outside the process and for energy production. Table 4.1 lists typical heating values of waste.

Waste recovery is one of the biggest problems that industry and public bodies have to tackle. Resolution of this problem, which began to be important in the 1980s after the previous decade of energy shortage, is evolving only slowly because of many technical, economic and environmental constraints.

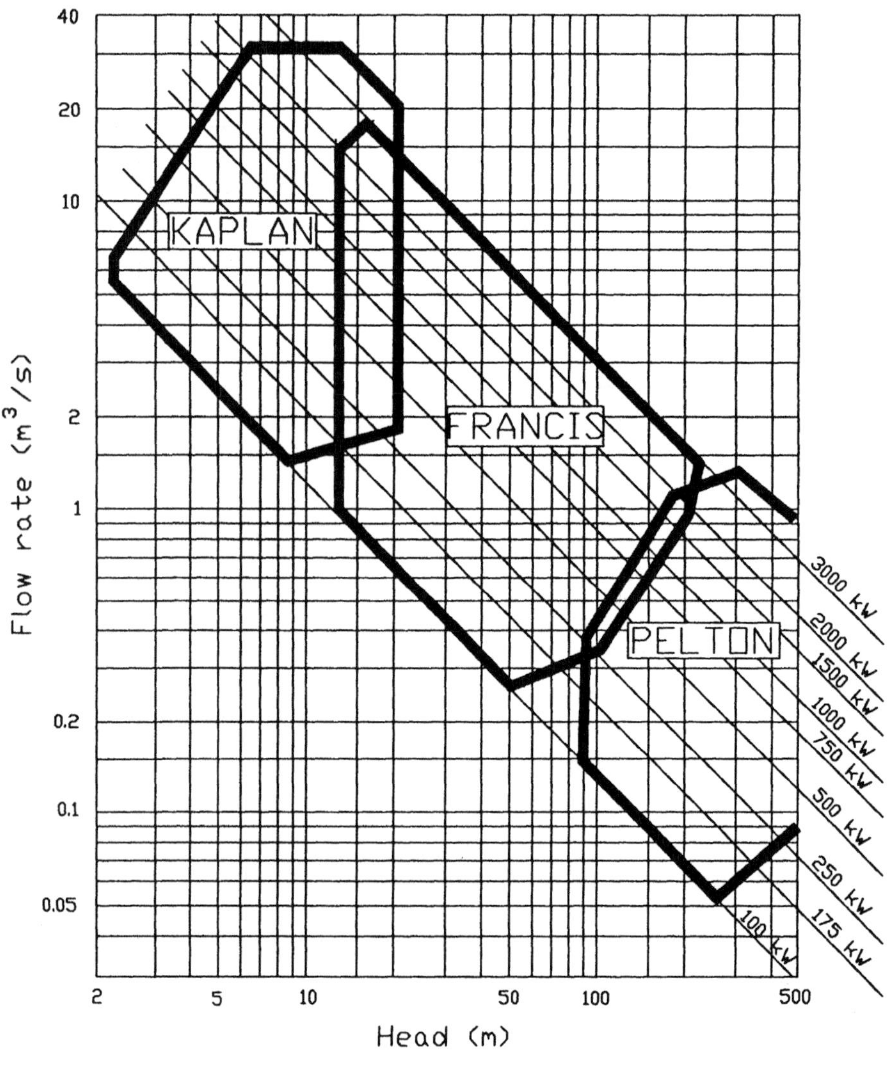

Fig. 4.3 Ranges of application of various types of hydraulic turbine units (100-3000 kW)

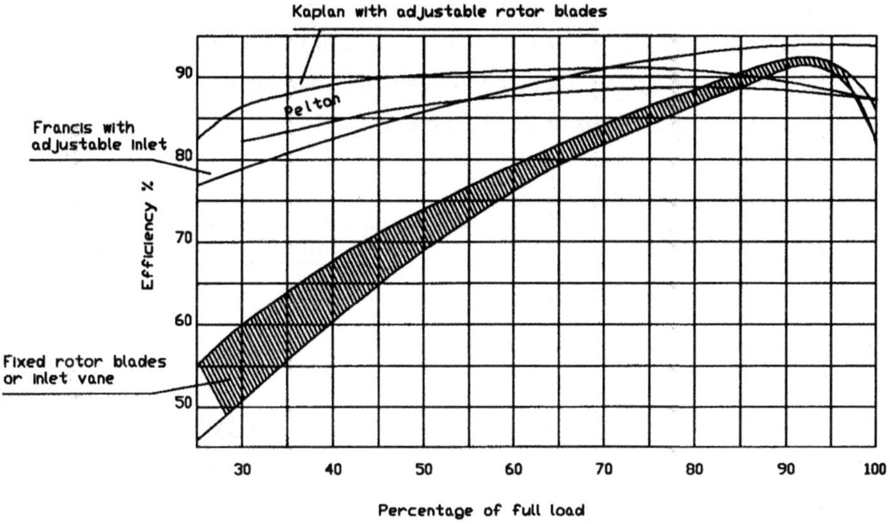

Fig. 4.4 Effect of load on the efficiency of hydraulic turbines

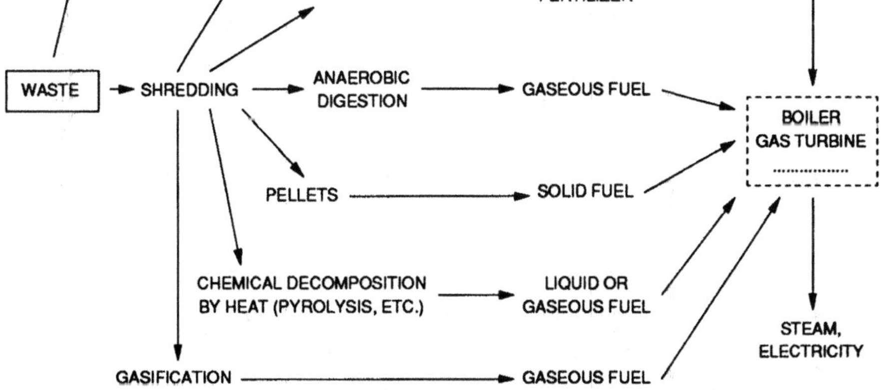

Fig. 4.5 Possible ways of converting waste into energy

Mass burning facilities where the refuse is burned as received and processed-fuel or Refuse-Derived-Fuel (RDF) where the solid waste is processed before burning are the two options that technology makes available.

A more detailed analysis of these problems can be found in Chapter 16.

Table 4.1 Typical heating values of waste

DESCRIPTION OF WASTE	GROSS HEATING VALUE	
	kJ/kg	Btu/lb
Industrial waste		
Leather scrap	23000	9888
Cellophane	27750	11930
Waxed paper	27500	11823
Rubber	28500	12253
Tires	41850	17992
Oil, fuel oil residue	41850	17992
Polyethylene	46000	19776
Agricultural		
Bark	11000	4729
Rice hulls	13500	5804
Corn cobs	19000	8169
Composite		
Municipal	10000-15000	4299-6449
Industrial	15000-17500	6449-7524
Agricultural	7000-14000	3009-6019

4.7 ENERGY STORAGE: HYDRO, MECHANICAL, ELECTRIC, THERMAL

Storage of energy after its transformation from primary energy into various forms is a deeply felt but rarely satisfied need. For both utility plants and end users, the storage of energy should be a way of improving energy management and reducing energy costs, by smoothing the profile of the energy demand, by exploiting low-rate opportunities in purchasing energy and by freeing energy recovery from user demand.

The main approaches to energy storage are hydro, mechanical, electric and thermal.

Energy stored as primary energy in combustibles has a very high mass-energy density with typical values of roughly 45000 kJ/kg (19350 Btu/lb) for liquid fuels. This value is vastly higher than the storage capacity of almost any other system available in which values of 100-200 kJ/kg (45-90 Btu/lb) seem to be the upper limit. The capital and operating costs of these systems make their extensive exploitation in industry difficult, except for hydro energy storage, which is the most attractive and which is widely used in both utilities and industries.

Notice that both hydro and mechanical storage are mainly electric storage because the energy available is generally electric which is transformed into mechanical energy at the electric motor shaft and then stored in different forms before being transformed back into electric energy.

Hydro storage

The principle of hydro storage follows the law of pumps (see Chapter 10.2) and of hydraulic source exploitation by turbines (see Chapter 4.5).

This system works, generally by using reversible machines, in the pumping mode when an excess of energy, usually electric, is available and in the turbine mode when energy is required back. Water is stored in an upper reservoir and it flows back through the turbine during periods of peak demand.

Remember that energy associated with 10 m^3 of water with 10 m of head is $9.81 \cdot 10^5$ J or 0.2725 kWh and that the efficiency of the whole system including energy transformations in pumping mode (electric motor, pump and flow losses) and in turbine mode (flow losses, turbine and electric generator losses) is generally 50-60%. It follows that the energy actually available is halved. A head of more than a hundred meters between the upper and the lower reservoir is usually required for an economic exploitation of the system.

Mechanical storage

The flywheel is a typical device for the storage of mechanical energy. The quantity of stored energy depends on the shaft speed, the mass and the radius of gyration of the flywheel (that is, the radius where the total mass is considered to be concentrated).

The energy stored in a flywheel is equal to the kinetic energy, that is:

$$E_{stored} = \frac{1}{2} \cdot \frac{1}{g_c} \cdot M \cdot (R \cdot \Omega)^2$$

where

E_{stored} = energy stored (J)

g_c = conversion factor $= 1 \frac{kg \cdot (m/s^2)}{N} = 32.17 \frac{lb \cdot (ft/s^2)}{lbf}$

M = mass of the flywheel (kg, lb)

R = radius of gyration (m, ft). In the case of a disc of uniform density, with uniform thickness and outer radius R_o, $R = 0.7 \cdot R_o$

Ω = revolution speed (rad/s); typical values (200-1000 rad/s)

The energy absorbed or released from a flywheel between two rotational speeds (Ω_1 and Ω_2) is

$$\Delta E = \frac{1}{2} \cdot \frac{1}{g_c} \cdot M \cdot R^2 \cdot (\Omega_1^2 - \Omega_2^2)$$

The specific energy referred to the mass of the flywheel depends on the revolution speed, on the stress-to-density ratio and on the geometry of the flywheel. Typical maximum values for high-speed composite-material flywheels are 100 kJ/kg; lower values such as 20-30 kJ/kg are generally reached for low-speed flywheels with isotropic materials and constant-stress disc geometry.

Electric storage

Electric energy can be stored directly by using electric batteries, among which the lead-acid type is the commonest.

The use of electric batteries for storage depends on many parameters such as charge and discharge cycle life, energy-mass and power-mass ratios and of course energy-cost ratio.

Table 4.2 lists typical operating parameters of industrial batteries such as lead-acid, nickel-iron, nickel-cadmium, sodium-sulfur, zinc-bromine.

The specific energy values range between 72-108 kJ/kg (20-30 Wh/kg) for lead-acid batteries and 360 kJ/kg (100 Wh/kg) for sodium-sulfur batteries; the specific power values range between 100 W/kg for lead-acid batteries and 200 W/kg for zinc-bromine batteries; the cycle life does not exceed 400 cycles for lead-acid batteries against 1000 cycles for nickel-cadmium batteries. The charge-discharge efficiency can reach 60-70% depending on the operating conditions and on the recharge equipment.

These values although higher than those of flywheel storage, are still remote from the energy storage capacity of combustibles, thus limiting this kind of storage to particular applications for short periods of demand.

In addition to industrial applications, electric storage affects electric vehicle diffusion. With typical consumption per kilometer and per metric ton of 490 kJ/t · km (0.15 kWh/t · km) if measured at the utility delivering node, a wide utilisation of these vehicles is strictly dependent on the availability of high-performance electric storage systems.

With an equal division of the gross weight among payload, vehicle body and traction equipment, and batteries, the urban range is about 70-80 km; this range could be improved by using higher energy-density batteries and by reducing vehicle weight through a proper design of the body and of the traction equipment.

Fuel cells, operating on non-petroleum fuels such as hydrogen, might provide an alternative energy source for electric traction and for other applications.

Table 4.2 Operating parameters of various batteries

	Specific energy Wh/kg	Specific power W/kg	Cycle life to 80% cycles
Lead-acid	35	80	400
Nickel/Iron	50	100	>1000
Nickel/Zinc	60	100	250
Zinc/Bromine	50	90	200
Nickel/Cadmium	25	350	>1000
Lithium Aluminium/ Iron sulfide	80	70	200
Sodium/Sulfur	100	120	250
Iron/Air	70	90	150

Heat and cold storage

Heat and cold storage can be classified as sensible and latent energy storage.

The first group includes systems where the storage is accomplished by increasing or decreasing the temperature of the material (water, organic liquid, solid); the storage energy density depends on the temperature change and on the specific heat of the material. The evolution of the temperature transient to reach the desired temperature is roughly exponential and the time constant, that is the time needed to reach 63.2% of the final temperature in a simplified model, depends on the product of the heat capacity of the mass involved multiplied by the thermal resistance (see Chapter 8.3) of the system.

The energy stored as sensible heat in the material at steady state, when the temperature transient has been completed, is:

$$E_{stored} = c \cdot M \cdot (t_2 - t_1)$$

where

E_{stored} = energy stored (J, Btu)
c = specific heat of the material to be heated (kJ/kg · K, Btu/lb · °F)
M = mass to be heated (kg, lb)
$c \cdot M$ = heat capacity of the mass to be heated (kJ/K, Btu/°F)
t_1, t_2 = temperatures of the mass to be heated or cooled (K, °C, °F).

If mineral oil is used for heat storage, the storage energy density associated with a temperature increase of 50 K (50°C; 83.8°F) is roughly 100 kJ/kg; the storage energy density with water for the same temperature increase is roughly 200 kJ/kg.
If water is used for cold storage, a temperature drop of 4-5°C (7.2-9°F) is usually obtained, so that the storage energy is 20 kJ/kg.

The second group includes systems where the energy is mostly stored in the form of the latent heat due to a phase change, such as melting a solid (the opposite in cold storage) or vaporizing a liquid. In releasing energy, liquids solidify (or solids liquefy) and vapors condense. The storage energy density per kg, which derives mainly from the latent heat, is greater than in the sensible energy storage systems. An additional advantage of this group is that it works at constant temperature during the phase change.

In addition to proper transition temperatures and high latent heat, materials for storage purposes must possess other physical and chemical properties such as thermal conductivity, stability, nontoxicity in the operating conditions. Eutectic salts such as NaF-FeF_2 and $ZnCl_2$ with a density of roughly 2000 kg/m^3, may reach energy densities of 400-1500 MJ/m^3 or 200-750 kJ/kg ($10.7 \cdot 10^3$ - $40.3 \cdot 10^3$ Btu/ft^3, 90-340 Btu/lb) with an overall efficiency of not more than 50%. Capital and operating costs are quite high and do not allow a wide exploitation of this system in industry.

If ice is used for cold storage, an average storage capacity is reached of 350 kJ/kg, of which 335 kJ/kg comes from the latent heat and 15 kJ/kg from the sensible heat. The electric energy required for producing the ice depends on the COP of the refrigerating system; if COP equals 3 (which can be considered as a typical value), the consumption is 0.032 kWh/kg of ice produced in the storage system.

4.8 APPLICABILITY IN INDUSTRY

The factory demand for both electricity and thermal energy is quite often satisfied by means of utility plants and related distribution lines. Generally, no alternative source of thermal energy is available. On the other hand, electric energy can be produced inside the single factory from either traditional fuels or refuse-derived fuels.

Cogeneration plants and waste-recovery plants seem to be the most attractive means of reducing primary energy consumption and the cost of the energy supply, but the capital and operating costs must be investigated together with technical aspects such as the quality, reliability and flexibility of the energy production.

Renewable sources can provide significant support only in particular conditions. Nevertheless, energy management programs must always verify these possibilities.

CHAPTER 5

ELECTRICAL SUBSTATIONS

5.1 INTRODUCTION

Electric energy is commonly delivered to industrial users by three-phase networks at medium or high voltage. The values of medium and high voltage are defined by international standards; broadly, medium voltages range from 5 to 25 kV and high voltages range from 50 to 500 kV. Higher and lower values can be found according to local conditions.

Individual industrial users have to cope with the reduction from the levels indicated above to lower voltages inside factories. Typical low voltage values range from 260 to 380 V.

Transformers at factory boundaries and related electrical equipment constitute the electrical substations, through which flows all the purchased electric energy, and they are responsible for transformation losses, as are boiler plants with thermal energy (see Chapter 6).

Although the transformer efficiency is quite high (generally not less than 96%) it is worth reducing these losses as much as possible because their quantity is related to the total electric energy consumed inside the factory.

5.2 BASIC PRINCIPLES OF TRANSFORMER LOSSES AND EFFICIENCY

Transformer losses may be no-load or load losses:

- no-load losses are almost exclusively iron losses. They are a squared function of voltage and they occur whenever the transformer is connected to a voltage source, independently of the load value;
- load losses are a squared function of current (or of apparent power load).They are mainly copper losses and they occur only when a load is connected to the transformer, through which a current is flowing.

The electric machine theory shows that the maximum efficiency is reached at a load level at which no-load and load losses are equal:

$$\text{Efficiency} = \frac{\text{Total output power}}{\text{Total input power}} = \frac{\text{Total output power}}{(\text{Total output power} + \text{Total losses})}$$

It follows that:

$$\text{Efficiency} = \frac{x \cdot P_n}{(x \cdot P_n + P_o + x^2 \cdot P_{cn})}$$

where

x = load factor = ratio between total actual output and rated output power
P_o = no-load losses (kW)
P_{cn} = load losses at rated power (kW)
Total losses = $P_o + P_{cn} \cdot x^2$ (kW)
$r = P_{cn}/P_o$
P_n = rated output power (kW at fixed power factor; $P_n = A_n \cdot \cos \varphi_n$)
Total actual output = $x \cdot P_n$ (kW)

Maximum transformer efficiency is reached if:

$$x^2 \cdot P_{cn} = P_o \qquad x^2 = \frac{P_o}{P_{cn}} \qquad x = \sqrt{1/r}$$

Then, the load value at which maximum efficiency can be reached is a function of the ratio r between load losses at rated power and no-load losses, which depends on the design parameters of the transformer. Let us consider a group of transformers with the same rated power and total losses, but with different values of ratio r; it follows that the maximum efficiency for each transformer corresponds to a specific load factor (x), that is, to an operating point where load losses ($x^2 \cdot P_{cn}$) equal no-load losses (P_o).

This means that maximum efficiency is reached if, as r increases, the load factor x decreases. Figure 5.1 shows transformer efficiency versus load factor for different values of ratio r for such a group of transformers. The dotted line represents maximum efficiency operating points. Notice that maximum efficiency at

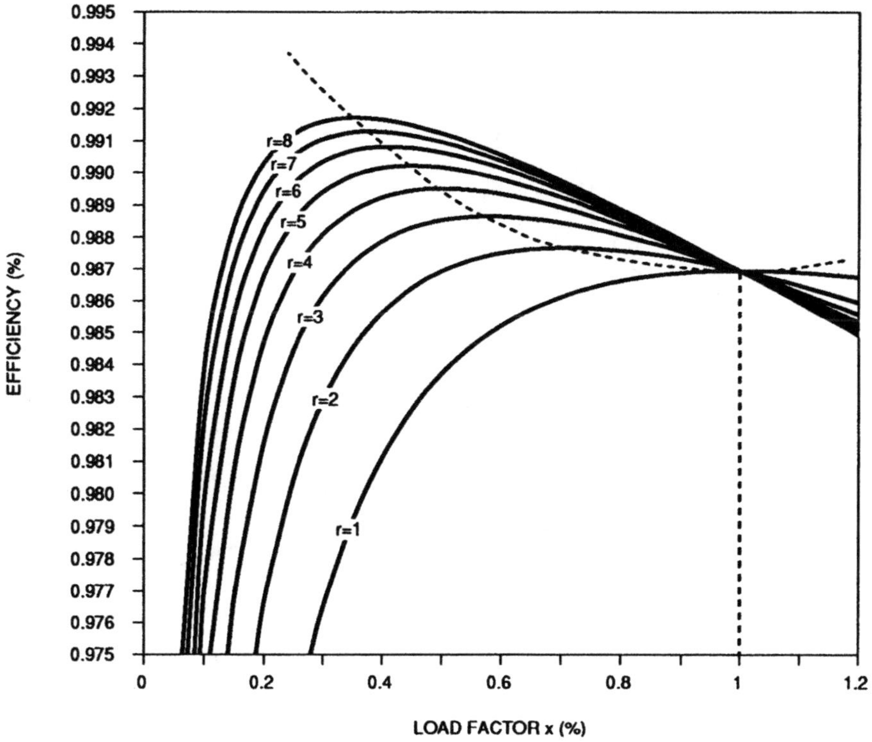

Fig. 5.1 Transformer efficiency versus load factor x for a group of transformers with the same rated power and rated total losses, but with different values of r ratio

rated power (x=1) is reached only if the transformer ratio r equals 1. If r is greater than 1, which is a typical design criterion, the maximum efficiency can be reached only when the transformer is partially loaded.

Typical values for industrial applications are: r equal to 5-7; maximum efficiency equal to 98.5-99%; load factor equal to 35-55%. This means that typical industrial transformers must be loaded well below nameplate values to obtain good efficiency and oversized machines must be installed.

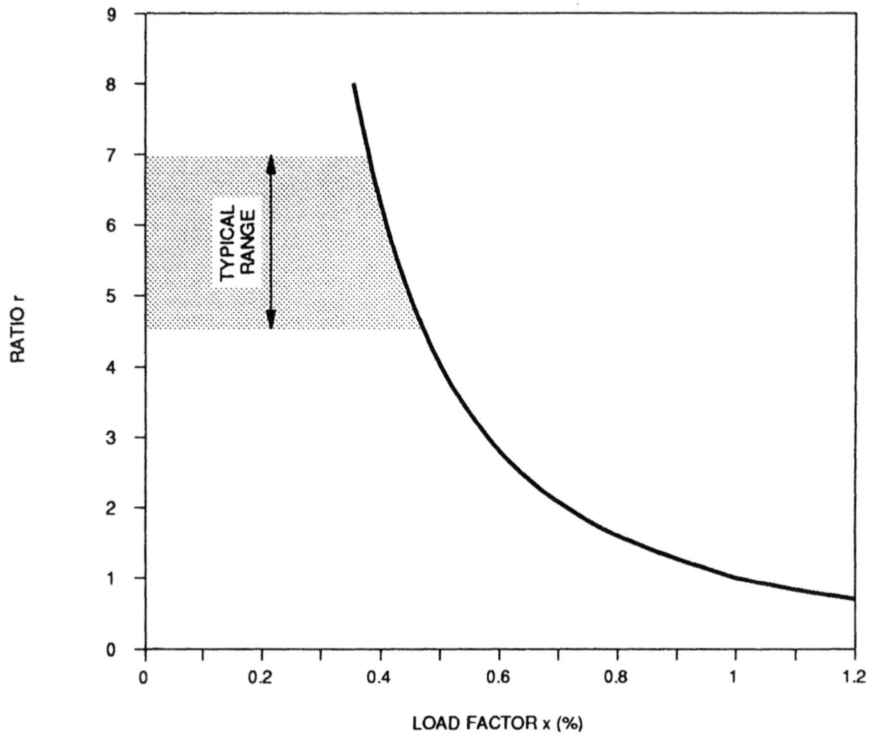

Fig. 5.2 Ratio r versus load factor x if the maximum transformer efficiency condition is imposed

As a general energy-management rule it is not advisable to allow transformers to operate constantly at full load.

Fig. 5.2 shows the ratio r versus load factor when the maximum efficiency condition is imposed. Typical values for load factor x (35-55%) and for ratio r (4-7) are also indicated.

5.3 HOW TO CHOOSE THE MOST SUITABLE TRANSFORMER

Transformer selection is based on many parameters such as capital cost, operating and maintenance costs, energy losses costs, predictable overload conditions, network short-circuit duties, large-motor starting, etc..

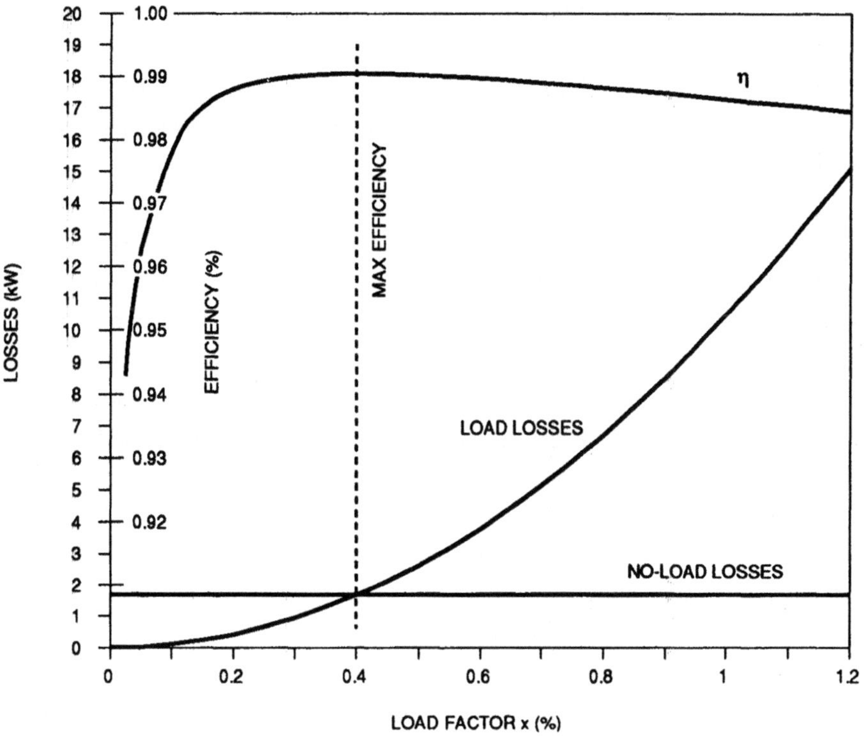

Fig. 5.3 Operating parameters for a standard 1000 kVA transformer, as specified in Table 5.2

Energy losses cost is one of the evaluation parameters although not the most crucial, but if the nameplate power is selected correctly, considering also energy saving, worthwhile economies should be obtained.

Firstly, it is necessary to assume as reference the average load during a prefixed period, i.e. the load to which correspond the same losses in the real or predicted cyclic conditions of operation. For a low average load, a transformer with a higher r ratio value should be selected. Finally, provided that the other selection parameters are satisfied, the nameplate power needed for maximum efficiency can be determined, taking account of the load average power value.

Table 5.1 lists standard transformer parameters. Notice that standard type transformers with standard losses are to be preferred; only for very large machines or for special applications should non-standard machines be considered.

Table 5.2 and Fig. 5.3 show the influence of no-load and load losses when the load is changing and the value of r ratio is defined. The point of balance between losses corresponds to the maximum efficiency.

Economic evaluations for transformer selection based on energy saving can be made as explained in Chapter 20. As a general rule, notice that in normal conditions the annual-losses cost is equal to roughly 10-30% of the capital cost of the transformer itself. It is worth remembering that in sizing transformers attention must be paid to correlated costs, such as control and protection switchgear, voltages level, cables, foundations, space occupied, etc., which are approximately equal to the cost of the transformer itself so that the incidence of annual losses is halved. Oversizing transformers may thus be more expensive with respect to switchgear and cables shortcircuit rating than with respect to the capital cost of the transformer itself. Energy optimization reduces the annual-losses cost but does not change substantially its incidence on the transformer and related equipments capital cost, which still remains quite considerable.

Roughly speaking, a 10% reduction of losses, to which corresponds an increase of 1% in transformer efficiency, should justify more than 10-15% of additional capital costs.

Table 5.1 Standard parameters for oil-insulated transformer MT/BT (normal and reduced losses)

Normal losses					
Rated Power A_n (kVA)	No-load losses P_o(W)	Load losses P_{cn}(W)	$r = P_{cn}/P_o$	No-load current %	Vcc %
50	190	1100	5,8	2,9	4
100	320	1750	5,5	2,5	4
160	460	2350	5,1	2,3	4
250	650	3250	5,0	2,1	4
400	930	4600	4,9	1,9	4
600	1300	6500	5,0	1,8	4
1000	1700	10500	6,2	1,5	6
1600	2600	17000	6,5	1,3	6
2000	3200	22000	6,9	1,2	6
2500	3800	26500	7,0	1,1	6
Reduced losses					
Rated Power A_n (kVA)	No-load losses P_o(W)	Load losses P_{cn}(W)	$r = P_{cn}/P_o$	No-load current %	V_{cc} %
50	150	850	5,7	1,9	4
100	250	1400	5,6	1,5	4
160	360	1850	5,1	1,3	4
250	520	2600	5,0	1,1	4
400	740	3650	4,9	0,9	4
630	1040	5200	5,0	0,8	4
1000	1300	9000	6,9	0,7	6
1600	2000	13000	6,5	0,5	6
2000	2400	16000	6,7	0,5	6
2500	2900	21000	7,2	0,5	6

Table 5.2 No-load and load losses for a normal-losses transformer at different load factor values (x ranges from 1.2 to 0.1)

Rated power 1000 kVA; power factor equal to 0.9; r = 10500/1700 = 6.2			
Load factor %	No-load losses P_o (W)	Load losses P_{cc} (W)	Efficiency %
120	1700	15120	98.47%
110	1700	12705	98.57%
100	1700	10500	98.66%
90	1700	8505	98.76%
80	1700	6720	98.84%
70	1700	5145	98.93%
60	1700	3780	99.00%
50	1700	2625	99.05%
40	1700	1680	99.07%
30	1700	945	99.03%
20	1700	420	98.84%
10	1700	105	98.03%

Table 5.3 Efficiency of different normal-losses transformers at the same load power

Average load power 400 kW; equal to power factor 0.85					
Transformer rated power A_n (kVA)	Ratio r	Load factor x %	No-load losses P_o (W)	Load losses P_{cc} (W)	Efficiency %
400	4,9	117.65%	930	6367	97.62%
600	5	78.43%	1300	3998	99.07%
1000	6,2	47.06%	1700	2325	99.45%
1600	6,5	29.41%	2600	1470	99.32%

5.4 PRACTICAL EXAMPLES

In what follows an example is given of choice based on energy optimization.

Example 1

Sizing of transformers based on maximum efficiency operation all the year round

Input data

— load average power 400 kW
— load average power factor 0.85 leading
— operating hours at full load 3200 h/year
— operating hours at no-load 5560 h/year
— oil-insulated transformers with technical characteristics as detailed in Table 5.1
— the load is supplied through only one transformer

Questions

1) Find the transformer rated power A_n (kVA) that gives the maximum efficiency at full-load operation;

2) calculate the efficiency that corresponds to differently rated power transformers with the same 400 kW load;

3) calculate the losses that correspond to different transformers at 400 kW load operation and the energy saved if the maximum efficiency operation is chosen instead of the minimum efficiency operation transformer.

Answers

1) 2) By reference to Table 5.1 it is possible to calculate the efficiency that corresponds to 400 kW load power at power factor 0.85 for differently rated power transformers:

$$x = \text{load factor} = \frac{400\,\text{kW}}{\text{kVA Rated power} \cdot 0.85}$$

$$\text{Efficiency} = \frac{x \cdot \text{kVA Rated power} \cdot 0.85}{x \cdot \text{kVA Rated power} \cdot 0.85 + P_o + P_{cn} \cdot x^2}$$

Table 5.3 gives the results of the calculations.

Notice that maximum efficiency corresponds to a transformer with a rated power equal to 1000 kVA.
The corresponding load factor is equal to 47.06% and the maximum efficiency is equal to 99.45%. The minimum efficiency, equal to 97.62%, corresponds to a transformer with a rated power equal to 400 kVA (rated powers less than this value are not considered, to avoid any thermal problem).

3) Table 5.4 shows annual energy losses (kWh) for different transformers at 400 kW load operation (see Table 5.3 for values of P_o and P_{cc}).

Table 5.4 Total annual losses of different transformers (as in Table 5.3)

No-load hours 5560 h/year; load hours 3200 h/year				
Average load power (kW)		400	power factor	0.85
Transformer rated power kVA	Annual losses at no-load operation kWh/year	Annual losses at average load power kWh/year	Total annual losses kWh/year	ANNUAL ENERGY SAVING kWh/year
400	5171	23350	28521	
600	7228	16954	24182	
1000	9452	12880	22332	
1600	14456	13024	27480	
(a)	(b) = $P_o \cdot 5560$ see Table 5.3	(c) = $(P_o + P_{cc}) \cdot 3200$ see Table 5.3	(d) = (b) + (c)	
case (400) - case (1000)				6189

Example 2
For a given rated power, choose between a reduced-losses transformer and a normal-losses transformer

Calculate the annual energy losses of the two transformers and the annual energy saving.

Input data

— load average power 400 kW
— load average power factor 0.85 leading
— transformer rated power 1000 kVA

— operating hours at full load 3200 h/year
— operating hours at no-load 5560 h/year
— reduced-losses and normal-losses transformers with technical characteristics as detailed in Table 5.1
— the load is supplied through only one transformer

Table 5.5 shows the results of the calculations. The annual energy saving is 4566 kWh.

The economic evaluation is given in Table 20.3.

Table 5.5 Comparison between total annual losses of two 1000 kVA transformers (a reduced-losses and a normal-losses transformer)

Normal-losses transformer		1700 W(P_o, Table 5.3)		10500 W(P_{cn}, Table 5.3)	
Type of operation	Working hours h/year	Load factor x %	Annual no-load losses kWh/year	Annual load losses kWh/year	Total annual losses kWh/year
No-load	5560	0.00%	9452	0	9452
Load	3200	47.06%	5440	7441	12880
TOTAL (a)	8760				22332
Reduced-losses transformer		1300 W(P_o, Table 5.3)		9000 W(P_{cn}, Table 5.3)	
Type of operation	Working hours h/year	Load factor x %	Annual no-load losses kWh/year	Annual load losses kWh/year	Total annual losses kWh/year
No-load	5560	0.00%	7228	0	7228
Load	3200	47.06%	4160	6378	10538
TOTAL (b)	8760				17766
ENERGY SAVING (a) – (b)				kWh/year	4566

CHAPTER 6

BOILER PLANTS

6.1 INTRODUCTION

Most of the thermal energy that crosses the factory boundaries as fuel (solid, liquid, natural gas) is transferred, by means of combustion inside the boiler plants, to intermediate fluids such as steam, oil, air, water and other fluids which have the task of carrying energy to the end users at the lowest possible cost and at maximum efficiency.

Alternatively, two boiler classes, generally called unfired boilers because they do not require fuel consumption, can be introduced: heat recovery and electric/electrode boilers.

If natural gas is used, a station where pressure is reduced could be necessary upstream of the boiler plant, depending on local gas pipeline conditions and burner requirements. Losses due to this transformation are generally low.

End users such as furnaces, ovens, dryers, kilns, etc. generally burn fuel directly without any upstream transformation.

Industrial-boiler plant efficiency generally ranges from 80% to 90%; this means that not less than 10-20% of the total fuel purchased is lost in the boundary zones of the factory before distribution to the end users. This is a considerable quantity, and it makes the boiler plant one of the most dissipative facilities on an industrial site; therefore attention must be paid to this plant with regard to both capital investment and maintenance efficiency, in order to minimize losses.

One of the basic tasks of industrial energy management is undoubtedly to optimize the operation of facilities which are responsible for the greatest part of the energy losses between factory boundaries and end users. Boiler plants are one of the most important facilities because of the great quantity of energy which generally flows through them. Any capital investments in boilers and related equipment such as combustion control, water treatment, air economizer, etc. must be carefully studied with the aid, if necessary, of detailed engineering analyses, to evaluate the return on investments and the benefits to be obtained from both energy-saving and air-quality improvements.

6.2 BASIC PRINCIPLES OF COMBUSTION

Combustion is an exothermic reaction of carbon and hydrogen with oxygen to produce carbon dioxide (CO_2) and water vapor (H_2O).

Combustion starts when the ignition temperature (see Table 6.1) is reached and ends when one element of the reaction is exhausted. This means that incomplete combustion will occur if the quantity of oxygen available in the combustion zone is insufficient for the reaction of all fuel molecula. In this case, the combustion will produce effluents which still contain combustibles, and the heat delivered will be less than expected. Secondary combustion products such as NO_x, SO_x, CO and solid particles are also released. Their influence on air pollution must be considered carefully.

Table 6.1 Ignition temperatures of different combustibles

Combustible	Ignition temperature		
	K	°C	°F
Dry wood	523	250	482
Coal	823	550	1022
Oil	603	330	626
Gasoline	753	480	896
Natural gas	918	645	1193
Carbon monoxide	883	610	1130

The main reactions for each element are outlined below:

— CARBON

complete reaction

$$C + O_2 = CO_2 + 34032 \frac{kJ}{kg\ Carbon}$$

if this reaction is not completed

$$C + \frac{1}{2} O_2 = CO + 10465 \frac{kJ}{kg\ Carbon}$$

$$CO + \frac{1}{2} O_2 = CO_2 + 10172 \frac{kJ}{kg\ CO}$$

— HYDROGEN

$$H_2 + \frac{1}{2} O_2 = H_2O\ \text{liquid} + 141486.8 \frac{kJ}{kg\ Hydrogen}$$

$$H_2 + \frac{1}{2} O_2 = H_2O\ \text{vapor} + [141486.8 - (9 \cdot 2511.6)] \frac{kJ}{kg\ Hydrogen} =$$

$$H_2O\ \text{vapor} + 118882.4 \frac{kJ}{kg\ Hydrogen}$$

(combustion of 1 kg of hydrogen produces 9 kg of liquid water which can be vaporized by roughly 2511.6 kJ/kg)

— SULFUR

$$S + O_2 = SO_2 + 10883.6 \frac{kJ}{kg\ Sulfur}$$

Most of the oxygen used for combustion comes from atmospheric air: in volume 21% of oxygen (O_2), 78% of nitrogen and traces of argon and carbon dioxide can be assumed as typical values. Notice that because of the nitrogen in the air, NO_x emissions always occur, depending mainly on flame temperatures.

Emission control in order to minimize the formation of NO_x, as well as of CO and SO_2, is a task of boiler-plant maintenance that must be performed with the same care as energy saving. Quite often the two problems are closely correlated and they can be solved together with considerable economies if they are tackled at the same time.

As shown in the reactions described above, water always forms through the combination of hydrogen and oxygen. The water may be in the liquid or the vapor state, so two different calorific heating values (gross and net values, otherwise named higher and lower heating values) are introduced (see Chapter 2.2). In industrial applications, effluents flow from the boiler when they are still at a temperature of more than 373.15 K (100°C; 212°F), so that the water is always in the vapor state.

The theoretical amount of air needed for complete combustion can be calculated as follows:

— the mass M_o (kg) of oxygen which realizes the complete combustion of 1 kg of solid or liquid combustibles is determined by their composition (mainly C%; H_2%; S%; O_2%). The composition of different combustibles is shown in Table 6.2;

— the related oxygen volume V_o is equal to M_o multiplied by the oxygen specific volume in standard conditions, that is:

$$V_o = M_o \cdot \frac{22.4}{\text{oxygen molecular weight}} = M_o \cdot \frac{22.4}{32} \qquad \left(\frac{\text{Sm}^3}{\text{kg comb}}\right)$$

where
22.4 m^3/kmol is the standard volume of 1 kmol of any gas at 273.15 K (0°C; 32°F) and 0.1 MPa (1.013 bar)
32 kg/kmol is the oxygen molecular weight;

— the theoretical amount of air, including the standard proportion of oxygen, already given as 21%, needed for the complete combustion of 1 kg of a solid or liquid combustible is:

$$V_{th} = \frac{100}{21} \cdot M_o \cdot \frac{22.4}{32} \qquad \left(\frac{\text{Sm}^3}{\text{kg comb}}\right)$$

The same procedure can be applied with gaseous combustibles.

Table 6.2 Solid, liquid and gaseous combustibles and combustion parameters

Fuel	Density	Net heating value	Oxygen for combustion	Air combustion volume			Exhausts at		
				Theoretical value	Practical excess air	Practical value	273 K / 0 °C / 32 °F	493 K / 220 °C / 428 °F	523 K / 250 °C / 482 °F
SOLID	kg/m^3	kJ/kg	Sm^3	Sm^3	%	Sm^3	Sm^3	m^3	m^3
Pitch lignite	650	18000	1.47	7	100.00%	14	15	27.1	28.8
Standard coal	700	29300	1.26	6	80.00%	11	12	21.7	22.9
Charcoal	750	31400	1.89	9	80.00%	16	17	30.7	32.6
Gas coke	400	26800	1.68	8	80.00%	15	16	28.9	30.6
(All values are referred to 1 kg of combustible with average values of C, N_2, S)									
LIQUID	kg/m^3	kJ/kg	Sm^3	Sm^3	%	Sm^3	Sm^3	m^3	m^3
Refined kerosene	800	43100	2.52	12	20.00%	14.5	15	27.0	29
Gasoil	825	42700	2.478	11.8	30.00%	15.3	16.2	29.3	31
Fuel oil No. 2	860	39000	2.352	10.3	40.00%	14.4	15	27	29
Fuel oil No. 6	1000	43000	2.352	11.2	40.00%	15.7	16.6	30.0	31.8
(All values are referred to 1 kg of combustible with average values of C, N_2, S)									
GASEOUS	$kg/1000 \cdot Sm^3$	kJ/kg	Sm^3	Sm^3	%	Sm^3	Sm^3	m^3	m^3
Natural gas	750	34325	1.995	9.5	20.00%	11.4	12.2	22.03	23.37
(All values are referred to 1 Sm^3 of combustible with average values of C, N_2)									

Notice that, practically, the amount of air needed to ensure complete combustion is higher than the theoretical one. If the real value is V, the ratio between $(V - V_{th})$ and V_{th} is called excess air. It is generally expressed as a percentage. Practical values range from 10% to 100% depending on the type of combustible (gaseous, liquid, solid).
Practical values of air combustion are: 10.5-11.5 Sm^3 for the combustion of 1 Sm^3 of natural gas; 12-15 Sm^3 for the combustion of 1 kg of oil.

6.3 COMBUSTION EFFICIENCY

It is customary to express combustion efficiency as the ratio of heat delivered by the combustion reactions to the fuel input.

Theoretical figures of heat delivered have already been expressed as calorific heating values; actual values may be lower because of lack or excess of oxygen in the combustion zone. To ensure complete combustion, fuels are always burned in excess of air and then of oxygen.

Excess air causes excess of flue gas from the combustion zone, so that the real is always lower than the theoretical combustion efficiency value, heat being spent on excess air. In spite of excess air, flue gas always contains some unburned combustibles and so efficiency is further reduced.

Another loss is the heat spent on evaporating and superheating the moisture in the fuel that enters the combustion zone, and on superheating the moisture in the combustion air. As shown in Chapter 6.2, moisture also comes from burning hydrogen in the fuel, and the heat spent on evaporating and superheating this is an additional loss.

Table 6.2 shows typical values of parameters able to qualify the combustion of different fuels (heating values, O_2, excess air, exhausts).

In conclusion, combustion efficiency can be improved by keeping the excess air as well as the unburned combustible content as low as possible. That seems to be contradictory, but valuable results can be obtained in this way by optimizing the design, operation and maintenance of burners and related auxiliaries, as well as of combustion control equipment.

Combustion losses are generally included among boiler losses, as discussed in Chapter 6.7. Notice that the main contribution to these losses is the heat going up the stack, which is closely related to both the excess air required by combustion and the temperature of the fluids required by the end users.

6.4 FUNDAMENTALS OF STEAM GENERATION

If boilers are used to produce steam, which is the most frequent case, the heat generated by combustion must be transferred to the water to allow it to reach boiling point and higher temperatures.

There are successive phases:

— heating the water.

The heat required to bring 1 kg of water from the reference temperature 273.15 K (0°C; 32°F) to boiling point is the enthalpy (h_i) or heat content of the liquid. The enthalpy increases with the temperature of the liquid and then with its pressure.

$$h_i = c \cdot (t - t_R)$$

where

c = specific heat of water = $4.186 \dfrac{kJ}{kg \cdot K} = 1 \dfrac{kcal}{kg \cdot °C} = 1 \dfrac{Btu}{lb \cdot °F}$

t = boiling point temperature (373.15 K, 100°C, 212°F)

t_R = reference temperature (273.15 K; 0°C; 32°F)

At atmospheric pressure at sea level:

$$h_i = 418.6 \frac{kJ}{kg} = 100 \frac{kcal}{kg} = 180 \frac{Btu}{lb}$$

If the heating starts at a temperature t_o above 273.15 K (0°C; 32°F), the heat needed to reach boiling point is reduced as follows:

$$h_i = c \cdot (t - t_o)$$

Liquid water can also be heated to over 373.15 K if the pressure inside the water circuit is kept higher than that of the atmosphere.

— Vaporization.

When the water boils and steam starts to be produced, they are both at the same temperature, that is the saturation temperature which has only one value for each boiling pressure. During this phase, the temperature remains constant and the heat converts water into steam. This heat is the evaporation enthalpy (h_v) which decreases as temperature and pressure increase.

Thus:

$$\text{enthalpy of saturated steam} = h_i + h_v$$

If saturation occurs at atmospheric pressure, the total amount of heat added to convert water at 273.15 K (0°C; 32°F) to 100% steam is equal to:

$$\text{enthalpy of saturated steam} = 418.6 + 2252 = 2670.6 \frac{\text{kJ}}{\text{kg}} = 638 \frac{\text{kcal}}{\text{kg}} = 1149 \frac{\text{Btu}}{\text{lb}}$$

When a mixture of liquid and vapor is in equilibrium, the liquid phase is a saturated liquid and the vapor phase is a saturated vapor.

This mixture can be classified by the quality index x which is the ratio between the mass of vapor present in the mixture and the total mass of the mixture itself.

The value of the quality index x ranges from zero to unity: at saturated liquid state $x = 0$, and at saturated vapor state $x = 1$. The quality index is frequently expressed as a percentage.

For saturated steam, typical values of enthalpy range from 2670 to 2770 kJ/kg (from 662 to 638 kcal/kg; from 1149 to 1192 Btu/lb).

— Superheating.

When the water is converted into 100% steam, the temperature will rise if heat is added. The enthalpy of the steam increases by the amount of heat added which can be calculated by multiplying the specific heat of the steam (average value at typical operating pressures is 2.1 kJ/kg · K, see Table 2.5) by the temperature increase above the boiling point. For a given pressure, the temperature depends on the heat added which is called the enthalpy of superheating (h_s). At a given pressure, any temperature above the saturation value can be reached.

Thus:

$$\text{enthalpy of superheated steam} = h_i + h_v + h_s$$

— Condensation.

During this phase, which is the opposite of the superheating and vaporization phases, vapor is transformed into liquid water by the extraction of heat. At a given pressure, the superheated steam temperature decreases until the saturation temperature is reached; then, liquid water and steam remain at constant temperature until 100% liquid water is produced. The temperature depends on the pressure.

Notice that the heat to be extracted is equal to the enthalpy of the evaporation phase at the same pressure and temperature.

Fig. 6.1, Fig. 6.2, Table 6.3 and Table 6.4 show steam properties which are generally obtainable from a Mollier chart.

6.5 INDUSTRIAL BOILERS

Boilers can be considered as containers into which water or other fluids are fed and there heated continuously by the application of the heat released during combustion. More frequently water is evaporated continuously into steam and the state of the fluid is then changed.

Industrial boilers can be classified as either watertube or firetube depending on the position of the hot combustion gases in relation to the fluid being heated (water, oil). Oil can be used as the heating medium both to carry thermal energy directly to the end users, and to produce steam by means of a heat exchanger acting as an evaporator inside or outside the boiler plant.

So-called unfired boilers such as heat recovery boilers and electric or electrode boilers, can also be installed, depending on technical and economic factors, in order to transfer thermal energy to end users without combustion inside the boiler itself.

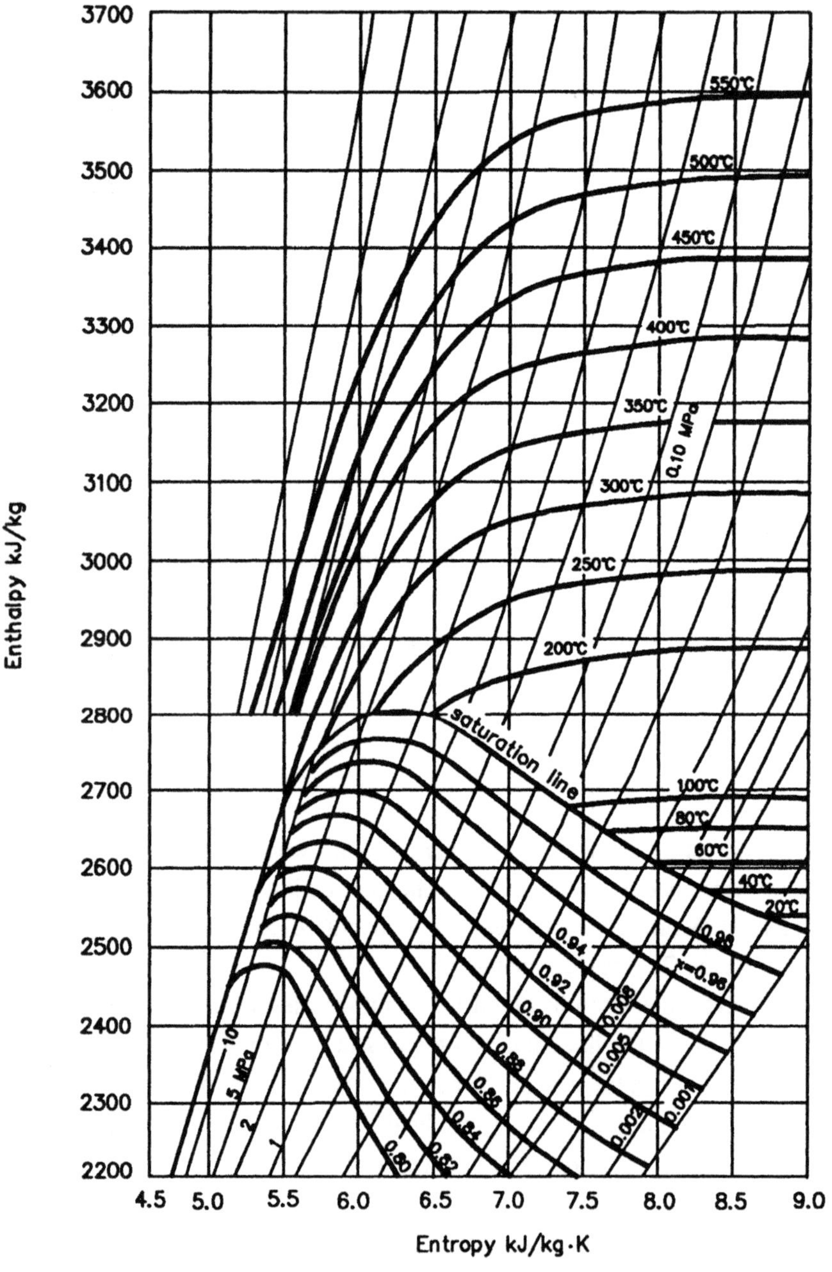

Fig. 6.1 Enthalpy-entropy diagram for water (SI units)

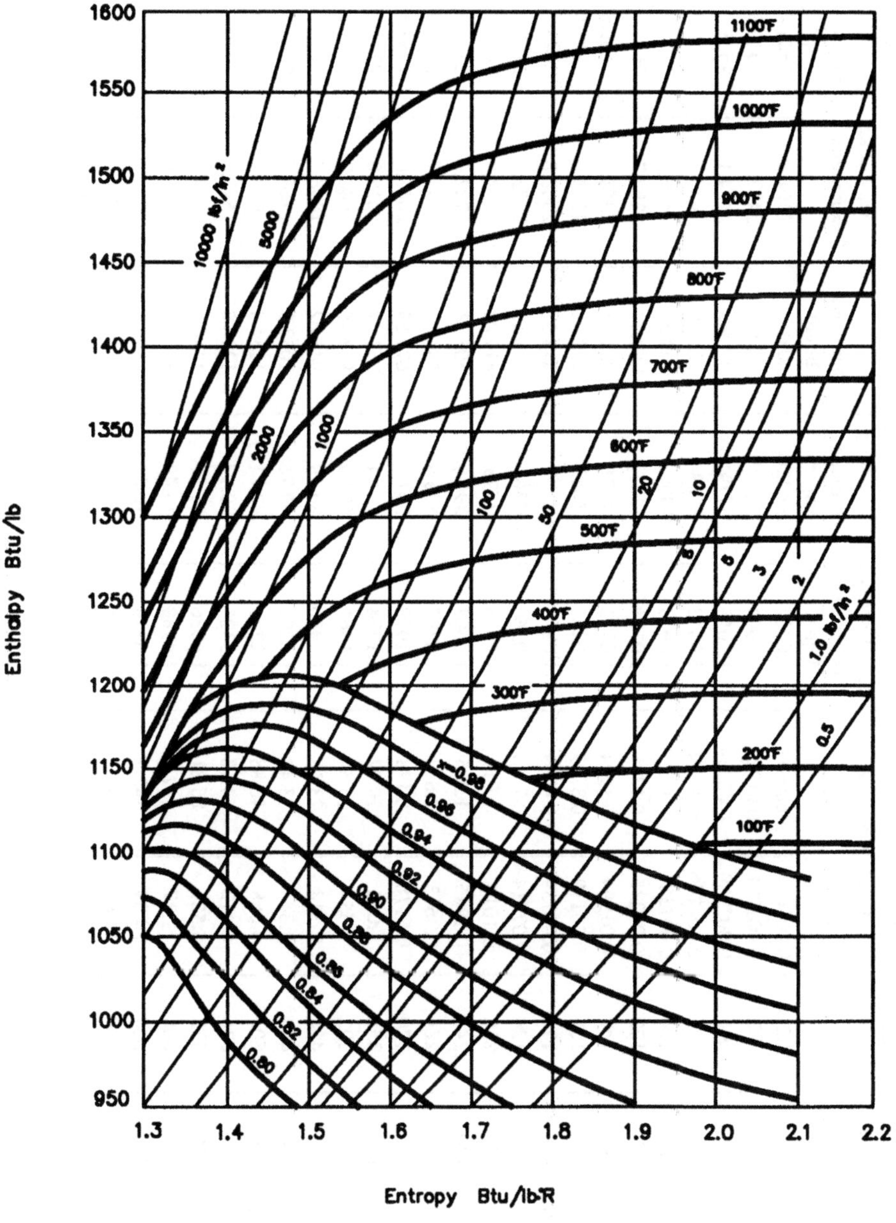

Fig. 6.2 Enthalpy-entropy diagram for water (English units)

Table 6.3 Main properties of saturated water (liquid-vapor)

Temperature			Absolute pressure		Specific volume m^3/1000 kg		Enthalpy kJ/kg		
K	°C	°F	MPa	bar	Saturated liquid	Saturated vapor	Saturated liquid	Evaporation	Saturated vapor
273.16	0.01	32.018	0.000611	0.00611	1.0002	206136	0.01	2501.3	2501.4
283.15	10	50	0.001228	0.01228	1.0004	106379	42.01	2477.7	2519.8
293.15	20	68	0.002339	0.02339	1.0018	57791	83.96	2454.1	2538.1
303.15	30	86	0.004246	0.04246	1.0043	32894	125.79	2430.5	2556.3
313.15	40	104	0.007384	0.07384	1.0078	19523	167.57	2406.7	2574.3
323.15	50	122	0.01235	0.1235	1.0121	12032	209.33	2382.7	2592.1
333.15	60	140	0.01994	0.1994	1.0172	7671	251.13	2358.5	2609.6
343.15	70	158	0.03119	0.3119	1.0228	5042	292.98	2333.8	2626.8
353.15	80	176	0.04739	0.4739	1.0291	3407	334.91	2308.8	2643.7
363.15	90	194	0.07014	0.7014	1.0360	2361	376.92	2283.2	2660.1
373.15	100	212	0.1014	1.014	1.0435	1673	419.04	2257.0	2676.1
393.15	120	248	0.1985	1.985	1.0603	891.9	503.71	2202.6	2706.3
413.15	140	284	0.3613	3.613	1.0797	508.9	589.13	2144.7	2733.9
433.15	160	320	0.6178	6.178	1.1020	307.1	675.55	2082.6	2758.1
453.15	180	356	1.002	10.02	1.1274	194.1	763.22	2015.0	2778.2
473.15	200	392	1.554	15.54	1.1565	127.4	852.45	1940.7	2793.2
493.15	220	428	2.318	23.18	1.1900	86.19	943.62	1858.5	2802.1
513.15	240	464	3.344	33.44	1.2291	59.76	1037.3	1766.5	2803.8
533.15	260	500	4.688	46.88	1.2755	42.21	1134.4	1662.5	2796.6
553.15	280	536	6.412	64.12	1.3321	30.17	1236.0	1543.6	2779.6
573.15	300	572	8.581	85.81	1.4036	21.67	1344.0	1404.9	2749.0
593.15	320	608	11.27	112.7	1.4988	15.49	1461.5	1238.6	2700.1
613.15	340	644	14.59	145.9	1.6379	10.80	1594.2	1027.9	2622.0
633.15	360	680	18.65	186.5	1.8925	6.945	1760.5	720.5	2481.0
647.29	374.14	705.452	22.09	220.9	3.155	3.155	2099.3	0	2099.3

Table 6.4 Main properties of superheated water vapor

Temperature			Specific volume	Enthalpy	Temperature			Specific volume	Enthalpy
K	°C	°F	$m^3/1000 \cdot kg$	kJ/kg	K	°C	°F	$m^3/1000 \cdot kg$	kJ/kg
			p = 0.1 MPa = 1 bar Tsat = 33.63 °C = 211.33 °F					p = 0.5 MPa = 5 bar Tsat = 151.863 °C = 305.35 °F	
363.78	99.63	211.334	1694	2675.5	425.013	151.863	305.3534	374.9	2748.7
373.15	100	212	1696	2676.2	473.15	200	392	424.9	2855.4
433.15	160	320	1984	2796.2	673.15	400	752	617.3	3271.9
593.15	320	608	2732	3114.6	773.15	500	932	710.9	3483.9
673.15	400	752	3103	3278.2	873.15	600	1112	804.1	3701.7
773.15	500	932	3565	3488.1	973.15	700	1292	896.9	3925.9
Temperature			Specific volume	Enthalpy	Temperature			Specific volume	Enthalpy
K	°C	°F	$m^3/1000 \cdot kg$	kJ/kg	K	°C	°F	$m^3/1000 \cdot kg$	kJ/kg
			p = 1 MPa = 10 bar Tsat = 179.91 °C = 355.84 °F					p = 2 MPa = 20 bar Tsat = 212.42 °C = 414.36 °F	
453.06	179.91	355.838	194.4	2778.1	485.57	212.42	414.356	99.6	2799.5
473.15	200	392	206.0	2827.9	513.15	240	464	108.5	2876.5
593.15	320	608	267.8	3093.9	593.15	320	608	130.8	3069.5
673.15	400	752	306.6	3263.9	673.15	400	752	151.2	3247.6
773.15	500	932	354.1	3478.5	773.15	500	932	175.7	3467.6
873.15	600	1112	401.1	3697.9	873.15	600	1112	199.6	3690.1
					973.15	700	1292	223.2	3917.4

Table 6.4 Main properties of superheated water vapor (continued)

Temperature			Specific volume	Enthalpy	Temperature			Specific volume	Enthalpy
K	°C	°F	$m^3/1000 \cdot kg$	kJ/kg	K	°C	°F	$m^3/1000 \cdot kg$	kJ/kg
			p = 4 MPa = 40 bar					p = 8 MPa = 80 bar	
			Tsat = 250.4 °C = 482.72 °F					Tsat = 295.06°C = 563.1°F	
523.55	250.4	482.72	49.78	2801.4	568.21	295.06	563.108	23.52	2758.0
553.15	280	536	55.46	2901.8	533.15	320	608	26.82	2877.2
673.15	400	752	73.41	3213.6	673.15	400	752	34.32	3138.3
773.15	500	932	86.43	3445.3	753.15	480	896	40.34	3348.4
873.15	600	1112	98.85	3674.4	873.15	600	1112	48.45	3642.0
973.15	700	1292	111.0	3905.9	973.15	700	1292	54.81	3882.4
Temperature			Specific volume	Enthalpy	Temperature			Specific volume	Enthalpy
K	°C	°F	$m^3/1000 \cdot kg$	kJ/kg	K	°C	°F	$m^3/1000 \cdot kg$	kJ/kg
			p = 16 MPa = 160 bar					p = 32 MPa = 320 bar	
			Tsat = 347.44°C = 657.39°F						
640.59	347.44	657.392	9.31	2580.6	673.15	400	752	2.36	2055.9
633.15	360	680	11.05	2715.8	793.15	520	968	8.53	3133.7
673.15	400	752	14.26	2947.6	873.15	600	1112	10.61	3424.6
753.15	480	896	18.42	3234.4	973.15	700	1292	12.73	3732.8
873.15	600	1112	23.23	3573.5	1073.15	800	1472	14.60	4015.1
973.15	700	1292	28.08	3833.9	1173.15	900	1652	16.33	4285.1

Steam Watertube Boilers

In watertube boilers water circulates inside the tubes, and flue gas outside. The basic principle is illustrated in the schematic circuit shown in Fig. 6.3. Water circulation is generally caused by the difference of density between the mixture of hot water and steam which is formed in the heated side of the circuit, generally called a riser, and the cooler water in the unheated side. Inside the drum, placed in the upper part of the circuit, steam is released. In normal operating conditions, water flows continously from the drum through the cooler water downcomer and up the riser to the drum where steam is released. Industrial boilers have many parallel riser circuits carrying the steam and water mixture and few larger downcomers forming a loop.

Watertube, like other industrial boilers, can be classified by tube shape and arrangement (horizontal, inclined, etc.), by the number of drums, by capacity or by pressure.

Watertube units typically have capacities between 5 t of steam per hour and a hundred or more and they are able to produce high-pressure steam.

The basic and auxiliary elements of a watertube boiler are shown in Fig. 6.4. The following points can be made:

— in the combustion chamber fuel is introduced and burned while excess air is guaranteed by means of fans, if necessary. The heat exchange is obtained by irradiation. In this phase, where a mixture of water and vapor flows through the tubes, an amount of heat equal to 35-40% of the total heat exchanged inside the boiler is involved;

— in the convective section, 40-60% of the total heat exchanged inside the boiler is exchanged by convection. This section is the main part of the exchange surface;

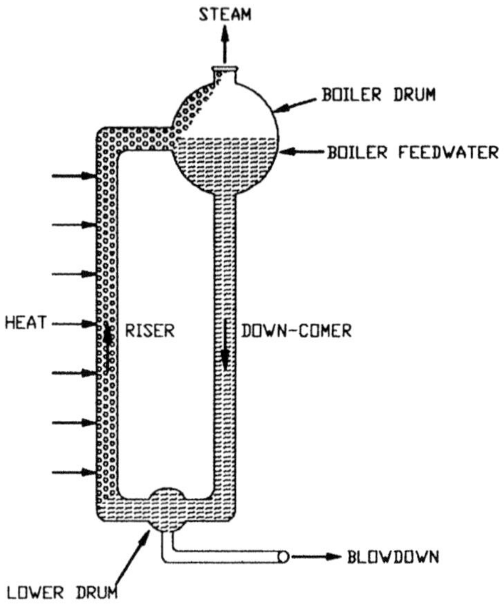

Fig. 6.3 Watertube boiler basic circuit

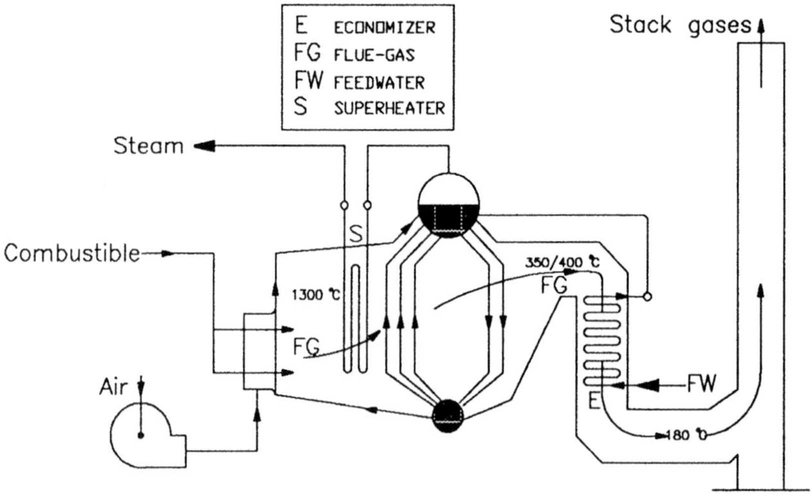

Fig. 6.4 Basic and auxiliary equipment for a watertube boiler

— the superheater (optional) is used to produce superheated vapor. Heat is exchanged by convection and irradiation and the amount of heat is roughly 20-30% of the total heat exchanged inside the boiler;
— the economizer (optional) is used to preheat the cold feedwater. The exchange is mainly by convection with exhaust flue gas at temperatures ranging from 673 K to 523 K (400°C to 250°C; 752°F to 482°F). It constitutes roughly 5% of the total heat exchanged inside the boiler;
— the preheater of combustion air (optional) preheats the combustion air exchanging low-temperature heat through the stack.

Notice that:

— water circulation is generally provoked by the difference of density variation between cold feedwater and the mixture of hot water and steam in the riser;
— a high speed of the mixture inside the boiler tubes should be maintained in order to improve the heat exchange. In some applications a forced circulation is used, particularly for high-pressure boilers (pressure higher than 15 MPa) and for instantaneous-vaporization boilers. The latter can be convenient when the load requires discontinuous peaks of steam, generally less than 1-3 t / h;
— the circulation of hot flue gases is induced by creating either a depression inside the combustion chamber with aspiration fans or an overpressure with inflation fans (pressurized boilers).

Steam Firetube Boilers

In firetube boilers flue gas flows inside the tubes and water is outside in a shell. They are usually available with a capacity of less than 20 t of steam per hour and for pressures below 1.5 MPa.

The basic and auxiliary elements of a firetube boiler are shown in Fig. 6.5. Notice that gases make several passes, generally from 2 to 4, to increase the surface area exposed to the hot gases and to distribute steam formation more

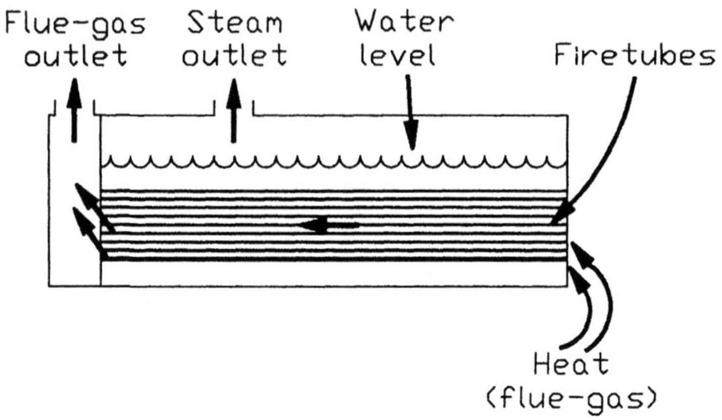

Fig. 6.5 Firetube boiler basic circuit

uniformly throughout the mass of water inside the shell. The main advantages of a firetube boiler include: ability to meet great and sudden load changes rapidly because of the large amount of water at the saturation temperature inside the shell, low capital cost and low maintenance costs. Disadvantages are the thermal inertia that prolongs startup and shutdown operations and the inherent inability to superheat.

Heat recovery boilers

Heat recovery boilers belong to the class of so-called unfired industrial boilers and are suitable for plants where process streams produce high-temperature gases which may be used to produce steam or hot water. A reliable supply of gas at 573-673 K (300-400°C; 572-752°F) or higher is required. Of course, the fluid temperature can be raised, if necessary, by burning fuel in the heat recovery system or outside. In both cases, combustion products from the auxiliary burners are generally mixed with the hot gases before entering the convective section of the boiler; if the gas contains enough oxygen, as gas-turbine exhaust does, the auxiliary burners operate directly in the gas stream.

Watertube and firetube boilers are used as heat-recovery boilers.

Firetube boilers are suitable for low-pressure exhaust from diesel engines, kilns, furnaces etc. and for low-capacity boilers with a capacity of less than 20 t of steam per hour.

Watertube boilers can be designed for almost any saturated or superheated steam pressure and temperature and for any capacity required. Natural or forced circulation can be used as in conventional units. Watertube boilers are generally more efficient than firetube boilers in similar conditions.

With heat recovery boilers, attention must be paid to the difference between waste gas temperature at the boiler outlet and saturated steam temperature: a small differential maximizes heat recovery but requires a larger heat transfer area and consequently higher capital costs.

Electric/electrode boilers

Boilers fed by electricity to produce steam or hot water comprehend two basic types: i) resistance units where electric current flows through resistors submerged in water inside a protective tube; ii) electrode units where two electrodes are submerged in water and electric current flows through the water which has its own resistance. In this case water conductivity control is one of the main keys to good, safe operation.

Resistance units are generally designed for small systems, both hot water at low temperature and steam at low pressure.

Electrode boilers can be divided into two groups, depending on the voltage: low-voltage types which operate at 600 V or lower with a maximum output of a few megawatts and medium-voltage types which operate at 10-20 kV with a maximum output of 50 MW.

Both groups are available with submerged electrodes where output power is controlled by moving insulating shields or by adjusting the immersion level of the electrodes. For the medium-voltage group, spray-electrode units are available where output power is regulated by varying the amount of water sprayed upon the electrodes.

Many factors such as absence of emissions, minimal space requirements, simple maintenance, relatively low cost may contribute to the attractiveness of these boilers, but the cost of electricity remains the main factor governing the choice.

Availability of electric power must also be taken into account because of the relatively high power needed, in comparison with that required by the whole factory (roughly 1 t of steam per hour corresponds to 0.8 MW of electric power).

Oil boilers

Oil and other organic fluids can be used as the heating medium to transfer thermal energy to end users for particular process demands (presses, dryers, ovens, etc.). Steam can also be generated outside the boiler by means of a hot oil/water vapor exchanger.

Forced circulation of oil is effected by electrically driven pumps inside the tubes while flue gas is outside. Care must be taken to limit temperature peaks throughout the circuits, particularly if there is an electricity black-out and pumps are switched off. Maximum temperatures of heating medium fluids are roughly 573-623 K (300-350°C; 572-662°F). Steam generation, usually at low pressure, is effected by heat exchangers outside the boiler itself.

Local regulations may accept for this system a lower control level, in both personnel and instrumentation, because there is no direct contact between the flame and pressurized equipments.

6.6 TECHNICAL PARAMETERS OF INDUSTRIAL STEAM BOILERS

The main parameters used to qualify any industrial boiler are follows:

— fuel type and rated consumption per hour;
— temperature of cold feedwater and condensate;
— temperature of combustion air;
— flow of steam expressed in kg/h or t/h;
— output steam pressure;
— temperature of saturated or superheated steam;
— temperature of flue gases from the stack;

— design pressure which all parts of the boiler must resist. This value is generally 10% higher than the steam pressure;

— useful output (generally in the form of steam or hot water) expressed in kW (or practically in tons of steam per hour).

It follows that:

$$\text{useful output (kW)} = m \cdot (h - h_o)$$

where

m = steam flow-rate in kg/s

h = enthalpy of the output steam in kJ/kg

h_o = enthalpy of the input water or condensate in kJ/kg.

Typical values of $(h - h_o)$ for saturated steam output are 2300-2600 kJ/kg. Notice that the same evaluation must be carried out for a hot water or intermediate oil boiler;

— specific capacity defined as the ratio between useful output and heat transfer area. It is expressed in kW/m^2 or kg steam/$m^2 \cdot h$;

— **boiler efficiency is expressed as the ratio of useful output (i.e. heat in the steam) to heat input in the fuel, based on either the gross or the net heating value, depending on local standards (gross value in the USA; net value in many European countries). Efficiency calculation procedures are expounded in Chapter 6.7;**

— **amount of output steam produced by the unit of input fuel in rated operating conditions. Typical values for saturated steam are 12-13 kg of steam/kg of oil, 10 kg of steam/Sm^3 of natural gas.**

6.7 BOILER LOSSES AND EFFICIENCY

Boiler efficiency can be defined as the ratio of useful output to heat input in the fuel. Depending on local usage, the latter can be either the gross or the net heating value.

A proper calculation of efficiency requires a definition of the boiler boundary which separates the elements to be considered part of the boiler from those that are excluded. Equipment is generally considered outside the boundary when it requires an outside source of energy (heat or electricity) or when the heat exchanged is not returned to the boiler generating system. Of course, boiler efficiency is lower than combustion efficiency (see Chapter 6.3), particularly because of additional heat losses from the exterior boiler surfaces through the insulation, generally called radiation losses.

A wider definition of boiler plant efficiency, instead of the single boiler, can take into account all kinds of energy consumption and recovery, which are not considered for the efficiency of the boiler itself.

Two basic procedures to calculate efficiency are discussed:

i) input-output or direct method;
ii) heat-loss or indirect method.

Of course, the two methods would give the same results if all the data required could be measured without error.

Input-output method

$$\text{Efficiency}(\%) = \left(\frac{\text{output}}{\text{input}}\right) \cdot 100 = \left(\frac{\text{useful output or heat absorbed by the fluid}}{\text{heat input in the fuel}}\right) \cdot 100$$

This method, also called the direct method, requires the direct measurement of both the input energy, that is the fuel flow rate, and of the useful output energy. In order to quantify the output energy it is necessary to measure the temperature, pressure and flow rate of generated steam and boiler feedwater. Difficulties in making these measurements, the lack of instrumentation at most industrial boiler plants, and the possibility of significant errors, make large-scale use of this method impracticable.

It is important to note that the choice of the heat input equal to either the net or the gross heating value determines a great difference in the related efficiency values. If the ratio between the gross and the net value, which roughly equals

1.12 for natural gas and 1.065 for oil, is called α and the boiler output power has the same value, the ratio between the two efficiency values is:

$$\frac{\text{Efficiency(\% gross reference)}}{\text{Efficiency(\% net reference)}} = \frac{1}{\alpha}$$

Heat-loss method

$$\text{Efficiency (\%)} = 100 - \left(\frac{\text{losses}}{\text{heat input in the fuel}}\right) \cdot 100$$

These losses are mainly the combustion losses already discussed in Chapter 6.3. They are listed below with some comments on their origin:

— waste heat going up the stack. The greater part is the heat carried out by dry flue gases (sensible heat). Another part of the stack loss is the moisture loss (latent heat), that is, the heat used to evaporate and superheat water vapor moisture resulting from the combustion of hydrogen in the fuel (this must be considered only if gross heating value is assumed as heat input in the fuel), the humidity of the combustion air and the water contained in the fuel. Industrial boilers not equipped with waste heat recovery equipment such as air preheaters or economizers have very large flue gas losses because of the high stack-gas temperature (473-573 K; 200-300°C; 392-572°F);

— losses due to incomplete combustion. These consist mainly of combustible material in the flue gas such as carbon monoxide, hydrogen, hydrocarbons. Additionally, refuse may contain unburned solid fuels and other solid combustibles;

— radiation losses through the exterior surfaces of the boiler. Approximate evaluation of these losses by charts developed by the ABMA (American Boiler Manufacturers Association) or similar ones is shown in Fig. 6.6;

— additional losses.

The indirect method is based on the determination of the above-mentioned losses, by first evaluating singly the heat losses per unit of fuel and then converting these values to a percentage loss by means of the heating (gross or net) value. A test procedure, the so-called short-form computation, has been devised by ASME (American Society of Mechanical Engineers).

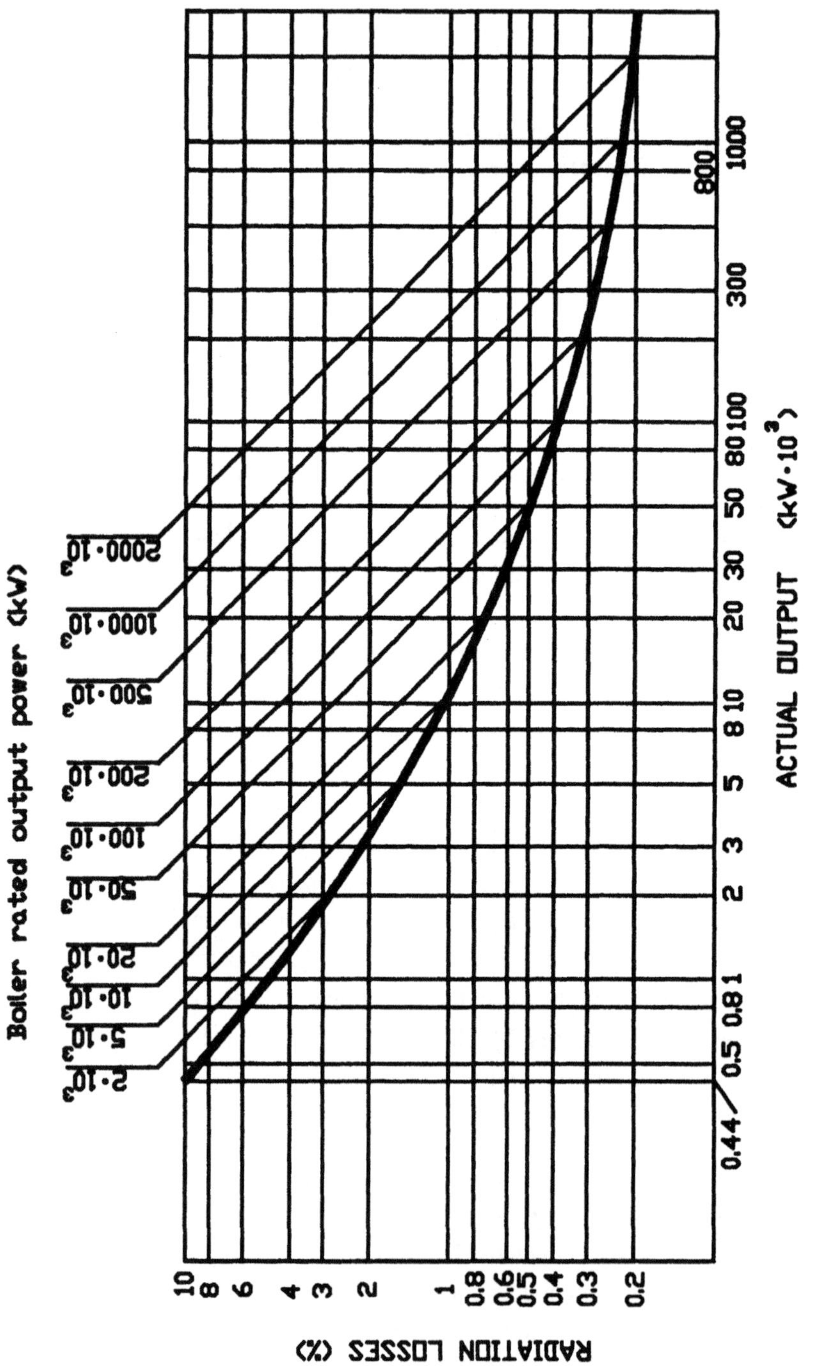

Fig. 6.6 Example of radiation losses chart (each curve corresponds to a boiler rated power value)

The waste heat going up the stack is evaluated as the sum of three components:

i) heat loss (sensible heat) due to dry gas is equal to the kilograms of dry gas per kilogram of fuel multiplied by the specific heat of the combustion gases (roughly 1 kJ/kg · K or 0.24 Btu/lb · °F) multiplied by the difference in temperature between the stack exit gas (t_s) and the inlet air for combustion (t_a). The stack exit temperature t_s is closely related to the recovery equipment temperature such as economizers and air preheaters;
ii) moisture loss due to the water contained in the fuel is equal to the kilograms of water per kilogram of fuel multiplied by the enthalpy difference between the water vapor mixture in the stack exit gas and water at ambient temperature;
iii) moisture loss due to the combustion of the hydrogen in the fuel equals the weight fraction of hydrogen in the fuel multiplied by 9 (as stated in Chapter 6.1 where it is pointed out that the combustion of 1 kg of hydrogen produces 9 kg of liquid water) multiplied by the enthalpy difference between the water vapor mixture in the stack and liquid water at ambient temperature. This loss is considered only if the gross heating value is assumed as heat input in the fuel.

The losses due to incomplete combustion are totally attributed to CO in stack gases and to the combustible in the refuse. In the latter case they are evaluated as the kilograms of dry refuse per kilogram of fuel multiplied by the heating value of the refuse determined by laboratory tests.

The radiation losses are estimated by reference to the ABMA charts mentioned above or equivalent charts.

Additional losses, generally ranging between 0.5 and 1.5 percent, are introduced to takc into account losses neglected in the short-form computation.

Graphic solutions were developed by using the above-mentioned or similar procedures in order to estimate stack gas losses (dry flue gas losses; all the moisture losses together) and losses due to incomplete combustion.

The heat-loss method requires the measurement of the stack flue gas parameters (temperature; O_2 or CO_2 and CO concentrations as % by volume or as ppm; combustible flow-rate; combustion air temperature).

The measurement of O_2 is generally preferred to that of CO_2 because simpler instrumentation achieves the same accuracy. Portable analyzers are available for both O_2 and CO_2 measurements.

Carbon monoxide is generally measured by means of hand-held chemical absorbing analyzers (Orsat or similar analyzers) and length of stain detectors.

Stack opacity or smoke density is assumed as an index of the combustion conditions. It can be measured by means of hand pump filter paper or a similar tester where the colour assumed by the paper is compared with a standard scale. Generally, index 0 means no opacity. Optimum values range between 0 and 3 (Bacharach index).

The relationships between excess air and stack gas concentration of O_2 and CO_2 for different fuels are shown in Fig. 6.7.

Fig. 6.8 shows a typical relationship between total stack gas losses and stack temperature for different values of stack excess O_2 and for a specific fuel (natural gas in this case). Notice that stack temperature increases with the excess of O_2 because complete combustion is achieved. Similar curves are available for different fuels. Fig. 6.9 shows the relationship between losses due to unburned combustibles and stack excess O_2 for different values of CO emissions. Sets of values referring to both net and gross heating values are reported.

Other methods, based on the evaluation of total stack losses and unburned losses through curves and standard coefficients are widely used:

— total stack gas losses $\% = K_s \cdot \dfrac{t_s - t_a}{CO_2\%}$ where K_s is called Hassenstein coefficient (see Table 6.5 where only the net heating value is assumed as reference), t_s is the stack gas temperature and t_a is the ambient temperature (typical values of $CO_2\%$ less than 10%). This formula is acceptable if incombustibles are kept low;

— losses for unburned combustible $\% = K_c \cdot \dfrac{CO\%}{CO_2\% + CO\%}$ where K_c equals 50.5 for fuel oil, 37.9 for natural gas, 59 for coal if the net heating value is assumed as reference (typical values of CO% less than 0.1%).

Examples of evaluation are given in Chapter 6.12.

In Fig. 6.10 the main combustion parameters are correlated by a set of curves called the Ostwald triangle. In particular, if CO_2% and O_2% are known, it is possible to check whether the combustion is complete, and if it is not, to determine the CO%. Otherwise, if only CO_2% or 02% is known, by introducing CO% (measured or estimated), it is possible to determine the other. Other sets of correlation curves also exist for use according to local practice.

Note that efficiency is generally defined according to national rules to which reference must be made. As already shown for the input-output method, particular attention must be paid to the heat input in the fuel, taken as gross or net heating value.

Table 6.5 Hassenstein coefficient values for different combustibles (K_s coefficient referred to net heating values of input fuels and to °C)

	Hassenstein coefficient values (1)			
CO_2 % by volume	Gasoil	Oil	Natural gas	Coal
4	0.523	0.543	0.418	0.683
5	0.530	0.550	0.427	0.684
6	0.536	0.556	0.437	0.685
7	0.543	0.563	0.447	0.686
8	0.550	0.570	0.457	0.687
9	0.557	0.576	0.466	0.688
10	0.564	0.583	0.476	0.689
11	0.571	0.590	0.486	0.690
12	0.578	0.596		0.691
13	0.585	0.603		0.692
14	0.592	0.610		0.693
15				0.694

(1) if °F is used, multiply K_s by 5/9.

The relationship between percentage losses, corresponding to the same output power, can be formulated as follows:

total losses (as % of gross heating value) =

= total losses (as % of net heating value)/α + 100 · (α – 1)/α

For natural gas, ratio α = 1.12; thus

total losses (as % of gross heating value) =

= total losses (as % of net heating value)/1.12 + 10.7

For fuel oil the ratio, α = 1.065; thus

total losses (as % of gross heating value) =

= total losses (as % of net heating value)/1.065 + 6.1

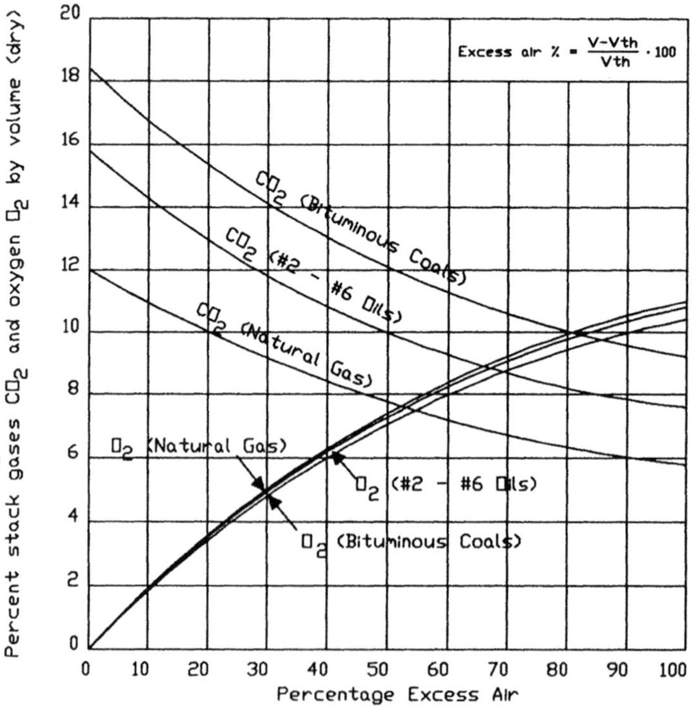

Fig. 6.7 Typical relationship between stack gas concentrations of CO_2(%) and O_2(%) and excess air (%)

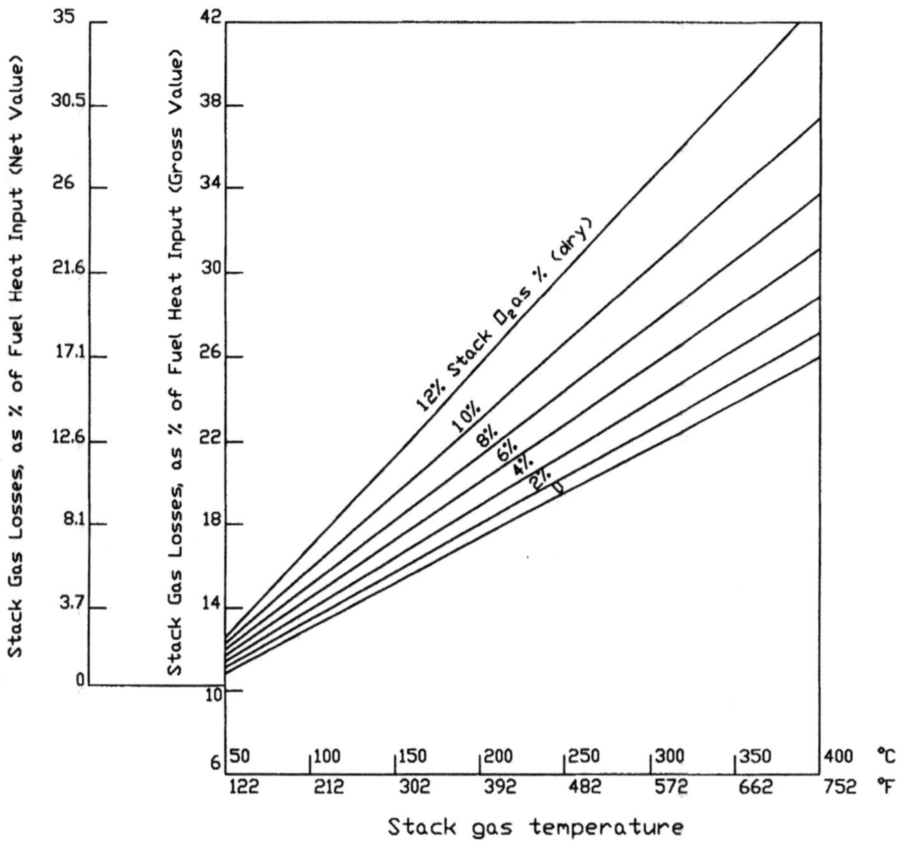

Fig. 6.8 Stack Gas Losses (natural gas)

6.8 HOW TO IMPROVE BOILER EFFICIENCY

Boiler efficiency is influenced by many factors. As shown in Fig. 6.11, efficiency varies according to the fuel, to the load and to the existence of stack gas recovery systems such as economizers and air preheaters.

Average efficiency values range between 75% and 91% and they can be obtained by a proper control of the operating parameters (the higher values within

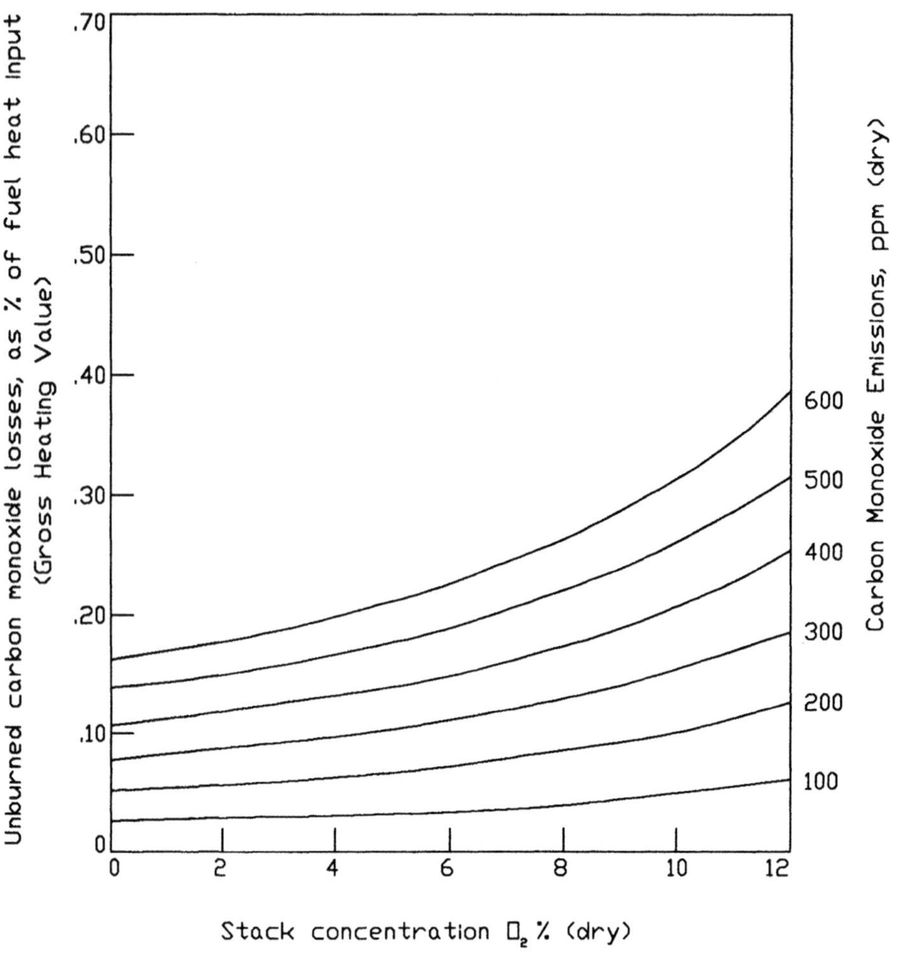

Fig. 6.9 Unburned CO losses versus O_2 concentration by volume for different values of CO emissions for natural gas (CO losses are here referred only to the Gross Heating Value because of their low value; CO concentration equal to 100 ppm or 0.01% in volume corresponds to 125 mg/Sm^3)

this range if recovery systems are installed and if the net heating value is used as input in the fuel).

Of course, efficiency is improved when losses are kept as low as possible. This happens mainly when excess air and flue gas stack temperature are reduced.

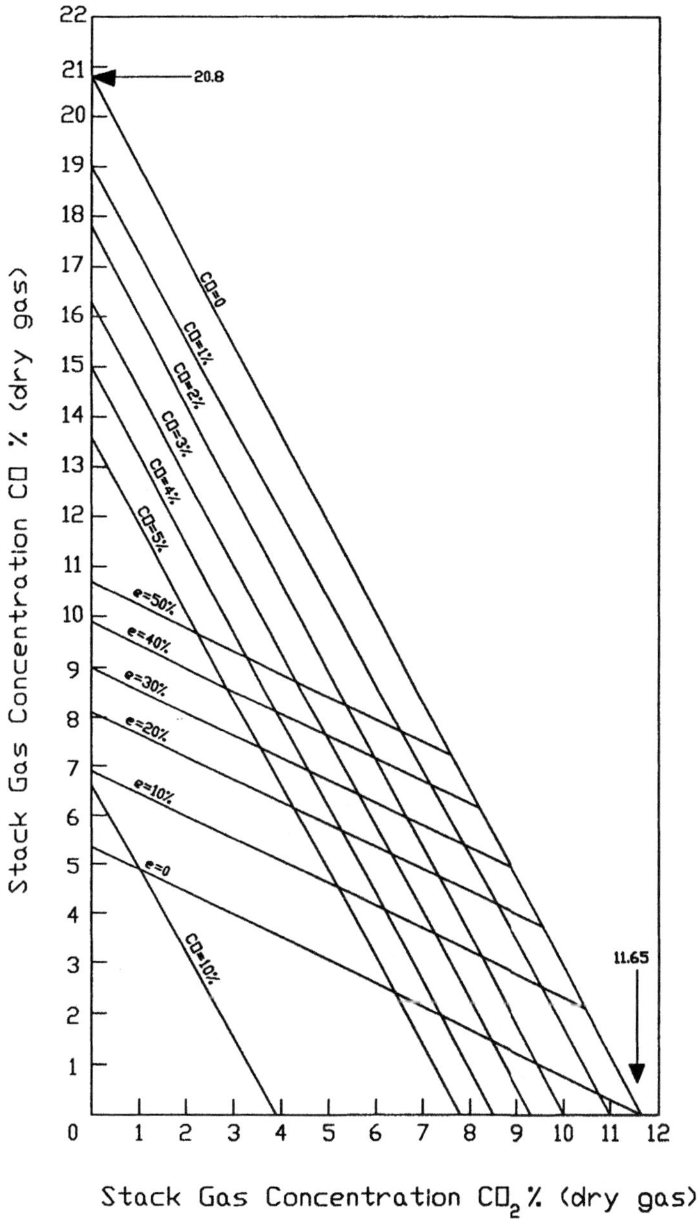

Fig. 6.10 Ostwald triangle for natural gas

$$e\% = excess\,air = \frac{V - V_{th}}{V_{th}} \cdot 100;\ concentration\ by\ volume\ (dry)$$

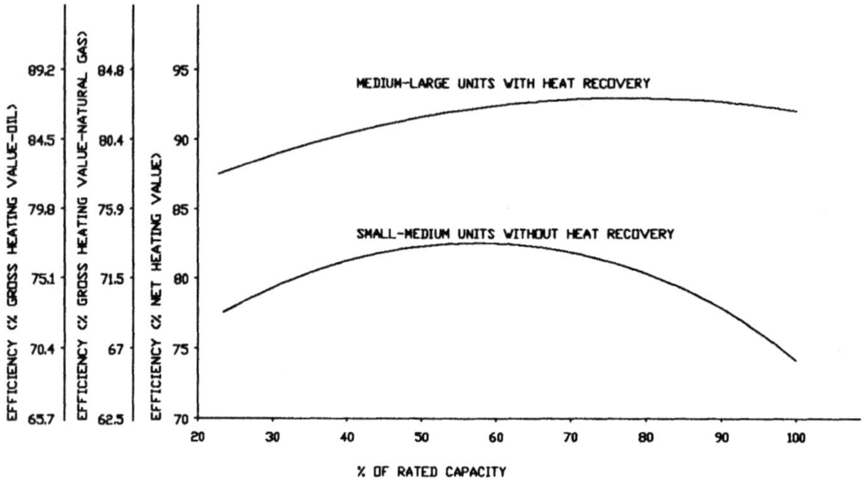

Fig. 6.11 Ranges of boiler operating efficiencies

In what follows, suggestions are made about how to keep industrial boiler losses to the minimum.

Excess air, ranging between 10% and 60%, but necessary to insure complete combustion and safe operation of the boiler, and high temperature of stack flue gas (473-573 K; 200-300°C; 392-572°F) are responsible for a significant part of the stack losses. The minimum values of sensible heat percentage losses represent less than half of total losses.

Losses due to the latent heat of stack water vapor mixture generally amount to at least half of the total losses, depending on the fuel and operating conditions. These losses cannot be reduced because no means are available to permit the water vapor mixture to condense before passing from the boiler into the stack.

Excess air can be reduced by improving the combustion control system which regulates the supply of air and fuel to meet load demand variations. Notice that a CO_2 content in stack flue gas higher or lower than standard values indicates anomalous operation of the boiler due to lower or higher excess air values.

Minimum levels of flue gas temperature are limited by potential corrosion and sulfuric acid condensation in the cold part of the stack. The minimum temperature depends on the sulfur content of the fuel and on the amount of moisture in the flue gas. As a general rule, natural gas allows lower stack flue gas

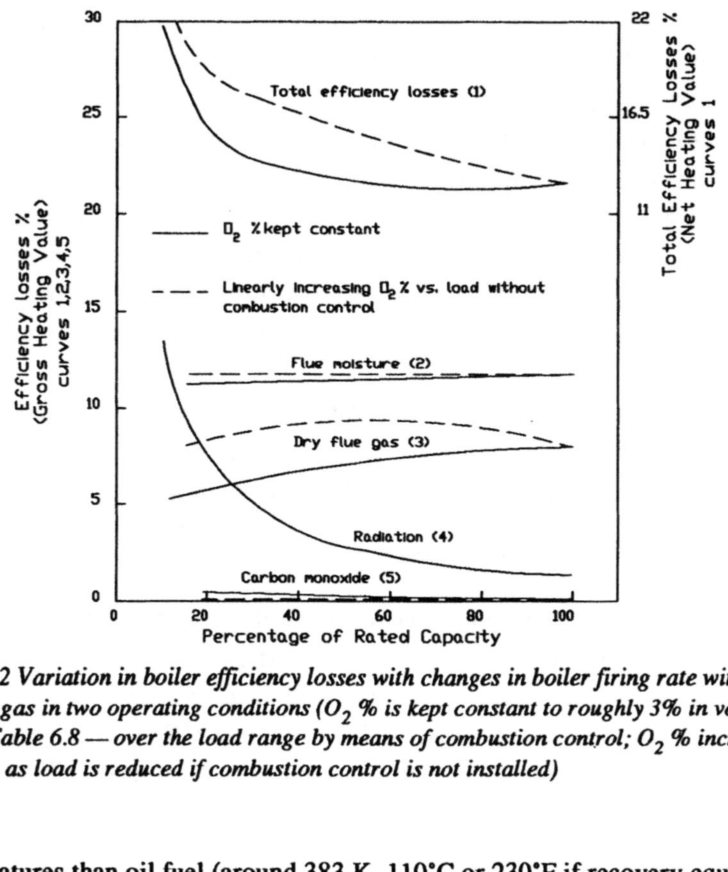

Fig. 6.12 Variation in boiler efficiency losses with changes in boiler firing rate with natural gas in two operating conditions (O_2 % is kept constant to roughly 3% in volume — see Table 6.8 — over the load range by means of combustion control; O_2 % increases linearly as load is reduced if combustion control is not installed)

temperatures than oil fuel (around 383 K, 110°C or 230°F if recovery equipment is installed) because of the absence of sulfur content.

The minimum flue gas temperature is higher than that given above if recovery equipment is not installed. In this case the minimum value depends on the boiler operating pressure which determines the saturation steam temperature. A difference of about 70-80 K or °C (126-144°F) between the two temperatures is generally suggested. Lower values would require a larger convective surface area and consequently additional cost.

Table 6.6 shows minimum stack temperatures and minimum losses for different fuels. If stack gas temperature and excess air are higher than these values, higher losses will occur.

Losses due to incomplete combustion are generally low if the right amount of excess air is maintained. The carbon monoxide content in stack flue gas must be kept approximately equal to zero.

Boiler firing rate or output considered as a percentage of rated capacity affects efficiency, particularly at low load. Fig. 6.12 shows boiler losses versus boiler firing rate in two operating conditions: i) O_2 % is kept constant over the load range; ii) O_2 % increases linearly as load is reduced (a common condition if combustion control is not installed). In the first situation, efficiency remains constant over a large range of load changes; in the second, efficiency diminishes as the load decreases.

In conclusion, to improve the efficiency of a boiler, excess air and stack temperature must be kept as low as possible at each load. Reduction of stack temperature is generally due to a reduction of excess air, that is, a reduction of excess O_2.

Improvement of efficiency depends on all the previous operating conditions; an improvement of about 1-2% is generally possible even if the boiler is operating near the maximum efficiency; of course, greater improvement is possible if the previous efficiency is very low. Any improvement of efficiency at constant output power will lead to energy saving as follows:

$$\text{power saved} = \left(\frac{1}{\eta_c} - \frac{1}{\eta_i}\right) \cdot P_{out}$$

where

η_c = current efficiency

η_i = improved efficiency

P_{out} = constant output power

Table 6.6 Minimum stack temperatures and losses for different combustibles without recovery equipment

Fuel	Burning equipment	C % by weight	H % by weight	H_2O % by weight	S % by weight	OTHERS % by weight	Minimum values of STACK TEMPERATURE			Minimum values of EXCESS O_2 in stack gases	LOSSES					
											Dry gas		Moisture		Total stack gas	
							K	°C	°F	% by volume	(a) %	(b) %	(a) %	(b) %	(a) %	(b) %
Natural gas	All burners	75.7	23.3	0	0	1	377	104	220	1	2.9	3.2	10.1	0.	13.0	3.2
Oil N.2 (*)	All burners	87	11.9	0	0.5	0.7	439	166	330	2	5.1	5.4	6.4	0.	11.5	5.4
Oil N.6 (**)	All burners	86.6	10.8	0	1.50	0.7	472	199	390	3	6.6	7.0	6.2	0.	12.6	7.0
Bituminous coal	Pulverized	71.3	5.0	4.5	2.80	16.4	416	143	290	4	4.8	5.0	4.5	0.	9.3	5.0
Bituminous coal	Stoker	71.3	5.0	4.5	2.80	16.4	416	143	290	6	5.5	5.7	4.5	0.	10.0	5.7

(a) with reference to the gross heating value

(b) with reference to the net heating value

(*) oil N.2 according to American Society for Testing and Materials (ASTM). It is a distillate oil

(**) oil N.6 according to American Society for Testing and Materials (ASTM). It is a residual oil

6.9 AUXILIARY EQUIPMENT

Auxiliary equipment can be divided into heat recovery systems and water treatment systems.

Heat-recovery systems

Flue gas going up the stack contains a considerable amount of thermal energy, part of which can be recovered by installing air preheaters and economizers.

Air preheaters are used to transfer heat from the stack flue gas to the incoming combustion air. In this way boiler efficiency is improved because of: i) reduction of stack gas temperature; ii) reduction of excess air due to improved burner combustion conditions if preheated air is supplied.

The total efficiency will increase significantly: roughly 2% for every 50 K or °C (90°F) decrease in temperature of flue gas and 2% for every 50 K or °C (90°F) increase in that of combustion air.

A high combustion air temperature seems to increase the emission of nitrogen oxides, particularly with natural gas. The reduction of excess air, however, may moderate the NO_x increase.

The maximum combustion air temperature does not generally exceed 200°C (392°F), depending on fuel, type of burner and boiler condition. Higher temperatures are required if pulverized coal is burned. The lowest flue-gas temperature at the outlet of the preheater is limited by potential corrosion, as mentioned in Chapter 6.8.

Air preheaters can be classified as either recuperative or regenerative.

In recuperative units, heat is transferred directly from the hot medium such as hot flue gas on one side of the stationary surface to air on the other side.

Regenerative units transfer heat indirectly from the hot flue gas to combustion air through an intermediate heat storage medium. They are used mainly in retrofitting existing boiler plants.

An alternative to the above-mentioned preheaters is the heat-pipe heater (see Chapter 15.5). They are not currently in use in boiler units but they are expected to become competitive with the traditional ones.

Economizers provide additional heating to the feedwater by recovering heat from stack flue gas. In this way the boiler firing rate needed to produce the same amount of steam is reduced and the overall efficiency is increased. Economizers can also be installed to heat hot water for users apart from the boiler water cycle.

The advantage of economizers is closely related to the existing feedwater temperature and condensate return system. Roughly, an increase in feedwater temperature of 5 K or °C (9-10°F) will result in an efficiency increase of 1%.

Water treatment systems

Water treatment is one of the main factors in boiler reliability. It is necessary because impurities in the feedwater can lead to scale deposits on the boiler heating surfaces and to corrosion. Suspended solid matter, colloids, dissolved minerals, atmospheric gases and impurities coming from the steam cycle are always present in varying degrees in boiler feedwater.

The main equipment for water treatment can be summarized as follows: demineralization, pretreatment systems to remove suspended solids such as clarifiers, softeners and media and precoat filters.

Demineralization removes dissolved salts from pretreated water, where they are present in the form of positive and negative ions (cations and anions). The goal is reached by means of ion-exchange resins exchanging harmless ions with impurities such as sodium cations and chloride anions. When resins become spent they can be regenerated for reuse; after a certain number of regenerations the resin bed must be discarded. The two-column system, which is one of the commonest installations, includes an acid cation exchanger followed by a base anion exchanger with a degasifier interposed to strip out the CO_2 produced in the first exchanger. For boilers operating at high pressure and temperature a feedwater of very high purity is required (roughly 5-10 parts per billion of sodium and silica). In this case other systems such as mixed cation-anion-resin beds and semipermeable membranes are generally installed.

Filters and clarifiers are used to remove solid particles and some dissolved substances. Clarifiers work with chemical coagulants which help to gather finely divided matter into larger masses. In addition, calcium and magnesium can precipitate depending on the chemicals added. Calcium and magnesium hardness can also be removed by separate softeners.

Softeners are used to remove hardness and also to remove silica and to reduce carbonate alkalinity. They generally work at ambient temperature and they are effective for low-pressure boilers that do not require complete demineralization of the feedwater. Hot-process softeners operating at above 373 K (100°C; 212°F) can be used for higher-pressure boilers.

Notice that blowdown is commonly used to remove boiler water impurities. Its amount depends on many factors, particularly on the boiler water treatment performances, and it can be as much as 5-10% of the total boiler steam. It is always possible to use heat recovery from blowdown water to preheat the incoming make-up water. Low-pressure boilers generally use heat exchangers; high-pressure and high-capacity boilers use flash systems where part of the blowdown waste water is flashed into high-purity steam, then condensed and recycled into the feedwater system. The remaining waste water exchanges heat with the make-up water and then is discarded.

Table 6.7 synthesizes the main features of the commonest types of water treatment equipment.

Condensate return systems

The recovery of condensate from process steam furnishes a great amount of usable energy and clean feedwater. The degree of recovery and its economic validity depend on many factors such as contamination from process, layout of steam end users and pipelines. The quantity of energy contained in the saturated liquid can be estimated as roughly 20% of the original energy contained in the steam at the same pressure.

Table 6.7 Effect of treatment techniques on makeup water contaminants under ideal operating conditions

Treatment Techniques	SUSPENDED SOLID	ALKALINITY	HARDNESS	DISSOLVED SILICA	DISSOLVED SOLIDS
Filtration	Excellent	No change	No change	No change	No change
Cold-process softening	Good	Fair	Fair	Good	Fair
Hot-process softening	Good	Good	Good	Good	Fair
Sodium-cycle cation exchange	Excellent	No change	Excellent	No change	No change
Demineralization	Excellent	Excellent	Excellent	Excellent	Excellent
Reverse osmosis	Excellent	Excellent	Excellent	Excellent	Excellent

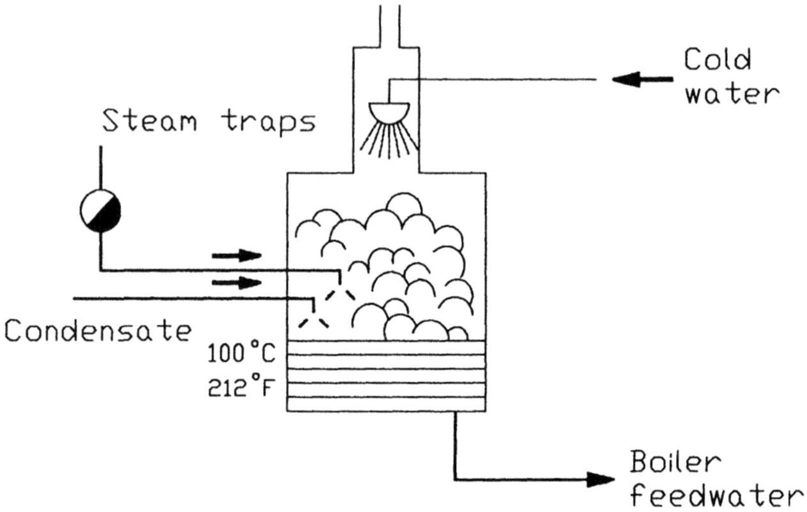

Fig. 6.13 Flash steam recovery in spray tank

The main systems used to recycle the condensate into the boiler are: the atmospheric system, pressurized systems and deaerators.

The atmospheric system consists of an open tank where condensate is received downstream of steam traps or drains. At atmospheric pressure the condensate temperature is reduced to 373 K (100°C; 212°F): a portion of the condensate, about 10-15%, will flash to steam and will be lost in the atmosphere; the remaining hot water will feed the boiler together with as much treated make-up water added as the boiler requires.

The semi-pressurized and fully-pressurized systems work with a receiver at intermediate pressure below the lowest process steam pressure or at process steam pressure. The flashed steam is piped to a low-pressure steam main where it can be utilized for process purposes; the flashed steam lost is conveyed to the outlet through relief valves or automatically operated vents which are installed to prevent abnormal pressure surges. This loss, too, can be reduced if some additional provisions are made for steam recovery.

The condensate is generally cooled by the make-up water to at least 283 K (10°C; 18°F) below the saturation temperature to prevent vapor generation in the boiler feedwater pump system.

Notice that although the use of flash steam in low-pressure mains seems to offer a significant heat recovery, its practical application involves a number of problems, essentially economic, that must be carefully considered. In addition to inherent factors which must be taken into account when planning a condensate recovery system, the quantity of flash steam must match potential needs at any time since steam cannot be stored economically for later use.

Spray deaerators, also called barometric condensers, consist of a tank where steam rises through a cold water spray which condenses it (see Fig. 6.13). This system, which achieves complete flash steam recovery to heat boiler feed-water, has the additional advantage of producing a deaerating effect: if the temperature of the tank is kept above about 363 K (90°C; 194°F) by a proper metering of the spray, dissolved gases in the condensate and feedwater, such as oxygen and CO_2, will come out of solution and will be released through the atmospheric vent.

6.10 COMPUTER CONTROL OF BOILER COMBUSTION

Control in the boiler plant means ensuring a correct balance of energy and mass entering and leaving the boiler.

Fuel is the major energy input, combustion air is the major mass input. The only useful energy output is steam (or hot water or other intermediate fluids), although heat from flue gas is often recovered.

A combustion control system must regulate the quantity of fuel and combustion air to meet the load requirements automatically by ensuring at the same time minimum fuel consumption, minimum pollution and safe operation. Flue-gas analyzers and measurements of the input combustion air and fuel flow, of the energy output parameters (steam pressure, temperature and flow) give all the information about boiler performance that a boiler control system may need.

The complexity of the control system is generally related to the capacity of the boiler plant and to the legal requirements for safe operation and air pollution.

The simplest system, once extensively applied to industrial boilers, consists of a single actuator. When the steam pressure changes in consequence of a change in the load demand, this actuator moves both the fuel and combustion air input devices, mechanically linked, to a preset position indicating a preset value of the fuel/air ratio. This system cannot compensate any shifting from design conditions, so that high excess-air levels are generally preset to guarantee safe boiler operation. Thus abnormal excess air may be present throughout the boiler's life.

Separate actuators for the fuel valve and the air combustion fan, together with an individual manual adjustment of the fuel/air ratio, can lower the margin of excess air and consequently improve boiler efficiency. An operating guide, based on the availability of oxygen analyzers or the steam flow/air flow ratios, must be accessible to the operator so that he can optimize the amount of excess air.

Metering of fuel and air flow, by means of pressure measurements or directly by means of flow metering, is a further improvement in boiler control. If any change in fuel flow is limited through a control logic to the air flow actually available, an additional refinement is introduced. This system is called the cross-limited metering control system.

A cross-limited metering control by continuous monitoring of flue gas O_2 levels can eliminate the effects of variations in fuel heating values and combustion air conditions. A preset O_2 level corresponding to a minimum fuel consumption can also be introduced into the control system. Oxygen analyzers using zirconium oxide sensors and computers are currently used for reliable operation.

6.11 BOILER EMISSION AND ENVIRONMENTAL POLLUTION

Emission control is undoubtedly one of the main requirements in boiler plant operation. Depending on fuels and boiler equipment, exhaust gases contain, in different quantities, byproducts of which some are air pollutants. Small and medium-sized boiler plants, which cannot economically accommodate abatement systems, must cope with this problem by a correct choice of fuel and/or of burners and auxiliary equipment.

Although every country has its own environmental protection regulations, a general classification of major pollutants can be made as follows: sulfur dioxide, nitrogen oxides, carbon monoxide, particulate matter and hydrocarbons. In addition, other pollutants depending on the particular fuel and operating conditions, can always be present. Typical ranges of values of permitted pollutant concentrations in the air emission from boiler plants are reported in Table 6.8 for the pollutants listed above. The permitted values vary from country to country but boiler plants must be conducted so as to ensure values lower than those in the table.

Sulfur dioxide

Sulfur dioxide SO_2 is formed during combustion (see Chapter 6.2) by the combination of sulfur contained in the fuel with oxygen from the combustion air.

Except for sulfur compound particles, most of the sulfur contained in the fuel is converted into SO_2, but a small percentage of the total oxides is converted into sulfur trioxide SO_3. Sulfur oxides combine with the moisture present in stack flue gas to form sulfuric acid, some of which can condense and corrode metallic parts of the boiler. Similarly, in the atmosphere a portion of SO_2 is converted into SO_3 and then into sulfuric acid and sulfur compounds.

The quantity of sulfur oxides formed during combustion is closely related to the sulfur content of the fuel and does not depend on the boiler operating conditions. Thus, reduction of sulfur oxides can be obtained by using low-sulfur fuel, natural gas in particular, or by introducing abatement devices such as stack gas scrubbers which remove SO_2 from stack flue gas.

The maximum SO_2 concentration in stack gas must be kept lower than 1500-2000 mg/Sm^3 or 525-700 ppm (see Table 6.8).

Emulsifying fuel oil with water or other compounds and atomization at the burner are other techniques quite widely applied to solve problems connected with the use of heavy oil.

Nitrogen oxides

Nitrogen oxides, NO_x, are formed during combustion by the combination of oxygen and nitrogen at high temperature. Both components are naturally present in the combustion air and nitrogen in the fuel itself.

NO_x emissions are mainly NO (95% in mass) and NO_2 (5% in mass). NO, when in the atmosphere, combines with oxygen to form NO_2.

NO_x formed during the combustion process can be classified as Thermal, Fuel or Prompt:

— Thermal NO_x is due to the oxidation of atmospheric nitrogen at temperatures higher than 1273 K (1000°C, 1832°F);
— Fuel NO_x is due to the oxidation of the nitrogen in the fuels (its value ranges between 0.05 and 1.5% in mass depending on the kind of fuel; lower values for natural gas and light oil, higher values for heavy oil and coal). The oxidation is accelerated if the excess air value is high;
— Prompt NO_x derives from the reaction of atmospheric air with hydrocarburic radicals.

The concentration of NO_x in stack gases is related mainly to the quantity of nitrogen in the air; typical values are 500-600 mg/Sm^3 or 244-292 ppm (see Table 6.8).

Although the reduction of nitrogen in the fuel is a technique that should be practised, most technologies applied to reduce NO_x are based on lowering peak flame temperature and on reducing the amount of oxygen available in the flame. Low excess-air burners, fuel-rich or staged-firing burners with separate primary and secondary combustion zones and flue gas recirculation are the main means used to reduce the production of NO_x inside the boiler. In the case of coal, fluidified-bed boilers are suitable for this purpose. Steam injection, too, in the combustion chamber is used to reduce NO_x.

Post-combustion NO_x can be reduced by means of abatement devices generally installed downstream of the combustion zone but above air preheaters such as selective-catalytic-reduction units (SCR) and selective-non-catalytic-reduction units (SNR). The choice of the best solution, both technically and economically, must be made individually for each plant in the light of local regulations and boiler operating modes.

Table 6.9 lists the main techniques used and comments on their capability of reducing NO_x emissions and on how they influence combustion efficiency.

Carbon monoxide

Carbon monoxide, CO, is always produced with incomplete combustion (see Chapter 6.2) and its concentration in stack flue gas is closely related to the boiler operating conditions. CO concentration at the stack must be very low; concentrations higher than 0.5% indicate poor combustion conditions and generally low excess air. CO emitted from the stack is dispersed in the atmosphere where it adds to that from other sources of CO concentration such as internal combustion engine vehicles.

The maximum CO concentration in stack gas must be kept lower than 100 mg/Sm^3 or 80 ppm (see Table 6.8).

Particulate matter

Particulate matter includes a wide variety of materials such as unburned fuel, sulfur compounds, carbon, ash and non-combustible dust that enters the combustion chamber with the combustion air.

The quality and quantity of particulate matter are influenced mainly by the type of fuel, the boiler operating mode and the type of burner. Natural gas, widely used in Europe, and some oil fuels produce little solid matter and ash. Most coals and heavy oils produce a great quantity both of particulates in stack flue gas and of ash, some of which remains in the stack flue gas.

In the case of coal, ash can amount to 20% or more of the total weight of the fuel. Particular attention must be paid during both the design and the maintenance of the boiler to avoiding the accumulation of ash on internal boiler surfaces.

To prevent particulate concentration from exceeding emission standards and to reduce damage to health, various abatement techniques are used, such as filtration, mechanical separation and electrostatic precipitation. These techniques can be applied to both coal and heavy oil boiler plants.

The maximum particulate matter concentration in stack gas must be kept lower than 100-150 mg/Sm^3 (see Table 6.8).

Hydrocarbons

Hydrocarbons can be grouped as either unburned fuel components or compounds from chemical reactions which occur during combustion.

Hydrocarbons can be reduced by proper combustion, but traces of their compounds will always be present in stack flue gas.

Other pollutants

Other pollutants can be present in stack flue gas depending on the type of fuel.

With heavy fuel oils, in addition to the pollutants already mentioned, asphaltenes can form a finely dispersed colloid producing tar particles which deposit in storage tanks and travel through the burner without volatilizing. Asphaltenes can be stabilized by chemical additives; the same technique can be applied to prevent formation of vanadium-pentoxide which is known to promote sulfuric acid reactions.

When refuse-derived fuels (RDF) are used, the stack flue gas contains particulates, complex hydrocarbons, trace-metal emissions and chloride as HCL, depending on both the nature of the refuse and the kind of treatment upstream of the combustion, for separating and/or recycling part of the waste.

Table 6.8 Typical ranges of values of permitted pollutant concentrations from combustion

Pollutants	Level of concentration C_R (1)	
	mg/Sm^3	ppm (3)
Sulfur dioxide	1500-2000	525-700
Nitrogen oxides (2)	500-600	244-292
Carbon monoxide	100	80
Particulate matter	100-150	

(1) referred to dry gas and to a concentration of O_2 equal to 3% in volume.
To convert from the actual pollutant concentration C_x (with an oxygen concentration W_x) to the concentration referred to 3% value of oxygen concentration, use the following relationship: $C_R = C_x\ (18/21-W_x)$.
(2) expressed in NO_2
(3) 10000 ppm = 1% in volume

Table 6.9 Main techniques used to reduce NO_x emissions

Techniques		EFFECTS ON	
		NO_x emissions	Combustion efficiency
Excess air	(+)	(+)	(–)
Flame temperature	(+)	(+)	(+)
Pre-heating of air	(+)	(+)	(+)
Steam injection	(+)	(–)	
Gas recirculating	(+)	(–)	
Combustion chamber size	(+)	(–)	
Combustion chamber charge	(+)	(–)	
Heat production rate	(+)	(+)	
Heat production exchange	(+)	(–)	
Duration of combustion	(+)	(+)	
Burners integration	(+)	(+)	
Nitrogen in fuel	(+)	(+)	
Type of fuel			
Coal		(+)	
Oil		(+)	
Natural gas		(–)	

Note
(+) means INCREASE (–) means DECREASE

6.12 PRACTICAL EXAMPLES

Two examples of efficiency evaluation for industrial boilers are appended with reference to both the net and the gross heating value of fuel.

Example 1

Calculate the efficiency of an industrial watertube steam boiler by means of the heat-loss method

This concerns an industrial watertube steam boiler, fed by natural gas, with the technical characteristics stated below. The efficiency evaluation has been conducted by the indirect method and by measuring O_2% and CO%. Correlation curves and relationships introduced in Chapter 6.7 are here used to determine CO_2% and losses (Ostwald triangle, K_s and K_c coefficients). The radiation losses are estimated by a chart similar to the chart in Fig. 6.6.

Nominal boiler operating conditions

— steam flow	18 t/h
— final saturated steam pressure	7 MPa = 70 bar
— condensate return temperature	70°C = 158°F
— rated output $m \cdot (h - h_o)$	12350 kW

Test operating conditions

— steam flow (estimated)	16 t/h
— stack temperature (measured)	148°C = 298°F
— air temperature (measured)	28°C = 82°F
— O_2% (measured)	4.6%
— CO% (from Ostwald triangle, see fig. 6.10)	0.013%
— CO_2% (measured)	9.2%

Efficiency evaluation (net heating value as reference 34325 kJ/Sm^3)

— total stack gas losses

$$K_s \cdot \frac{t_s - t_a}{CO_2\%} = 0.468 \cdot \frac{148 - 28}{9.2} = 6.1\%$$

— losses for unburned combustible

$$K_c \cdot \frac{CO_2\%}{CO_2\% + CO\%} = 37.9 \cdot \frac{0.013}{9.2 + 0.013} = 0.05\%$$

— radiation losses (see Fig. 6.6)

1.2%

EFFICIENCY = 100 – (6.1 + 0.05 + 1.2) = 92.65%

Efficiency evaluation (gross heating value as reference 38450 kJ/Sm³)

—gross heating value/net heating value = 38450/34325 = 1.12

—total losses referred to the gross heating value if the output power is kept constant

(6.1 + 0.05 + 1.2)/1.12 + (112 – 100)/1.12 = 17.29%

EFFICIENCY = 100 – 17.29 = 82.71%

Efficiency is quite high, so no improvement is practicable.

Example 2

For a given rated output power, compare the energy consumption of two steam boilers having different efficiencies

This concerns an industrial firetube steam boiler, fed by natural gas, with the technical characteristics detailed below.

The efficiency evaluation has been conducted by the indirect method and by measuring $O_2\%$, $CO_2\%$ and CO%. Correlation curves and relationships introduced in Chapter 6.7 are here used to determine losses (K_s and K_c coefficients). The radiation losses are estimated by a chart similar to the chart in Fig.6.6.

Nominal boiler operating conditions

— steam flow	10 t / h
— final saturated steam pressure	0.9 MPa = 9 bar
— condensate return temperature	80°C = 176°F
— rated output $m \cdot (h - h_o)$	6805 kW

Test operating conditions

— steam flow (measured)	8 t / h
— stack temperature (measured)	245°C = 473°F
— air temperature (measured)	21°C = 69.8°F
— O_2% (measured)	5.2%
— CO% (for Ostwald triangle, see Fig. 6.10)	0.01%
— CO_2%(measured)	8.7%

Efficiency evaluation (net heating value as reference 34325 kJ/Sm³)

— total stack gas losses

$$K_s \cdot \frac{t_s - t_a}{CO_2\%} = 0.467 \cdot \frac{245-21}{8.7} = 12.02\%$$

— losses for unburned combustible

$$K_c \cdot \frac{CO\%}{CO_{2\%} + CO\%} = 37.9 \cdot \frac{0.01}{8.7 + 0.01} = 0.043\%$$

— radiation losses (see Fig.6.6)

3.5%

EFFICIENCY = 100 – (12.02 + 0.043 + 3.5) = 84.44%

Efficiency evaluation (gross heating value as reference 38450 kJ/Sm³)

—gross heating value/net heating value = 38450/34325 = 1.12

— total losses referred to the gross heating value if the output power is kept constant

(12.02 + 0.043 + 3.5)/1.12 + (112 –100)/1.12 = 24.62%

EFFICIENCY = 100 – 24.62 = 75.38%

The efficiency can be increased up to 91-92% (referred to NHV) or to 81-82% (referred to GHV) by replacing the boiler (particularly if it is near the end of its useful life) or by installing proper equipment to improve the combustion efficiency or the heat recovery.

Table 6.10 shows energy-saving evaluations for different operating conditions over a year.

The economic evaluation is in Table 20.3.

Table 6.10 Comparison between two boilers with different efficiencies (Example 2) fed by natural gas (34325 kJ/Sm^3 as Net Heating Value))

Existing boiler-rated steam flow		10 t/h		2770 kJ/kg		Saturated steam
6805 kW		0.9 MPa		320 kJ/kg		Condensate
Output		Working hours	Efficiency (*)	Natural gas input		
% rated capacity %	kW	h/year	%	kW	Sm^3/h	Sm^3/year
80	5444	2000	84.44	6448	676.2	1352469
40	2722	2000	80	3403	356.9	713765
20	1361	2000	76	1791	187.8	375666
TOTAL (a)						2441900
New boiler-rated steam flow		10 t/h		2770 kJ/kg		Saturated steam
6805 kW		0.9 MPa		320 kJ/kg		Condensate
Output		Working hours	Efficiency (*)	Natural gas input		
% rated capacity %	kW	h/year	%	kW	Sm^3/h	Sm^3/year
80	5444	2000	91	5983	627.5	1254972
40	2722	2000	90	3025	317.2	634458
20	1361	2000	88	1547	162.2	324439
TOTAL (b)						2213869
ENERGY SAVING	(a) – (b)				Sm^3/year	228031
					TOE/year	186.984
(*) Net Heating value as reference 34325 kJ/Sm^3						

CHAPTER 7

ELECTRIC DISTRIBUTION SYSTEMS FROM FACILITIES TO END USERS

7.1 ELECTRIC NETWORK AND RELATED EQUIPMENT

Electric energy, which is generally delivered to factories by means of utility electric networks at high or medium voltages, is distributed to end users in the factory through medium- and low-voltage networks. Depending on the power demand of loads and the process layout, transformer substations (see Chapter 5), whose main task is to reduce the supply voltage downstream of the utility delivering node, are concentrated at the factory boundary or distributed around the factory itself.

The choice among different distribution systems, such as radial or loop-feeder systems, shown schematically in Fig. 7.1, is based on technical and economic evaluations which generally do not consider energy-saving targets, because of the insignificant amount of the energy losses involved.

As a general indication, only a small part of the total electric power travelling along the internal lines, not more than 2-3%, is lost as Joule and additional losses, both related to the square of the flowing current. This figure, if compared with the heat distribution losses that can reach 10-20% of the total thermal energy flowing along pipelines (see Chapter 8), shows that these problems can easily be regarded as of secondary importance in an energy management program.

Most electric losses occur in end users which can be grouped basically as (i) electrical machinery and drives; (ii) electrically heated users such as furnaces, ovens, boilers, induction heating equipment, resistors, microwave equipment,

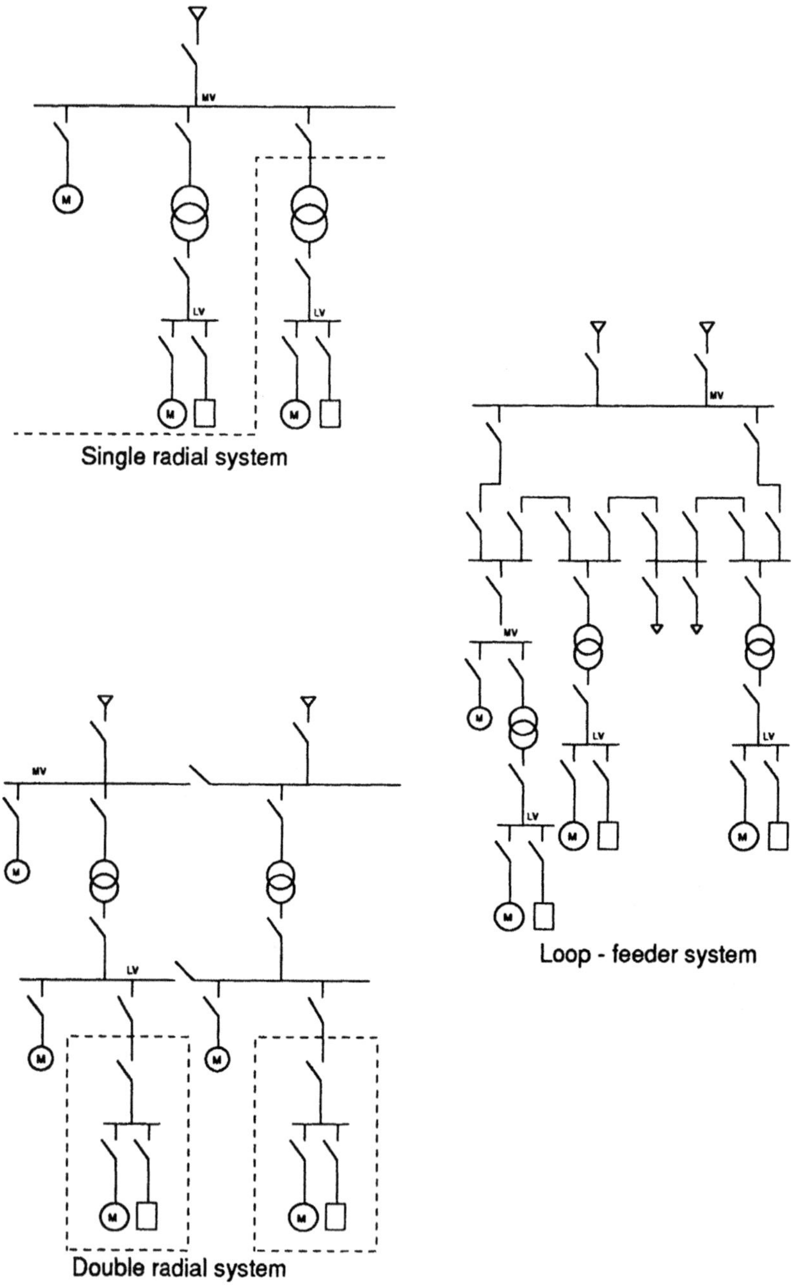

Fig. 7.1 Industrial electric distribution systems: radial, loop-feeder

etc; (iii) lighting; (iv) others such as electrochemical equipment, control and communication systems, etc..

Electrical drives and electrically heated users will be discussed in this chapter; lighting will be the topic of Chapter 14; for electrochemical and other applications, specialized technical literature should be consulted.

7.2 ELECTRIC DISTRIBUTION LOSSES

Electric distribution losses in the factory are due mainly to the Joule effect which depends on the square of the current and on the line resistance. The basic relationship can be expressed as follows:

$$P_{losses} = n \cdot R \cdot I^2 \qquad \text{(W)}$$

where

n = number of phase conductors (usually 2 or 3)

R = resistance of the phase conductor (Ω)

I = RMS or effective current (A)

If the current flows through a series of n electric conductors, the total resistance is expressed as the sum of the single conductor resistances:

$$R_{total} = \sum_{1}^{n}{}_{j}\, \rho_j \cdot \frac{l_j}{S_j} \qquad (\Omega)$$

where

ρ_j = resistivity of the conductive material which is defined as resistance per unit length of unit area at a given temperature (reference or operating temperature)

S_j = cross section of the j-conductor

l_j = length of the j-conductor

In the case of a line having n electric conductors connected in parallel which share the total current, the lower the resistance the higher the current flowing in each conductor. The total resistance is expressed as the reciprocal of the sum of the reciprocals of the resistances:

$$R_{total} = \frac{1}{\sum_{1}^{n} {}_j \frac{1}{\rho_j} \cdot \frac{1}{l_j} \cdot S_j}$$

If n conductors are in parallel, in practice, they have the same section; then, the resistance of the line is (1/n) multiplied by the resistance R of each phase conductor:

$$R_{total} = \left(\frac{1}{n}\right) \cdot R$$

Industrial distribution systems are mainly three-phase systems; here, too, the relationships shown above are used as follows:

$$\mathbf{P_{losses} = 3 \cdot R \cdot I^2}$$

where
R = resistance of a single-phase conductor (as above)
I = RMS or effective value of the current flowing in each line

The resistivity of conducting materials, such as copper and aluminium, varies with the temperature at a rate of about 0.4%/°C (0.22%/°F). The reference temperature for resistivity values is 75°C (167°F) for electrical machinery and 20°C (68°F) for cables. The temperatures at which cables can work in steady state conditions range from 50°C to 100°C (122-212°F); this range derives from the ability of various insulating materials to support high temperatures for a long time without unacceptable deterioration.

Additional losses occur if the frequency of the current differs from zero. In this case, which is the commonest, since losses vary according to the square of the current as do the basic Joule losses, an equivalent resistance is generally introduced to take these phenomena into account. In practice, this resistance is expressed as a percentage of the zero frequency or direct current resistance; values range from 3% to 100% and higher. As a general rule, the higher the frequency and the larger the conductor cross-section, the higher the percentage. Additional losses, by their nature, decrease as the temperature rises.

Values of resistivity for different materials, and other technical parameters, are reported in Table 7.1. Note that the resistivity is here expressed also in $\Omega \cdot mm^2/km$ in order to facilitate the calculation for line conductors.

The main parameters to consider in designing electrical systems are the current density and the voltage drop through lines and transformers, to which losses are related. Widely-accepted criteria for the design of electric power distribution systems in industrial plants are based on a predetermined voltage drop along the lines, generally fixed at less than 5% in normal operating conditions, to which corresponds less than 2-3% of losses along the same line. These values are related to the ratio resistance R/reactance X of the cable and to the power factor of the load. The maximum current density in this situation depends on the operating temperature and on the conductor cross-section. As a general rule, the larger the cross-section of the conductor the lower the acceptable current density (see Fig. 7.2).

For low-voltage cables (0.6-1 kV), and for various modes of installation, the following formula may be used to calculate the electric current capacity:

$$I = A \cdot S^m - B \cdot S^n \qquad (A)$$

where A, B, m, n are parameters defined by IEC Standard and S is the conductor cross-section.

A simplified formula for non-buried cables is:

$$I = a \cdot S^b \qquad (A)$$

where

a = current capacity of a conductor with a cross-section of 1 mm^2. Typical values for copper conductors are 13 A/mm^2 for PVC insulation and 17 A/mm^2 for EPR, XLPE insulation.

S = conductor cross-section in mm^2

b = 0.625 for the commonest cables and installation modes.

The electric current density is:

$$\delta = \frac{I}{S} = a \cdot S^{b-1} = a \cdot S^{-0.375} \qquad (A/mm^2)$$

For two typical cables made with conductors having cross-sections in the ratio 100 to 1 (e.g. 150 mm^2 and 1.5 mm^2), the ratio between the two values of the current density is:

$$\frac{\delta_{150}}{\delta_{1.5}} = \left(\frac{150}{1.5}\right)^{-0.375} = 0.17$$

This example shows that a given current density cannot be assumed as reference value to design cable lines, but that voltage drops and continuous operating temperatures are the main parameters to be considered.

Notice that the exploitation of the lines is generally lower than the designed level because the growth of the factory is often slower than originally foreseen; in consequence, line losses are much lower than the values given above.

In conclusion, because actual losses are comparatively small, it is not worthwhile to approach the industrial line design or retrofit with energy-saving design criteria as the most important.

Table 7.1 Technical parameters for different conducting materials

Parameter		Silver	Copper	Aluminium	Lead
Density (*)	kg/m^3	10523	8937	2707	11372
	lb/ft^3	657	558	169	710
Resistivity (*)	$\Omega \cdot m$	$1.60 \cdot 10^{-8}$	$1.76 \cdot 10^{-8}$	$2.83 \cdot 10^{-8}$	$2.10 \cdot 10^{-7}$
	$\Omega \cdot mm^2/km$	16.0	17.6	28.3	210
Temperature coefficient	%/°C	0.38	0.39	0.40	0.40
	%/°F	0.21	0.22	0.22	0.22
Specific heat (*)	kJ/kg · K	0.234	0.383	0.896	0.130
	Btu/lb · °F	0.056	0.092	0.214	0.031
Melting point	°C	961	1083	658	327
	°F	1762	1981	1216	621
Thermal conductivity (**)	W/m · °C	415	381	228	33
	Btu/h · ft · °F	240	220	132	19

(*) at 20 °C = 68 °F
(**) at 100 °C = 212 °F

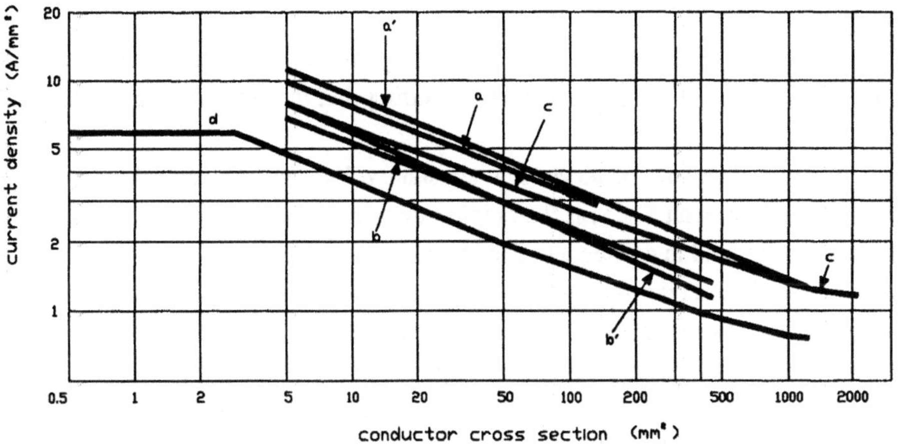

Fig. 7.2 Current density of different copper cables: (a) single-core paper insulated in air; (a') single-core paper insulated underground; (b) three-core paper insulated in air; (b') three-core paper insulated underground; (c) non-insulated conductor in air; (d) single-core rubber insulated cables

7.3 POWER FACTOR CONTROL

The power flowing through the electric lines has two components: (i) the active power (kW), which is the power available for conversion to mechanical, thermal, chemical, light or sound energy; (ii) the reactive power (kvar), which is used for exciting magnetic fields in transformers and electrical machinery or electric fields in capacitors.

The expressions for the active and reactive power are:

$$P = \sqrt{3} \cdot V \cdot I \cdot \cos\varphi \quad \text{(kW)}$$
$$Q = \sqrt{3} \cdot V \cdot I \cdot \sin\varphi \quad \text{(kvar)}$$

The power factor, i.e. the cosine of the angle between voltage and current in a specific section of an electric circuit, is an index of the ratio between the active and reactive power flowing in the same section:

$$\tan\varphi = Q/P$$
$$\cos\varphi = \cos \operatorname{atan} Q/P$$

This relationship can be used to calculate the power factor value in each section of the network.

In order to minimize losses in an existing line, the current flowing along it must be kept as low as possible while ensuring that the active power P required by the loads can still flow.

A constant amount of power P at constant voltage can be carried by keeping constant the product $I \cdot \cos\varphi$. The higher the power factor $\cos\varphi$, the lower will be the current and in consequence the line losses.

In electric circuits, this means that electric active power is transferred with the reactive power intrinsically needed by the majority of electric loads such as motors and transformers. If the reactive power is fed in nodes along the network, generally by means of capacitors, the excess of current, which flows downstream of the node itself because of the reactive power demand, will be reduced upstream.

Electric utilities usually charge for a part of the reactive energy consumed. This reactive energy E_Q must be paid for, since the utility has to provide plants of sufficient size to generate and transmit both the active and reactive power required by the end users and the active power corresponding to the line losses.

If $E_Q = 0.5 \cdot E_P$, then:

$\cos\varphi = 0.9$

This value, $\cos\varphi = 0.9$, represents a situation that is generally accepted by the utility companies in the delivering node. At lower $\cos\varphi$ values, corresponding to $E_Q > 0.5\ E_P$, a charge must be paid for each kvarh exceeding 0.5 kWh; at higher $\cos\varphi$ values, corresponding to $E_Q < 0.5 \cdot E_P$, no charges are scheduled.

Values of $\cos\varphi$ exceeding unity, which may occur when capacitors remain in operation even with low load rates, are not accepted by utilities because of overvoltages along the lines upstream. Consequently, factories must install control equipment or manual procedures in order to avoid this problem. The same procedures must be used to protect internal distribution lines if capacitors are installed at the end-user nodes.

The capacitor power (kvar) needed to raise cos φ to a determined value can be calculated with reference to the mean reactive and active power over a determined period ($Q = E_Q$/operating hours, $P = E_p$/operating hours), as follows (see also Fig. 7.3):

— reference situation

$$Q_0, P_0, \cos\varphi_0 = \cos \operatorname{atan} \frac{Q_0}{P_0}$$

— final situation ($\cos\varphi_1 > \cos\varphi_o$)

$$Q_1, P_0, \cos\varphi_1 = \cos \operatorname{atan} \frac{Q_1}{P_0}$$

$\Delta Q = Q_0 - Q_1$ = power of capacitors to be installed

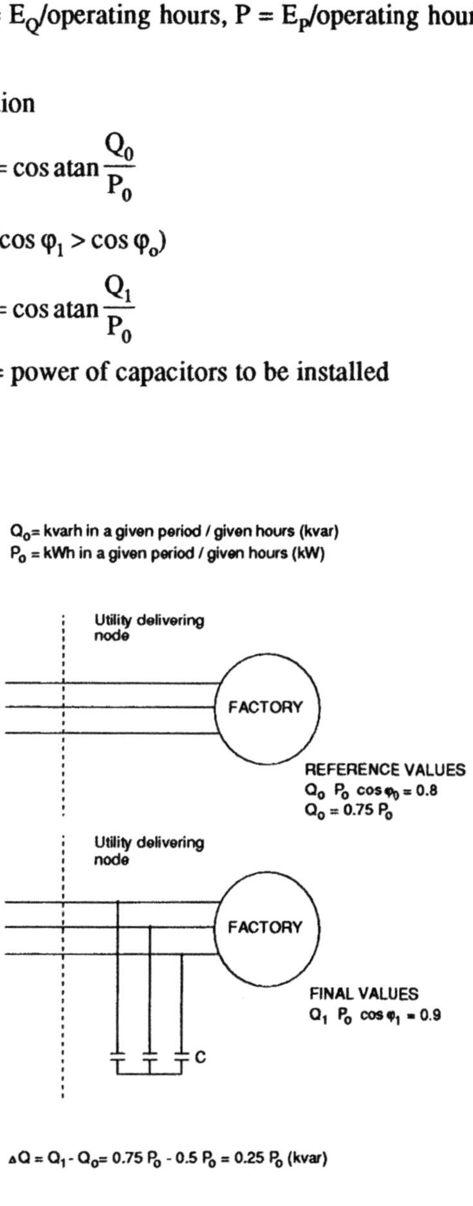

Fig. 7.3 Power factor control: a basic scheme of installation

— if the final situation must be $\cos\varphi_1 = 0.9$

$$\cos\varphi_1 = 0.9 \text{ corresponds to } \varphi_1 = \text{atan}\frac{Q_1}{P_0} = \text{atan } 0.5$$

$$Q_1 = 0.5 \cdot P_0 \quad \Delta Q = Q_0 - Q_1 = Q_0 - 0.5 \cdot P_0$$

If the section in the utility delivering node is considered, and capacitors are installed in this section, cost rather than energy is saved because only utility rates for the excess of reactive power are reduced. In consequence, energy is saved only in utility systems upstream of the delivering node, since these carry the same active power with a lower current.

The foregoing considerations apply also to internal sections of the power distribution systems, in which capacitors must be installed, generally near the loads demanding most reactive power (high-power motors, transformers, inductive loads, etc.), if energy saving is desired for the factory too.

7.4 ELECTRICAL DRIVES

Electrical drives can be defined as systems where electric energy is transformed into mechanical and vice versa by controlling electrical (voltage, frequency and current) and mechanical (torque, speed) parameters. The main components of electrical drives are shown in Fig. 7.4: electrical rotating machinery, power electronic devices and related controls with transducers of electrical and mechanical quantities (current, voltage, frequency, magnetic flux, speed, shaft position, torque, etc.).

Electrical rotating machinery may operate as a motor or as a generator depending on the power flow direction, from the shaft to the electric network or vice versa. Motors and drives transform electric energy into the mechanical energy needed at fixed speed or at variable speed to operate process and facility equipment of which pumps, compressors and fans are the commonest. Downstream are performed other energy transformations, e.g. from mechanical energy into cold fluid or compressed air.

A rated power or nameplate power is introduced as the maximum output power which can be delivered as mechanical or electric power in specific conditions (voltage, current, frequency, speed, shaft profile demand and service, type of cooling) without exceeding the maximum temperature allowed for each class of insulating materials used inside the electrical machinery. These temperatures depend on the class of insulating materials, but they do not exceed 453 K (180°C; 356°F) for the highest class (class H in IEC standard); if operating temperatures exceed these limits, insulating materials are damaged to the extent of losing all their useful properties, thus provoking short circuits and machine faults. Attention must be paid to the cooling system (natural ventilation, forced ventilation, intermediate fluid cooling, etc.) in order to keep temperatures below the rated values.

Efficiency of the rotating machinery (that is: the ratio between mechanical output power and electric input power for motor operation; the ratio between electric output power and mechanical input power for generator operation) is generally in the range 85-95%, the higher values for medium- and high-power machines (medium-power machines between 50 kW and 500 kW; high-power above these values). Efficiency also depends on the percentage of the rated power as shown in Fig. 7.5. Smaller machines usually have lower values of efficiency.

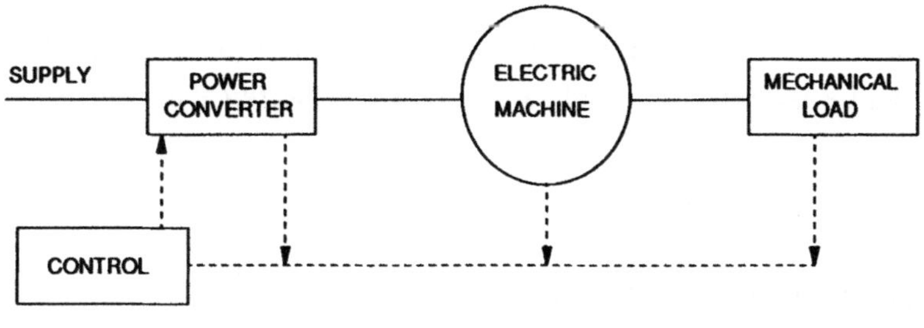

Fig. 7.4 Basic scheme of electrical drives

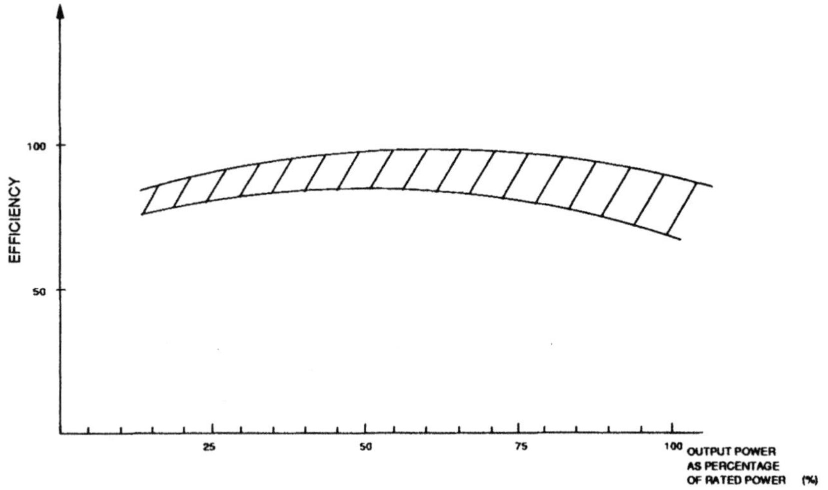

Fig. 7.5 Efficiency versus rated power for electric motors

Efficiency can be improved (i) in the design phase, with the so-called high-efficiency motors where the efficiency is 2-3 percentage points higher than in normal-efficiency motors, by using lower-losses materials, in particular low-losses iron; (ii) in the operating phase by allowing the machine to operate around the maximum efficiency point (see Chapter 5 for transformers).

Efficiency of the entire drive is lower because of the power converter efficiency which however is about 95%. Overall efficiencies of 85-90% can be assumed as actual values for many electrical drives.

Basic principles and operating modes of electrical machines.

Electrical machines are traditionally classified as alternating current machines (synchronous machines, asynchronous or induction machines, commutator machines) and direct current machines (commutator machines) according to the voltage of the feeding line. The stationary structure (stator) and the rotary structure (rotor) are both generally provided with windings. One structure has to generate a magnetic flux by means of a particular set of windings (field windings in commutator machines and synchronous machines) or permanent magnets, the other structure has the main windings (armature windings) through which power flows from the electric line to the shaft or vice versa. Asynchronous machines (also called induction machines) have polyphase stator windings (generally two- or three-phase) which generate a rotating flux and carry the electric power fed by the line; rotor windings are usually polyphase squirrel-cage windings.

A.c. (alternating current) machines can be fed directly from the line at medium or low voltage depending on the power of the machines (medium voltage for high-power machines, more than 500 kW). Synchronous machines must rotate at a synchronous speed related to the line frequency (synchronous speed: in r/min = 60 f/pair of poles; in rad/s = 2 π f/pair of poles where the frequency f is expressed in Hz) to effect the energy transformation, so they have to rotate at a constant speed and are not self-starting. All the other a.c. machines (except the synchronous) may work at any speed and are self-starting.

D.c. (direct current) machines require a d.c. voltage line which is not generally available in industrial applications. Power converters from a.c. to d.c. voltage are therefore necessary.

As a general rule, shaft torque can be expressed as the product of magnetic flux (generated by a current flowing inside a field winding or inside a polyphase winding or by a permanent magnet in one structure) multiplied by current flowing inside the winding in the other structure (stator or rotor). This relationship is the basis of all torque control methods.

At steady state, torque-speed characteristics can be used to represent the output of the machines when operating as motors and the input when operating as generators. The working point is the intersection between the torque-speed curve

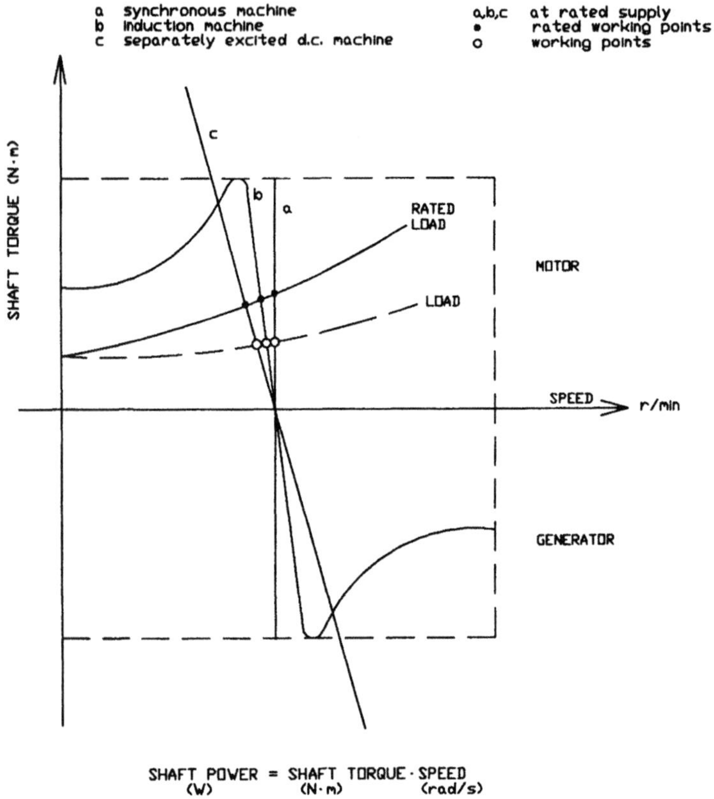

Fig. 7.6 Torque-speed characteristics for electrical machines at given input voltage and current

of the machine and the load curve as shown in Fig. 7.6 where the characteristics of the various classes of machine are represented.

If the machine is fed directly from the line, at constant voltage and frequency, only one torque-speed characteristic (that is, the locus of working points of the machine) exists and therefore only one working point is allowed for each load. Thus, at constant voltage and frequency, torque and speed values at the working point are imposed by the load and may change only if one of the load parameters varies.

Rheostatic control, which changes the speed of a motor by introducing a variable amount of resistance into the stator or into the rotor circuit, always involves waste of energy when the motor runs at reduced speed. In fact, a voltage drop occurs at the terminals of the rheostat, thus provoking Joule effect losses in it. Notice that these systems cannot change the synchronous speed already defined because they do not change the frequency of the supply. Once widely used for controlling d.c. and induction machines with wound rotors, rheostats have gradually become obsolete because of the introduction of power electronic converters which exercise a better control with lower losses.

In conclusion, energy-saving at the end-user level is possible only by a correct choice of the nominal power of the machine, so that the working point is situated in the maximum efficiency zone, and by installing high-efficiency machines instead of standard machines. Once the installation has been made, there is no flexibility to come into play if the load changes, unless more complex systems, such as power electronic converters, are introduced between the supply and the electrical machine.

Basic principles and operating modes of electrical drives

Electrical drives, instead of electrical machines alone, allow a set of working points inside a working area instead of one fixed point.

The electric input variables (voltage, current, frequency) can be modified within a given range by controlling the power electronic converter properly, so that the mechanical output quantities (torque, speed, shaft position) are regulated independently within defined boundaries. In this way, the external characteristics of the machines can be continuously modified, giving rise to a family of curves which allow operation of the drive in an entire region of the torque-speed plane. Any point can then be reached with an appropriate control law, by varying the motor voltage (or current), the frequency or both. The working point of the drive is no longer confined, for a given load, to the single intersection with the electrical machine characteristic resulting from a fixed-parameters source.

Two working regions are generally defined: (i) constant-torque region as a set of operating points where any torque can be delivered by the drive at any speed, within the thermal and insulation limits of the drive; (ii) constant-power region as an operating region where the torque is delivered only within the power limit of the drive (remember that shaft power = shaft torque · speed), so that its value decreases as the speed increases.

Operation at constant torque is normally performed with a constant value of the machine flux (by permanent magnets or by constant field current) and implies that any torque, within the motor's capabilities, can be delivered at any speed by action on the proper control variables, usually the current. Torque being flux multiplied by current, when the flux and the current are kept constant, a constant value of torque can be delivered at any speed by varying the voltage (and the frequency in a.c. machines) according to the control system requirements.

Operation at constant power occurs beyond the constant-torque region, and higher speeds are reached at the expense of the torque within the limits of the maximum power delivered by the machine. Operation is bounded by the hyperbola of the equation Power = Torque · speed = constant. In this region, voltage is kept constant (at its rated value or lower), while the machine flux is weakened. A definitive limit on the speed axis can be either a mechanical constraint related to the rotor type of construction, or an electrical limitation (converter or motor current). Because of the constant-voltage value, constant power at any speed corresponds to a constant current from the line (neglecting losses).

Fig. 7.7 shows constant-torque and constant-power regions for various drives; both motor and generator operating modes are illustrated.

Notice that electrical drives may work in a reversible mode by transforming mechanical energy into electric energy without significant modification of their structure. This capability is used mainly to effect regenerative braking by transforming energy stored inside rotating masses at the shaft into electric energy which is fed back to the network. Accurate braking control together with energy recovery are the main reasons for using this mode of operation both in industry and in traction.

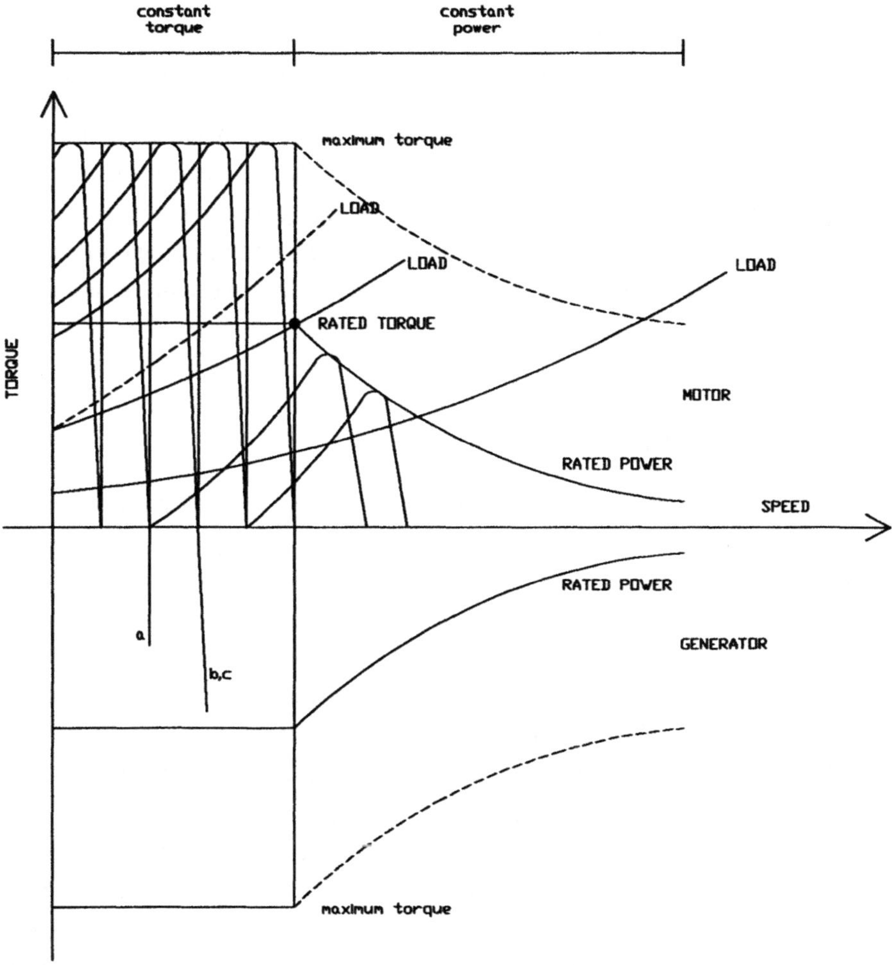

Fig. 7.7 Constant-torque and constant-power region for drives with synchronous machines (a), asynchronous (b) and d.c. machines (c)

Electrical drives play an important role in energy-saving when machines such as pumps, compressors and fans are controlled by speed variation as an alternative to other, dissipative, methods (see Chapters 10, 11, 12, 13).

In addition, electrical drives are among the main components of industrial automation systems and allow improvements in production quality and quantity, thus generally determining a decrease in specific energy consumption per unit of production. Interfacing with the plant centralized control, diagnostics and monitoring are other factors that make electrical variable-speed drives attractive for industrial applications.

Application problems

Electrical variable-speed drives may be a source of thermal, mechanical and electrical malfunctioning which must be investigated carefully in order to guarantee a safe operation of the drive itself and of the electric network.

The power factor at the drive feeding node changes according to the operating mode: generally, the lower the actual load, the lower the power factor. That depends on the internal structure of the power converter and on the current and voltage harmonics, but it can be assumed as a general rule. A control system is usually required to keep the power factor value within the accepted range.

Derating of electrical machines, when fed by power converters, is necessary because of the increase in losses due to the harmonics and because of the reduction in efficiency of cooling systems at low speed. A derating of 10-15% depends on the harmonics; higher derating, up to 50%, is due to thermal phenomena caused by the reduced speed range. In any case, an independent cooling system, generally a fan driven by an independent electric motor, is installed.

7.5 ELECTRICALLY HEATED END USERS

Industrial uses of electric energy for heating include metal treatments, boilers, induction heating, electric arc furnaces, heat pumps, electromagnetic wave heating, mechanical compression of steam, etc.. They are different types of application based on different principles.

Joule losses principle

The operating principle refers to the Joule losses (power = resistance · current^2).

In metal treatment applications, two situations may occur: (i) the current is fed from the line and flows through a resistive winding; (ii) the current is fed from the line and flows through an inductive winding thus determining induced voltage and current inside the metal to be heated which is generally inside the inductor winding.

Because of the losses in the power plant producing electric energy from fuels or other combustibles, electric heating seems to be unacceptable from the primary energy consumption point of view (roughly 9628-10465 kJ/kWh or 2300-2500 kcal or 9126-9919 Btu/kWh are required from a power plant instead of the 3600 kJ or 860 kcal or 3412 Btu made available as heat per 1 kWh at the end users).

Heating from combustibles thus consumes less primary energy and is quite often cheaper, except for special electric energy low rates (night, holiday, summer, others) charged by utilities.

Production requirements, in terms of constant quality and quantity, may impose the use of electric heating. Notice that the end-user efficiency of heating from combustibles (typical values 50-60%) is lower than that of electric heating (the ratio between the two efficiencies is roughly 0.7) because of the unavoidable flue gas losses due to the combustion, but that does not modify the considerations already made.

Boilers to produce steam or hot water by using the Joule losses principle have been discussed in Chapter 6.5.

Electromagnetic wave heating

An alternative method of heating is electromagnetic wave heating. This subjects the material under treatment to electric waves ranging from a few Hz to 1 GHz in frequency for radiofrequency systems and from 1 GHz to 300 GHz for microwave systems. Typical values are 13-27 MHz for the former and 2.5 GHz for the latter.

Advantages of these systems are the possibility of heating the material from the core to the surfaces instead of vice versa, as in traditional heating systems based on fuel combustion, the reduction of local pollution, the easy control of the energy flow and the reduction of the heating time to 1/3-1/20.

These systems are used in sterilization, defreezing, cooking and drying. Industrial drying applications are quite common in textile, food processing and other industries; in addition, both domestic and industrial cooking increasingly employ these systems.

Notice that these systems alone do not permit a significant energy saving in drying because the specific consumption, the energy consumed per kg of evaporated water, is quite similar to or higher than the consumption in the case of traditional drying based on combustion.

Typical values are 1.2-1.3 kWh/kg of evaporated water, which are similar to those obtained when fuels are used for standard drying; of course, specific consumption as primary energy needed by utility power plants is higher than with combustion-based dryers.

These systems are quite often installed as a second step after thermal or mechanical, e.g. centrifugal, drying, to eliminate water from the core of the material. In this way, the quality of production is improved by action on a phase of the process in which traditional systems would require higher energy consumption to achieve similar results.

Single modules generally size from 1 to 10 kW output for microwaves and from 20 to 100 kW for radiofrequency systems. They are assembled for higher power.

Heat pumps

A heat pump is a device which operates cyclically to transform low-temperature energy from a source (air or water) into high-temperature energy by the application of external work, mainly mechanical. The prime mover is either an electric motor or a fuel engine. Energy can be saved depending on the low-temperature source available and on the requirement of the end users at high temperature (see Chapter 12.9).

Mechanical recompression of steam

This system increases the pressure and thus the temperature of saturated steam by using mechanical energy of compression instead of thermal energy from a boiler. It serves to make small pressure changes, particularly when rejected steam from process (for instance liquid concentration or distillation) is available for re-use in the same process equipment.

Typical applications are in dairy industries, food processing and chemical industries.

In mechanical vapor recompression systems, typical values are 20-30 kWh for recompressing 1 t of steam with a temperature increase of 6-8 K or °C (11-15°F). Greater increases will require higher energy consumption.
For mechanical recompression in concentration plants the basic scheme is shown in Fig. 7.8: typical values are 20-30 kWh of electric energy plant consumption for evaporating 1 t of water.
Lower values can be achieved if mechanical compression is combined with multiple-effect evaporation plants.
If the compressor is driven by a turbine or a diesel engine with heat recovery (see cogeneration in Chapter 9), operating costs can be reduced.

An alternative to mechanical recompression is thermocompression (see Fig. 7.8) where the steam pressure (and thus the temperature) increase is realized by means of an ejector (see Chapter 11.1). In this case, high-pressure steam from boilers is used to transport the rejected steam into the ejector, thus converting the velocity of the mixture to pressure in a diffuser.

The higher the driving steam pressure in proportion to the suction pressure of the rejected steam, the lower the driving steam flow-rate will be.

Typical operating parameters of thermocompression are: 0.6-1.2 MPa for the driving steam pressure, 0.05-0.02 MPa for the suction pressure, 1.5-2 for the compression ratio of the ejector (between sections 1 and 2 in Fig. 7.8; 15-20 K or °C of temperature increase), 0.5-1 for the ratio between the driving steam flow rate and the rejected steam in the ejector. The energy consumption is ten times greater than with mechanical recompression systems.

This system can be used in the first stage of a multiple-effect evaporation plant; in this case, the plant requires less steam or fewer effects than a pure multiple-effect evaporation plant with the same evaporation rates, but still more than plants with mechanical evaporation.

7.6 END USERS AND LOAD MANAGEMENT

Figure 7.9 shows a basic scheme for an electric system where average values of efficiency for different end users are also indicated.

Note that the efficiency of the electric end users mentioned above is generally high, particularly in electrical drives, and no significant improvement can be made in the efficiency of the component alone. Nevertheless, high-efficiency motors, where the efficiency is 2-3 percentage points higher than in normal-efficiency motors, can be justified both technically and economically when working hours are more than 4000-5000 h/year. Better results can be obtained if steps are taken to improve the performance of the whole system, for example in pumps and fans (Chapter 10), in air compressors (Chapter 11), in industrial cooling systems (Chapter 12), in HVAC systems (Chapter 13) and in lighting (Chapter 14). Of course, all these considerations can be taken as general guidelines to determine priorities in an energy-saving program.

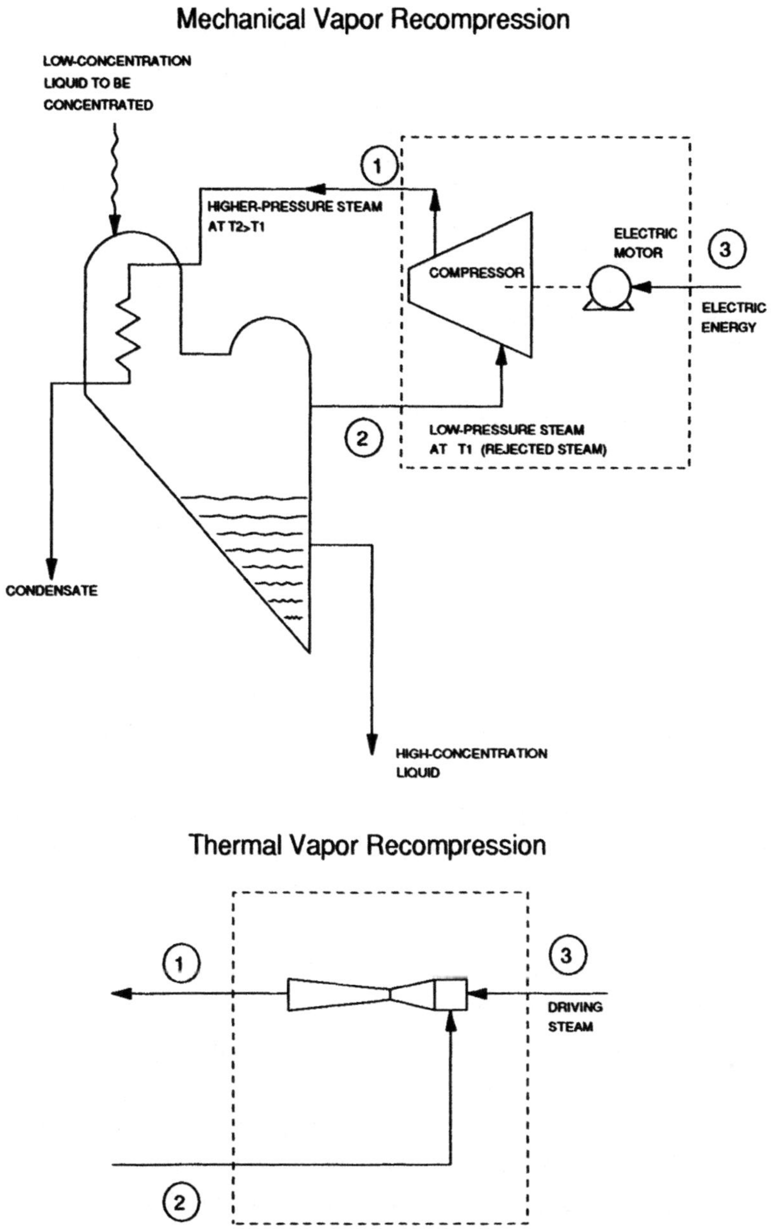

Fig. 7.8 Operating principles of MVR (Mechanical Vapor Recompression) and TVR (Thermal Vapor Recompression)

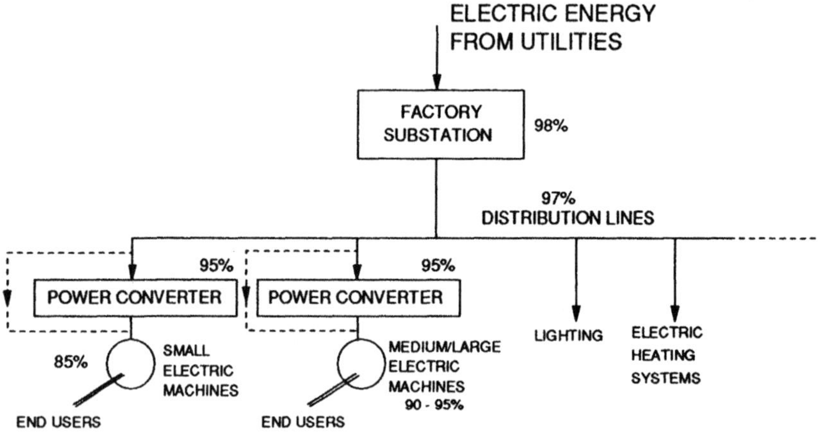

Fig. 7.9 Typical values of overall efficiency from factory boundaries to electric equipment terminals

In addition, actions such as the application of load-management techniques may give economic benefits on utility rates. This result is achieved by reducing power peaks without sacrifing production quality or quantity and by shifting energy consumption from peak hours to low-rate hours. A proper choice of the loads to be controlled and of the shut-down sequence can lead to appreciable cost-saving and also to effective energy-saving by reducing energy waste, particularly in facility plants.

Microcomputer or computerized energy management control systems are widely used in factories and in large buildings. They generally prove economical, depending on the complexity of the installation (number of loads, layout of the plant, power of the single controlled load); they are becoming more and more attractive for use in conjunction with non-energy functions such as monitoring and controlling pollution-producing equipment, managing preventive maintenance, recording production, and other non-engineering functions.

7.7 PRACTICAL EXAMPLES

Example 1

Calculate the losses in an electric industrial distribution line with different levels of power factor control

A practical example is given below of the calculation of losses in an industrial electric distribution line with different levels of power factor control at load nodes. This approach must be followed in order to evaluate the economic validity of installing capacitors near the loads instead of concentrating them at the utility nodes (see Chapter 7.3).

In reference to Fig. 7.10 and on the assumption that the load voltage V is kept constant, the formulae used for calculation are:

— line current

$$I = \frac{P}{\sqrt{3}\ V \cos\varphi}$$

— total line losses

$$PL = 3 \cdot RT \cdot I^2 = \frac{RT \cdot P^2}{(V \cdot \cos\varphi)^2}$$

— line voltage drop

$$\Delta V = \sqrt{3} \cdot I \cdot (RT \cdot \cos\varphi + XT \cdot \sin\varphi)$$

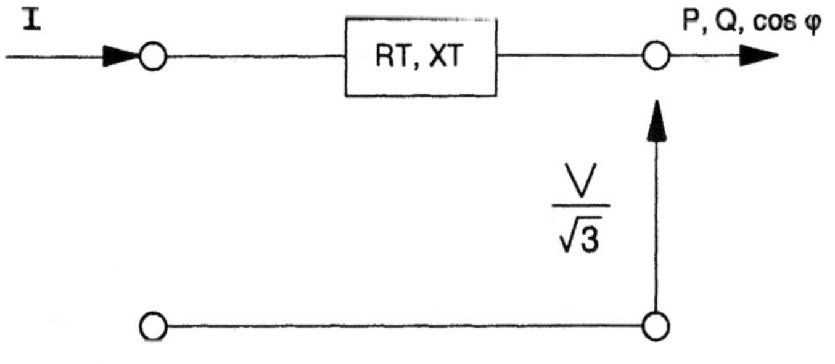

Fig. 7.10 Simplified equivalent circuit for a distribution line

— maximum loading power with a voltage drop of 5%

$$PMV = \frac{5}{100} \cdot \frac{V^2 \cdot \cos\varphi}{RT \cdot \cos\varphi + XT \cdot \sin\varphi}$$

— maximum loading power with rated current I_n

$$PMI = \sqrt{3} \cdot V \cdot I_n \cdot \cos\varphi$$

The different distribution lines examined are represented in Fig. 7.11.a and in Fig. 7.11.b together with their equivalent series circuit RT, XT.

The basic assumptions for the calculation are:

— the load voltage is kept constantly equal to the nominal value;

— the recorded active and reactive power upstream the distribution line (point 1) in the factory electrical substation are assumed as load powers. The load is taken to be totally concentrated in the point most unfavorable for losses and voltage drop. In a real situation the load is divided into many loads distributed along the line. The calculation used gives the maximum losses;

— voltage drop along the line calculated with the simplified formula given above.

Results for the circuits illustrated in Fig. 7.11.a and in Fig. 7.11.b are shown in Table 7.2.

CASE 1

Active and reactive load power are the reference values which have been recorded upstream the distribution line in normal operating conditions (point 1).

CASE 2

Active load power is kept equal to the reference value as in CASE 1 but a power factor equal to 0.9 is imposed because of the installation of capacitors at the load node.

CASE 3

Active load power equal to the maximum loading power with power factor equal to the reference value as in CASE 1 and a current corresponding to a voltage drop equal to 5%. Alternatively, a current equal to the line nominal value is imposed if the previous value of the current was higher.

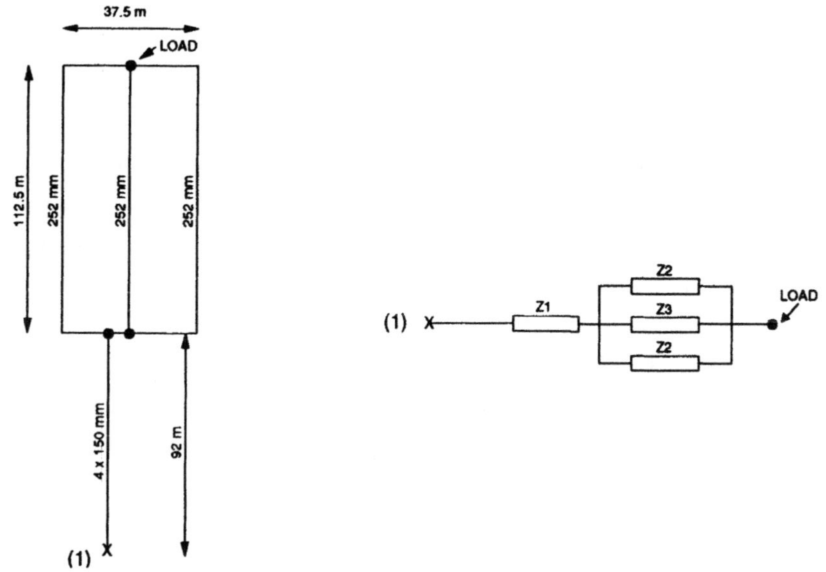

Fig. 7.11a An example of an industrial electric distribution line; see Table 7.2a for calculation

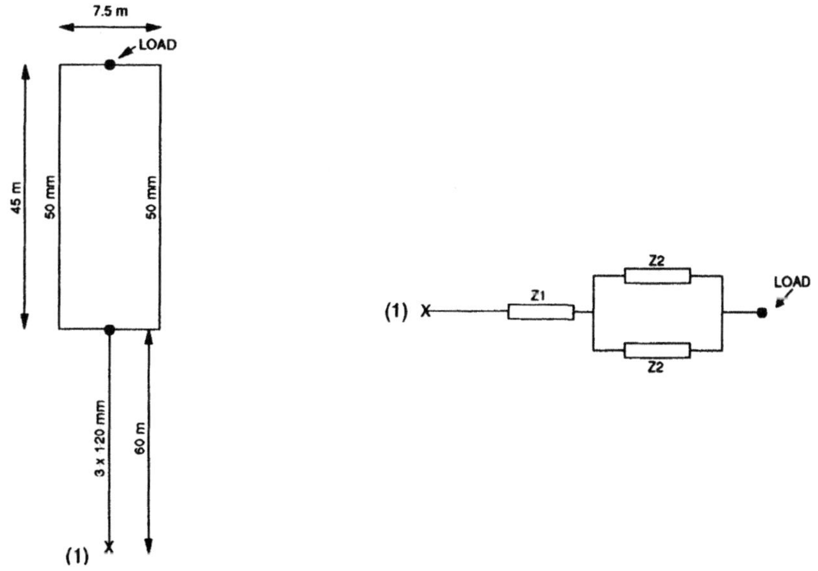

Fig. 7.11b An example of an industrial electric distribution line; see Table 7.2b for calculation

CASE 4

The same as in CASE 3, but a power factor equal to 0.9 is imposed because of the installation of capacitors at the load node.

Note that CASE 1 represents a typical operating condition, so that calculated losses are the maximum possible in this situation. CASE 2, with a power factor equal to 0.9, shows the maximum reduction of losses obtainable with a power factor control in the operating conditions shown in CASE 1.

CASE 3 represents the maximum exploitation of the line with a voltage drop equal or inferior to 5% or a line current equal or inferior to the nominal value. This case, which is different from the operating situation in CASE 1, shows the maximum losses obtainable with the maximum exploitation of the line.

CASE 4, with a power factor equal to 0.9, shows the maximum reduction of losses obtainable with a power factor control in the operating conditions shown in CASE 3.

For all cases, Table 7.2 shows the annual energy losses and possible energy saving if capacitors are installed as in CASE 2 and CASE 4. The power of the capacitors needed is also given for these cases, so that the investment can be evaluated. The cost of the capacitors alone represents a small part of the total investment which is determined mainly by installation costs dependent on many factors, such as the state of the electric line and related equipment and the distribution of loads along the line.

Economic evaluations for CASE 2 and CASE 4 are given in Table 20.3.

Table 7.2a Voltage drops and losses for different cases shown in Fig. 11a

Z1 (cable 4 · 150 mm^2; 92 m)		R1 = 0.0034 Ω	X1 = 0.0014 Ω	
Z2 (bus duct 252 mm^2; 150 m)		R2 = 0.0129 Ω	X2 = 0.0196 Ω	
Z3 (bus duct 252 mm^2; 112 m)		R3 = 0.0096 Ω	x3 = 0.0146 Ω	
RT	0.0073 Ω			
XT	0.0073 Ω			
Three-phase voltage	380 V, 50 Hz	P, Q in point (1)		
	CASE 1	CASE 2	CASE 3	CASE 4
LOAD POWER P, Q				
P (W)	260000	260000	459200	459200
Q (var)	300000	125000	529840	220770
φ = atan(Q/P)	0.86	0.45	0.86	0.45
power factor cos φ	0.65	0.90	0.65	0.90
POWER LOSSES PL				
PL (W)	7967	4207	24852	13124
LINE VOLTAGE DROP ΔV				
ΔV (V)	10.758	7.396	19.000	13.063
ENERGY LOSSES				
working hours (h/year)	3000	3000	3000	3000
losses (kWh/year)	23902	12622	74556	39372
ENERGY SAVING				
case 1-case 2 kWh/year		11280		
case 3-case 4 kWh/year				35184
CAPACITOR POWER				
case 1-case 2 kvar		175		
case 3-case 4 kvar				309

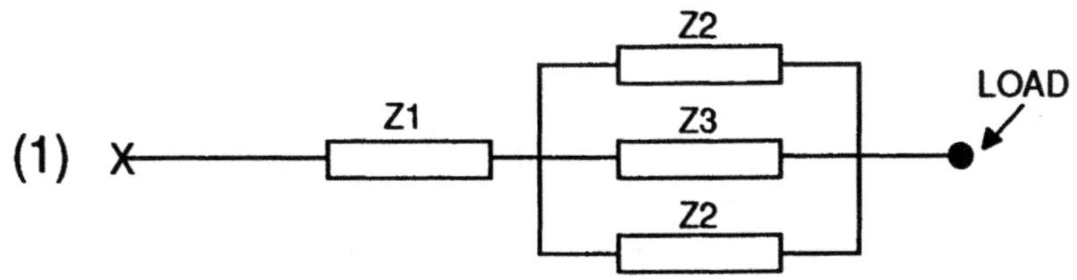

Table 7.2b Voltage drops and losses for different cases shown in Fig. 11b

Z1 (cable 3 · 120 mm^2; 60 m)		R1 = 0.0037 Ω	X1 = 0.0015 Ω	
Z2 (bus duct 50 mm^2; 52 m)		R2 = 0.025 Ω	X2 = 0.009 Ω	
RT	0.0162 Ω			
XT	0.006 Ω			
Three-phase voltage	380 V, 50 Hz	P, Q in point 1		
	CASE 1	CASE 2	CASE 3	CASE 4
LOAD POWER P, Q				
P (W)	190000	190000	122160	122160
Q (var)	265000	90000	171570	57870
φ = atan(Q/P)	0.95	0.44	0.95	0.44
power factor cos φ	0.58	0.90	0.58	0.90
POWER LOSSES PL				
PL (W)	11928	4959	4977	2050
LINE VOLTAGE DROP ΔV				
ΔV (V)	12.284	9.521	7.917	6.122
ENERGY LOSSES				
working hours (h/year)	3000	3000	3000	3000
losses (kWh/year)	35785	14876	14930	6150
ENERGY SAVING				
case 1-case 2 kWh/year		20909		
case 3-case 4 kWh/year				8780
CAPACITOR POWER				
case 1-case 2 kvar		175		
case 3-case 4 kvar				114

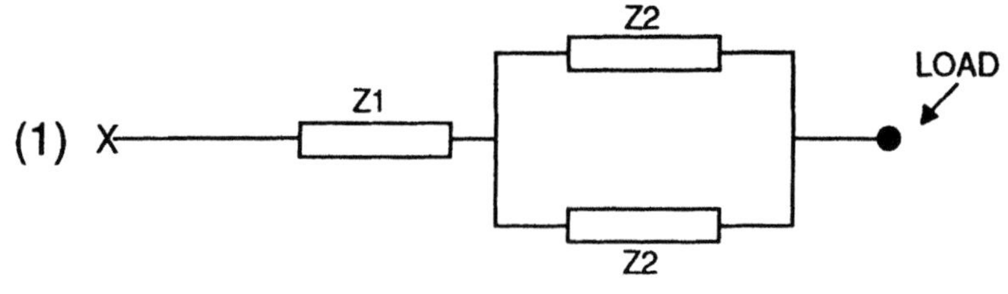

Example 2

Compare the energy consumption of a microwave and a thermal dryer system in the textile industry

An example is given here of a microwave heating system used for drying in the textile industry.

The drying line processes 250 kg/h of dry cotton cops for 6000 hours per year. The cotton entering the line has a water content equal to 250% by dry weight; that means 875 kg/h of total weight entering the line of which 625 kg/h is water. The water content must be reduced to 8% by dry weight at the end of the process by eliminating 605 kg/h of water in two steps: (step 1) an expulsion of 500 kg of water/h by mechanical centrifugation or by compressed air in an autoclave (water content at the exit of this step is 125 kg/h); (step 2) removal of 105 kg of water/h by a thermal dryer or a microwave system.

The two systems are compared below in regard to the second step of drying.

The thermal dryer

The dryer works with hot air produced by a steam/air exchanger and has a requirement of useful power for drying equal to 6500 kJ/kg of eliminated water. The electric consumption for ventilation is 120 kW for a rate of 105 kg of water/h.

The microwave system

The microwave dryer removes water from the core of the material.

The rated useful power of the microwave equipment is 90 kW (30 modules of 3 kW each) with an absorbed power of 128 kW for a rate of 105 kg of water/h.

The specific consumption is then 128/105 = 1.22 kWh/kg of water.

Table 7.3 lists the main data for the comparison and the resultant energy saving.

Economic evaluation is shown in Table 20.3.

Table 7.3 Comparison between microwave-based dryer and hot-air dryer

	Operating parameters		Hot air (II)	Microwave (I)
a	Working hours	h/year	6000	6000
	DRYING (step 2)			
b	Input dry cotton in cops	kg/h	250	250
c	Water content	kg/h	125	125
d	Water to be eliminated	kg/h	105	105
$e = 6500 \cdot d/3600$	Useful power for drying	kW	190	90
f	Absorbed electric power			
	for drying	kW	0	128
	for ventilation	kW	120	0
$g = e/0.87$	Absorbed thermal power (*)	kW	218	0
$h = g \cdot 3600/41860$		kgoil/h	18.8	
	SPECIFIC CONSUMPTION			
$i = f/d$	Electric	kWh/kg H2O	1.143	1.219
$l = h/d$	Thermal	kgoil/kg H2O	0.179	0
$m = l \cdot 41860$		kJ/kg H2O	7500	
	ANNUAL ENERGY CONSUMPTION			
$n = f \cdot a$	Electric	kWh/year	720000	768000
$o = h \cdot a$	Thermal	kgoil/year	112875	0
$o' = o/1000$		TOE/year	112.875	0
	ENERGY SAVING (I)-(II)			
$p = n(I) - n(II)$	Electric	kWh/year		–48000
$q = o'(I) - o'(II)$	Thermal	TOE/year		112.875

Note
(*) boiler efficiency equal to 0.87 (referred to Net Heating Value)

Example 3

For a given rated output power, compare a high-efficiency with a normal-efficiency electric motor

In many manufacturing industries (bricks, tiles, foundries, paper, etc.) and in HVAC systems, motor-driven fans and pumps work at constant load for the whole year.

A 55 kW rated power electric motor works at 80% of its rated power for 5000 h/year with 86% efficiency.

This motor can be replaced by an electric motor having the same rated power but with higher efficiency (90%).

Table 7.4 shows the calculation of the energy saved, which proves to be 11370 kWh/year.

The economic evaluation is in Table 20.3.

Table 7.4 Comparison between two electric motors with different efficiencies (Example 3)

Standard-efficiency motor			55kW	Rated output power	
Output		Working hours	Efficiency	Input	
% rated capacity %	kW	h/year	%	kW	kWh/year
80	44	5000	86	51	255814 (a)
Standard-efficiency motor			55kW	Rated output power	
Output		Working hours	Efficiency	Input	
% rated capacity %	kW	h/year	%	kW	kWh/year
80	44	5000	90	49	244444 (b)
ENERGY SAVING (a)–(b)				kWh/year	11370

CHAPTER 8

THERMAL FLUID DISTRIBUTION SYSTEMS

8.1 THERMAL FLUID PIPELINES

Thermal energy entering the factory as fuel is distributed to end users in different ways: pipelines downstream of the boiler plants may carry thermal energy by means of heating media such as steam, hot water, hot oil, hot air; alternatively, pipelines may carry natural gas or oil to furnaces and ovens where fuels are burned directly.

In most industrial applications, steam is certainly the least expensive and most effective heat-transfer medium if water is available. The conversion of water into vapor absorbs a large quantity of heat per kilogram of water. The resulting steam, which is easy to distribute around the factory, contains a relatively high amount of heat if compared with other heat-transfer media. Finally, the process of heat-transfer by condensation at end-user level, is generally very efficient because of the quantities transferred in a relatively short time by means of simple equipment.

Natural gas is generally delivered by utilities to factories at high pressure (1.5-5 MPa; 15-50 bar) and then brought down to lower pressure at the factory boundary or near the end users. Losses resulting from these transformations are relatively low.

8.2 COMPONENTS OF STEAM AND CONDENSATE SYSTEMS

Figure 8.1 shows a typical arrangement for a steam system in an industrial plant.

The boiler produces steam at a pressure, and consequently at a temperature, slightly higher than those required by the end users. The steam is carried from the boiler plant through steam main and secondary transfer lines to the process

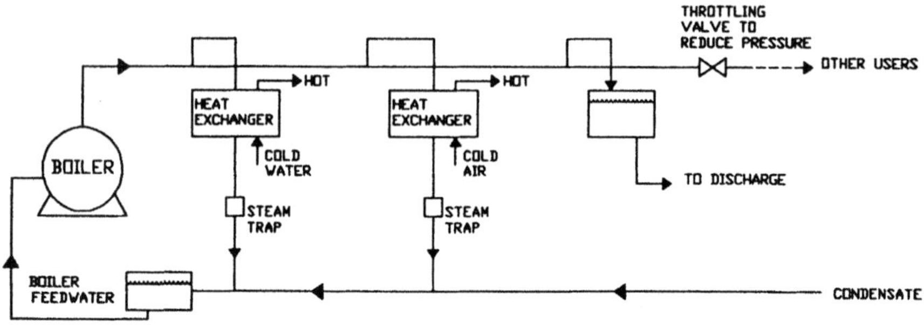

Fig. 8.1 Typical steam system circuit

equipment. If some equipment requires a lower temperature, the steam will be throttled to lower pressure by means of a pressure-regulating valve or, if large quantities of steam are involved, by means of a backpressure turbine.

In most applications, steam/water or steam/air heat exchangers are the part of the end-user equipment where heat-transfer by condensation takes place.

Depending on the process conditions, condensate can be discarded or recovered by steam traps, located near the steam end users, which allow condensate to drain into the condensate return line along which it flows back to a tank and then into the boiler as feedwater.

In addition to the basic system shown in Fig. 8.1, other components are used to ensure safe operation of the system.

Insulation of pipelines, tanks and also valves and flanges and other components, constitutes a potential energy saving that must always be investigated to check whether it is economically valid.

The steam trap is one of the most important components of steam and condensate systems. Its main functions are to allow condensate to be drained completely from end-user equipment as quickly as possible and to facilitate the removal of air from the steam pipelines. Condensate in the steam chamber reduces the heat-transfer effectiveness and in addition, if liquid inclusions are trapped in a steam line, they can be accelerated to high velocity and impelled against an obstruction such as valves or other components, with an impact similar to that obtainable by hitting the component with a hammer. This phenomenon is called water hammer.

Air and some gas, generally liberated from the water in the boiling process, when mixed with steam make a gas mixture in which each component contributes to the pressure proportionately to its share of the space available. In consequence, the pressure of the steam is reduced, as is the temperature. This means that the heat transfer is reduced or that more steam must be produced to obtain the same heating effect. In addition, the air very near to the heat-transfer surface will act as an insulating layer thus further reducing the heat transfer.

For steam distribution lines, temperatures generally range between 373 K (100°C; 212°F) and roughly 873 K (500°C; 932°F) for superheated steam. Higher temperatures can be found in particular applications.

Other fluid distribution lines work at different temperatures in a range from 233 K (–40°C; –40°F) to 373 K (100°C; 212°F): plumbing, HVAC and process refrigerating systems. If oil is used as the thermal medium for industrial applications, temperatures range between 523 K (250°C; 452°F) and 623 K (350°C; 662°F).

Typical values of velocity are 20-40 m/s for steam and 2-4 m/s for water. The volumetric flow rate can be calculated by multiplying the fluid velocity (m/s) by the pipe section (m^2). The mass flow-rate is equal to the volume flow-rate multiplied by the fluid density (kg/m^3). See Table 6.4 for steam density values.

8.3 BASIC PRINCIPLES OF PIPELINE LOSSES AND INSULATION

Insulation of any hot or cold system plays an important role in energy saving. Insulation can reduce by at least 90% the unwanted heat transfer, that is distribution losses, occurring with the bare surface.

To understand better the criteria for economic thicknesses of pipeline and flat surface insulation, the basic concepts of heat transfer are briefly reviewed below.

The same concepts can be applied to any problem concerning heat transfer and thermal energy balance.

Thermal energy is transferred in three main modes: conduction, radiation, convection.

Conduction: heat is transferred through both fluids (gas and liquid) and solids when the two sides of a volume of these materials are at different temperatures. The thermal power transfer is proportional to the temperature gradient through a proportionality factor k, called thermal conductivity, and the surface.

Thermal conductivity (or k-value in insulation technology) is the measure of the capability of a material to transmit heat. Substances with high values of thermal conductivity, such as copper, are good thermal conductors, and those with low conductivity, such as polystyrene foam or cork, are good insulators. Thermal conductivity is expressed as the quantity of heat that will be conducted through unit area of a layer of material of unit thickness with unity difference of temperature between the faces in unit time. The SI unit is W/m · K; other units commonly used are: (kcal/h · m · °C), (Btu · in/ft^2 · h · °F), (Btu/ft · h · °F).

The expression of the thermal conduction power transfer Q through a surface A is as follows:

$$Q = k \cdot A \cdot \Delta t_x \qquad \text{(W), (kcal/h), (Btu/h)}$$

where Δt_x is the temperature gradient between the two faces (K/m; °C/m; °F/ft). The sign of Q, minus in accordance with the laws of thermodynamics, is here omitted.

Thermal conductivities of various materials at room temperature or higher are reported in Table 8.1. If the temperature rises, thermal conductivity decreases with homogeneous metals and it increases in composite materials almost linearly in an interval range of 323-373 K (50-100°C; 122-212°F). With insulating materials, because of the composite or multilayer structure where also air can be present, the k coefficient must take into account the presence of all the heat transfer basic modes and its value is assessed on empirical bases. For detailed evaluations, manufacturers' specifications should be consulted.

Radiation: heat is transferred as radiant energy and it does not require any medium of diffusion. Radiation can take place even in a vacuum. Solid surfaces, gases and liquids: all emit, absorb and transmit thermal radiation in varying degrees.

The rate at which energy is emitted from a system with a surface A situated inside a large space is quantified by the basic relationship:

$$Q = \varepsilon \cdot \sigma \cdot A \cdot \left(T_s^4 - T_o^4\right) \quad \text{(W)}$$

where T_s and T_o are the surface and output surrounding absolute temperatures (K), ε $(0 < \varepsilon < 1)$ is an adimensional quantity that indicates how effectively the surface radiates ($\varepsilon = 1$ for the black body or the perfect emitter) and σ is the Stefan-Boltzmann constant equal to $5.67 \cdot 10^{-8}$ W/(m$^2 \cdot$ K^4). As thermal radiation is associated with the fourth power of the absolute temperature of the surface, the importance of this mode of heat transfer increases rapidly with the temperature.

Values of ε for different operating conditions (materials, different states of surfaces) are shown in Table 8.2. As with the conduction energy transfer, different units are commonly used alternatively to the SI units.

The above-mentioned relationship is valid if some basic assumptions are respected such as transparent medium, surface with particular spectrum emission behavior, etc.. For pipelines and tanks in industrial applications, these hypotheses are widely accepted and the relationship can be used with a proper choice of parameter.

Convection: heat is transferred from a solid surface at one temperature to an adjacent fluid, liquid or gas, at another temperature, when motion occurs. The thermal power transfer from the system with a flat A surface is expressed as follows:

$$Q = h \cdot A \cdot (t_s - t_f) \qquad \text{(W)}$$

where t_s and t_f are the surface and fluid temperatures and h is an empirical parameter, depending on operating conditions and surface geometry, called the heat transfer coefficient. The SI unit is W/m$^2 \cdot$ K; other units commonly used are: (kcal/h $\cdot$ m$^2 \cdot$ °C), (Btu/ft$^2 \cdot$ h $\cdot$ °F).

In most cases, heat is transferred by a combination of the above-mentioned basic modes through composite or multilayer systems, and the evaluation is generally made by means of empirical parameters and linearization of the heat-transfer relationships. A simplified approach, based on a monodimensional scheme, is generally introduced in industrial applications. As a rule, radiation is included in the heat-transfer coefficient in a wide range of temperature values.

Values of h for different operating situations with surface exposed to the air are given in Table 8.3.

Values of h for fluid flowing in pipelines depend on many factors (temperature, speed, state of pipe, etc.) and must be calculated by using special formulae. They are greater than those for air and they range between hundreds and thousands $W/m^2 \cdot K$ (see also Chapter 15).

In a multilayer system, heat is transferred from one fluid (fluid inside i) at higher temperature (t_i) through an n-layer slab to another fluid (fluid outside o) at lower temperature (t_o).

The fundamental equation for the heat-transfer calculation is:

$$\text{Heat flow through unit area} = \frac{\text{fluid temperature difference}}{\text{overall thermal resistance}} =$$

$$= \frac{t_i - t_o}{R_{th}} = (t_i - t_o) \cdot U \qquad \left(\frac{W}{m^2}\right) \qquad \left(\frac{Btu}{h \cdot ft^2}\right)$$

where

t_i = **inside fluid temperature** (K), (°C) (°F)

t_o = **outside fluid temperature** (K), (°C) (°F)

R_{th} = **overall thermal resistance** $\left(\frac{m^2 \cdot K}{W}\right) = \left(\frac{m^2 \cdot {}^\circ C}{W}\right) \quad \left(\frac{h \cdot ft^2 \cdot {}^\circ F}{Btu}\right)$

$U = 1/R_{th}$ = **overall heat transfer coefficient (see Chapter 15)**

Note that:

—for homogeneous materials, thermal resistance is the reciprocal of the system thermal conductance which represents the quantity of heat that will be conducted through unit area of a layer of material of a defined thickness with unity difference of temperature between the two fluids in unit time;

—for n-layer systems thermal resistance R_{th} is the reciprocal of the system thermal conductance which represents the quantity of heat conducted through unit area of a n-layer material of defined thicknesses with unity difference of temperature between the two fluids in contact with the end faces in unit time;

—the relationship given above is based on the assumption that the radiation heat transfer is negligible or that its effect is introduced into the R_{th} by linearizing the radiation law. In the same way, all heat-transfer models are linearized;

—at thermal steady state, the same amount of heat flows through each layer of the system. Thus, the temperature of each layer can be calculated.

Specialized handbooks on this subject should be consulted for more detailed evaluations.

Examples are given below of flat-surface and of pipeline multilayer systems with bare surfaces and with insulation.

Flat-surface multilayer system (see Fig. 8.2)

When a flat surface is heated on one side and cooled on the other, as shown in Fig. 8.2, heat flows from the hot side to the cold side.

The flow of heat is defined as follows:

$$Q = A \cdot \frac{t_i - t_o}{R_{th}}$$

where

$A =$	internal and external surfaces with the same value	(m^2)	(ft^2)
$R_{th} = \frac{1}{h_i} + \Sigma_j \frac{d_j}{k_j} + \frac{1}{h_o}$		$\left(\frac{m^2 \cdot K}{W}\right) = \left(\frac{m^2 \cdot °C}{W}\right)$	$\left(\frac{ft^2 \cdot h \cdot °F}{Btu}\right)$
$d_j =$	thickness of the j layer	(m)	(ft)

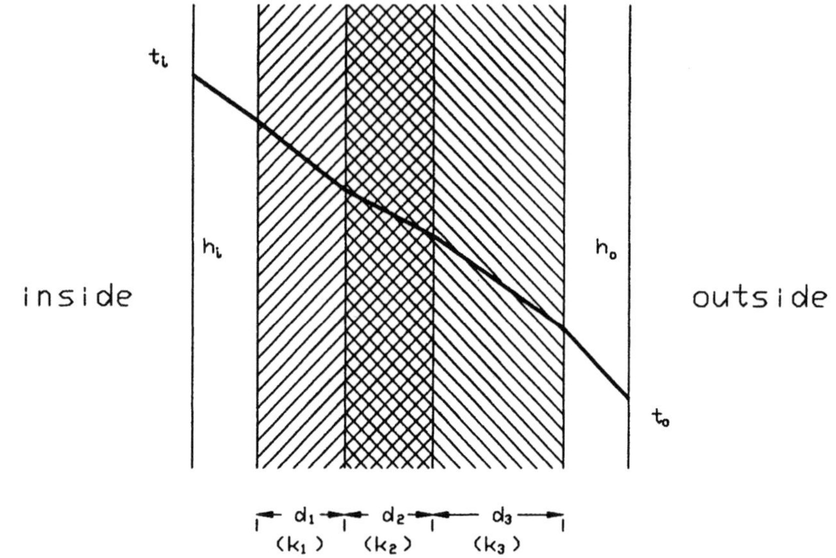

Fig. 8.2 Heat transfer through a multilayer wall

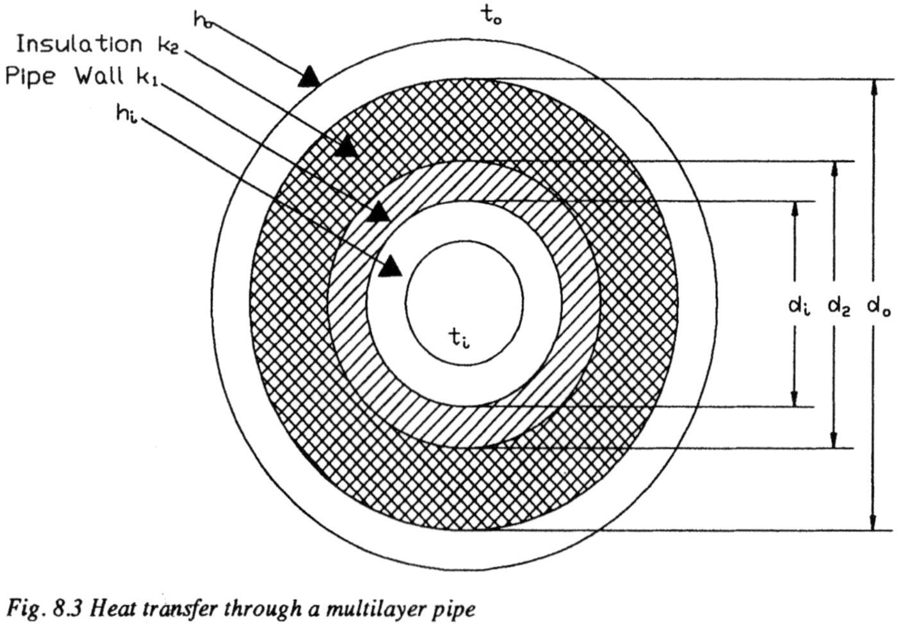

Fig. 8.3 Heat transfer through a multilayer pipe

k_j =	thermal conductivity of the j layer	$\left(\frac{W}{m \cdot K}\right)=\left(\frac{W}{m \cdot {}^\circ C}\right)$	$\left(\frac{Btu}{ft \cdot h \cdot {}^\circ F}\right)$
d_j/k_j =	thermal resistance of the j layer	$\left(\frac{m^2 \cdot K}{W}\right)=\left(\frac{m^2 \cdot {}^\circ C}{W}\right)$	$\left(\frac{ft^2 \cdot h \cdot {}^\circ F}{Btu}\right)$
h_i =	heat-transfer coefficient, radiation included, between the inside fluid and the internal surface of the first layer	$\left(\frac{W}{m^2 \cdot K}\right)=\left(\frac{W}{m^2 \cdot {}^\circ C}\right)$	$\left(\frac{Btu}{ft^2 \cdot h \cdot {}^\circ F}\right)$
h_o =	heat-transfer coefficient, radiation included, between the external surface of the last layer and the fluid outside	$\left(\frac{W}{m^2 \cdot K}\right)=\left(\frac{W}{m^2 \cdot {}^\circ C}\right)$	$\left(\frac{Btu}{ft^2 \cdot h \cdot {}^\circ F}\right)$

If one of the layers is made of insulating material, this layer has the highest thermal resistance (d_j/k_j) and the flow of heat is less than with a bare surface.

Table 8.4 shows heat losses from a composite wall. Notice that heat transfer through the layers is reduced by insulation to roughly 10%.

Cylindrical surface multilayer system (see Fig. 8.3)

This example is typical of pipelines for steam or other hot fluids. The same approach can also be followed in the case of cold fluids.

The flow of heat is:

$$Q = A \cdot \frac{t_i - t_o}{R_{th}}$$

where

A = the outside ($2\pi L r_o$) or the inside ($2\pi L r_i$) surface of the pipeline of length L. Note that the R_{th} expression will be modified by the choice between the outside and the inside surface

r_i = inside pipe radius

r_o = outside multilayer pipe radius

Thus:

outside surface

$A = 2 \cdot \pi \cdot L \cdot r_o$(outside surface) (m^2) (ft^2)

$$R_{th} = \frac{1}{h_i} \cdot \frac{r_o}{r_i} + \sum_{1}^{n} j \frac{r_o}{k_j} \cdot \ln \frac{r_{j+1}}{r_j} + \frac{1}{h_o} \qquad (m^2 \cdot K/W) \qquad (ft^2 \cdot h \cdot {}^\circ F/Btu)$$

inside surface

$A = 2 \cdot \pi \cdot L \cdot r_i$ (inside surface) (m^2) (ft^2)

$$R_{th} = \frac{1}{h_i} + \sum_{1}^{n} j \frac{r_i}{k_j} \cdot \ln \frac{r_{j+1}}{r_j} + \frac{1}{h_o} \cdot \frac{r_i}{r_o} \qquad (m^2 \cdot K/W) \qquad (ft^2 \cdot h \cdot {}^\circ F/Btu)$$

In the case of a bare tube, if the outside surface is taken as reference, the above-mentioned relationships (with n = 1) are modified as follows:

$$Q = A \cdot \frac{t_i - t_o}{R_{th}}$$

where

$r_2 = r_o$

$A = 2 \cdot \pi \cdot L \cdot r_o$

$$R_{th} = \frac{1}{h_i} \cdot \frac{r_o}{r_i} + \frac{r_o}{k_1} \cdot l_n \frac{r_2}{r_1} + \frac{1}{h_o} = \frac{1}{h_i} \cdot \frac{r_o}{r_i} + \frac{r_o}{k_1} \cdot l_n \frac{r_o}{r_i} + \frac{1}{h_o}$$

If the thickness ($r_o - r_i$) of the pipe is small ($r_o - r_i < 0.1 \cdot r_i$), this expression can be approximated, as for a flat surface, to the following:

$$R_{th} = \frac{1}{h_i} + \frac{r_o - r_i}{k_1} + \frac{1}{h_o}$$

Notice that $1/h_i$ is usually the lowest term in the case of pipelines, because of the moving fluid inside the pipe.

For insulated pipelines with one insulation layer, it follows that:

$$Q = A \cdot \frac{t_i - t_o}{R_{th}}$$

where $A = 2 \cdot \pi \cdot L \cdot r_i$

$$R_{th} = \frac{1}{h_i} + \frac{r_i}{k_1} \cdot \ln \frac{r_2}{r_1} + \frac{r_i}{k_2} \cdot \ln \frac{r_3}{r_2} + \frac{1}{h_o} \cdot \frac{r_i}{r_o}$$

$j = 2;\ r_1 = r_i;\ r_3 = r_o$

If $r_1 = r_i$ is assumed to be equal to r_2 (small thickness of the pipe) it follows that:

$$R_{th} = \frac{1}{h_i} + \frac{r_i}{k_2} \cdot \ln \frac{r_3}{r_2} + \frac{1}{h_o} \cdot \frac{r_i}{r_o}$$

If the thickness of the insulation layer is small in comparison with the bare pipe radius, that is $(r_3 - r_2) < 0.1 \cdot r_2 = 0.1 \cdot r_i$, the expression given above can be approximated as follows:

$$R_{th} = \frac{1}{h_i} + \frac{r_3 - r_2}{k_2} + \frac{1}{h_o} \cdot \frac{r_i}{r_o}$$

Table 8.5 shows heat losses from a composite metal pipeline to still air. Notice that insulation reduces heat transfer through the layers to roughly 10%.

Attention must be paid to the surface conditions such as film deposits, surface scaling, corrosion. A fouling factor can be introduced to take account of these phenomena as an additional term in the overall thermal resistance R_{th} (see also Chapter 15).

In all the foregoing relationships, average values of temperatures must be introduced in the following situations:

—when a fluid enters a pipe at one temperature (t_{in}) and leaves at another (t_{out}) an average temperature must be used equal to $(t_{in} + t_{out})/2$ at any point along the pipe (see also log-mean temperature in Chapter 15);

—when the value of the parameter k for insulation material varies significantly with the temperature, the mean temperature between the two faces of the insulation must be considered. The mean value is calculated as half the sum of the temperatures on either side of the insulation.

Since the same amount Q of heat flows through each layer of a multilayer wall or pipeline, the temperature difference for each layer is proportional to the thermal resistance of each layer ($(t_i - t_o) = Q \cdot R_{th}$).

Tanks, other equipment

In these cases specialized handbooks should be consulted to ascertain relationships and coefficients related to the shape of the structure to insulate and to the operating temperature.

8.4 THERMAL ENERGY END USERS

At the end of thermal distribution pipelines, end users can be grouped as follows:

—furnaces and ovens where fuels are burned directly;

—systems where steam or hot water is released to make direct contact with the products (steam to heat a water tank, steam to sterilize food products, etc.). In this case, because of possible contamination during the process, the condensate is generally discarded; potential recovery and thus potential energy saving must be considered carefully in each individual case;

—heat exchangers where the heat medium (steam, water, oil, etc.) transfers energy to a colder fluid (air, water or other fluids) by reducing its temperature or by changing its state, as in steam condensation (see Chapter 15).

As a general rule, the steam consumed in the final users can be calculated by using a few basic relationships:

Heating a mass by steam

$m_{steam} = (c \cdot m \cdot \Delta t)/h_v$

where
m_{steam} = steam mass flow-rate
c = specific heat of the material to be heated
m = mass flow-rate to be heated
Δt = temperature increase of the mass
h_v = evaporation enthalpy or latent heat of the steam

Exchanging heat between a fluid and steam (see also Chapter 15)

$$m_{steam} = A \cdot \frac{\Delta t}{R_{th}} \cdot \frac{1}{h_v} = A \cdot \Delta t \cdot U/h_v$$

where
m_{steam} = steam mass flow-rate
A = surface of exchange
Δt = the log-mean difference between the temperatures of the two fluids at the two sides of the exchanger (see Chapter 15)
R_{th} = overall thermal resistance = 1/U
U = overall heat transfer coefficient (see Chapter 15)
h_v = evaporation enthalpy or latent heat of the steam

Extraction of water in industrial dryers by evaporation

> **The necessary quantity of heat medium (steam, oil) is related to the quantity of water contained in the product and to the efficiency of the system. In industrial applications, at least 4000-5000 kJ/kg of water (950-1200 kcal/kg of water; 1710-2160 Btu/lb of water) are generally required, that is roughly 1.6-2 kg of steam, to evaporate 1 kg of water contained in the product.**

Operation of most process equipment can basically be derived from the relationships given above; low-consumption process equipment is designed to ensure recovery of heat inside the equipment itself and the minimum requirement of thermal energy (hot water, hot air, steam, hot oil).

8.5 PRACTICAL EXAMPLES

Examples are given below of insulation with reference to Fig. 8.2 and Fig. 8.3.

Example 1
Compare losses from a composite wall with different thicknesses of insulation (see Fig. 8.2 and Table 8.4)

A flat surface with air on both sides is heated on one side. The difference between inside and outside fluid temperatures is 100°C (180°F).

The thickness of the wall is 0.1 m and the thermal conductivity is 2.5 W/m · K. Insulation with fiberglass of different thicknesses is considered.

Calculate losses and energy saving with different thicknesses of insulation.

Details of input data and results of calculation are shown in Table 8.4.

In the case of 70 mm insulation, for example:

— losses without insulation are 526 W/m^2 (167 $Btu/h \cdot ft^2$)

— losses with 70 mm of insulation are reduced to 68 W/m^2 (22 $Btu/h \cdot ft^2$)

— reduction of losses is 458 W/m^2 (145 Btu/h · ft^2)
— if combustibles are used for heating and combustion efficiency is 85% (Net Heating Value as reference), the saving in input oil entering the boiler plant is $(458 \cdot 3600/1000)/(0.85 \cdot 41860) = 0.046\ kg_{oil}/h \cdot m^2$
— if the system works for 5000 h/year the energy saved is $0.046 \cdot 5000 = 230$ $kg_{oil}/year \cdot m^2$
— with a surface of 100 m^2 the energy saved is $230 \cdot 100 = 23000\ kg_{oil}/year = 23$ $t_{oil}/year$ = 23 TOE/year.

Example 2
Compare losses from a composite pipeline with different thicknesses of insulation (see Fig. 8.3 and Table 8.5)

A pipeline has steam inside (160°C; 320°F) and still air outside (10°C; 50°F). The difference between inside and outside fluid temperatures is 150°C (270°F).

The pipeline has an external radius of 0.05 m; the thermal conductivity is 50 W/m · K. The steam flow rate is 2 t/h. Insulation with fiberglass of different thicknesses is considered.

Calculate losses and energy saving with different thicknesses of insulation.

Details of input data and results of calculation are reported in Table 8.5.

In the case of 50 mm insulation, for example:

— losses without insulation are 3308 W/m^2 or 935 W/m
— with 50 mm of insulation losses are 254 W/m^2 or 72 W/m
— saving is 3054 W/m^2 (968 Btu/h · ft^2) or 863 W/m (895.5 Btu/h · ft)
— if combustibles are used for heating and if combustion efficiency is 85% (Net Heating Value as reference), the saving in input oil entering the boiler plant is $(3054 \cdot 3600/1000)/(0.85 \cdot 41860) = 0.309\ kg_{oil}/h \cdot m^2$
— if the system works for 5000 h/year the energy saved is $0.309 \cdot 5000 = 1545$ $kg_{oil}/year \cdot m^2$
— with a bare pipeline surface of 100 m^2 (353 m in length of the pipeline or less if 100 m^2 includes additional equipment insulation as equivalent surface) the energy saved is $1545 \cdot 100 = 154500\ kg_{oil}/year = 154.5\ t_{oil}/year$ = 154.5 TOE/year.

Economic evaluations for both examples are given in Table 20.3.

Table 8.1 Average values of thermal conductivity of different materials at room temperature

Material	Thermal conductivity k in the temperature range 273.15-373.15 K 0-100°C 32-212°F		
Pure aluminium	W/m · K	(a)	228
	Btu/ft · h · °F	(b) = (a)/1.731	132
	Btu · in/ft^2 · °F · h	(c) = (b) · 12	1581
	kcal/h · m · °C	(d) = (a) · 860/1000	196
Cast iron	W/m · K		48
	Btu/ft · h · °F		28
	Btu · in/ft^2 · °F · h		333
	kcal/h · m · °C		41
Fiber glass	W/m · K		0.058
	Btu/ft · h · °F		0.034
	Btu · in/ft^2 · °F · h		0.402
	kcal/h · m · °C		0.050
Ceramic fiber	W/m · K		0.055
	Btu/ft · h · °F		0.032
	Btu · in/ft^2 · °F · h		0.381
	kcal/h · m · °C		0.047
Cellular polyurethane	W/m · K		0.025
	Btu/ft · h · °F		0.014
	Btu · in/ft^2 · °F · h		0.173
	kcal/h · m · °C		0.022

Table 8.2 Total emissivity average values (ε)

Surface	Temperature range		ε
	°C	°F	
Aluminium			
commercial sheet	200-600	400-1100	0.09
heavily oxidized	100-550	200-1000	0.3
Iron			
steel polished	40-250	100-500	0.085
steel oxidized	250	500	0.8
Brick			
red	40	100	0.93
white refractory	1100	2000	0.29
Concrete	40	100	0.94
Glass	40	100	0.94

Table 8.3 Values of heat transfer coefficient (h) for hot service in case of air

Surface Emittance	Wind velocity			Typical values of h (a) $W/m^2 \cdot K$ (b) $Btu/ft^2 \cdot h \cdot °F$			
	km/h	Operating temperature	°C	100	200	300	400-550
			°F	212	392	572	752-1022
1.0	0	(a)		9.31	11.13	12.34	13.52
		(b)		1.64	1.96	2.17	2.38
	3			11.83	13.85	15.35	16.22
				2.08	2.44	2.70	2.86
	6			13.85	16.70	18.32	19.58
				2.44	2.94	3.23	3.45
0.8	0			8.23	10.32	11.13	11.59
				1.45	1.82	1.96	2.04
	3			11.13	13.21	14.94	15.35
				1.96	2.33	2.63	2.70
	6			13.21	15.77	17.75	18.32
				2.33	2.78	3.13	3.23
0.6	0			7.28	9.01	9.96	10.32
				1.28	1.59	1.75	1.82
	3			10.32	12.62	13.85	14.20
				1.82	2.22	2.44	2.50
	6			12.62	14.94	16.22	16.70
				2.22	2.63	2.86	2.94
0.2	0			5.56	7.14	7.69	8.33
				0.98	1.26	1.35	1.47
	3			8.48	10.32	11.13	11.59
				1.49	1.82	1.96	2.04
	6			10.32	12.91	14.20	14.94
				1.82	2.27	2.50	2.63

(a) SI units $W/m^2 \cdot K$
(b) British units $Btu/ft^2 \cdot h \cdot °F$
(b)=(a)·0.1761
In the case of low-temperature service, values are lower
In the case of higher wind velocity, values are higher

Table 8.4 Losses from a composite wall (reference Fig. 8.2)

Thickness of insulation (d_2)		Overall thermal resistance R_{th}		Heat losses Q		Energy saving		
mm	$10^{-3}\cdot$ ft	$m^2\cdot K/W$	$ft^2\cdot h\cdot °F/Btu$	W/m^2	$Btu/h\cdot ft^2$	W/m^2	$Btu/h\cdot ft^2$	$kgoil/h\cdot m^2$
0	0.0	0.190	1.079	526	167	0	0	0.000
10	32.8	0.372	2.112	269	85	257	82	0.026
20	65.6	0.554	3.144	181	57	346	110	0.035
30	98.4	0.735	4.177	136	43	390	124	0.039
40	131.2	0.917	5.209	109	35	417	132	0.042
50	164.0	1.099	6.242	91	29	435	138	0.044
60	196.9	1.281	7.274	78	25	448	142	0.045
70	229.7	1.463	8.307	68	22	458	145	0.046
80	262.5	1.645	9.340	61	19	466	148	0.047
90	295.3	1.826	10.372	55	17	472	149	0.048
100	328.1	2.008	11.405	50	16	477	151	0.048

INPUT DATA

wall conductivity k_1	2.500 W/m·K	1.444 Btu/ft·h·°F
fiberglass conductivity k_2	0.055 W/m·K	0.032 Btu/ft·h·°F
wall thickness d_1	0.100 m	0.328 ft
heat-transfer coefficient h_i	20 $W/m^2\cdot K$	3.522 $Btu/ft^2\cdot h\cdot °F$
heat-transfer coefficient h_o	10 $W/m^2\cdot K$	1.761 $Btu/ft^2\cdot h\cdot °F$
$t_i - t_o$	100°C = 1000 K	180°F

BASIC FORMULA

$$Q = A \cdot \frac{t_i - t_o}{R_{th}}$$

$$R_{th} = \frac{1}{h_i} + \frac{d_1}{k_1} + \frac{d_2}{k_2} + \frac{1}{h_o}$$

saving in $\frac{kg_{oil}}{h \cdot m^2} = \frac{W}{m^2} \cdot \frac{3600}{\eta \cdot 41860 \cdot 10^3}$

η combustion assumed 0.85 (Net Heating Value as reference)

Net Heating Value of oil 41860 kJ/kg

ANNUAL ENERGY SAVING
70 mm insulation
5000 h/ year
100 m^2 surface
23 TOE/year

Table 8.5 Heat losses from a composite steel pipeline to still air (see reference Fig. 8.3)

Thickness of insulation $(r_o - r_2)$	External radius (r_o)	R_{th}	Heat losses Q		R_{th} simplified	Energy saving		
mm	m	$m^2 \cdot K/W$	W/m^2	W/m	$m^2 \cdot K/W$	W/m^2	$Btu/h \cdot ft^2$	$kgoil/h \cdot m^2$
0	0.05	0.045	3308	935	0.045	0	0	0.000
10	0.06	0.187	802	227	0.220	2506	794	0.254
20	0.07	0.308	487	138	0.396	2821	894	0.285
30	0.08	0.413	363	103	0.574	2945	933	0.298
40	0.09	0.506	296	84	0.753	3012	955	0.305
50	0.1	0.590	254	72	0.932	3054	968	0.309
60	0.11	0.666	225	64	1.112	3083	977	0.312
70	0.12	0.735	204	58	1.292	3104	984	0.314

INPUT DATA

pipe conductivity k_1	50.000 W/m·K	28.885 Btu/ft·h·°F
fiberglass conductivity k_2	0.055 W/m·K	0.032 Btu/ft·h·°F
internal radius (r_i)	0.045 m	0.1476 ft
pipe thickness	0.005 m	0.016 ft
bare pipe external radius (r_2)	0.05 m	0.164 ft
heat-transfer coefficient h_i	4000 $W/m^2 \cdot K$	704.400 $Btu/ft^2 \cdot h \cdot °F$
heat-transfer coefficient h_o	20 $W/m^2 \cdot K$	3.522 $Btu/ft^2 \cdot h \cdot °F$
$t_i - t_o$	150°C = 150 K	270°F

BASIC FORMULA REFERRED TO INSIDE SURFACE

$$Q = A \cdot \frac{t_i - t_o}{R_{th}}$$

$$R_{th} = \frac{1}{h_i} + \frac{r_i}{k_1} \cdot \ln \frac{r_2}{r_i} + \frac{r_i}{k_2} \cdot \ln \frac{r_o}{r_2} + \frac{1}{h_o} \cdot \frac{r_i}{r_o}$$

SIMPLIFIED FORMULA

$$R_{th} = \frac{1}{h_i} + \frac{r_o - r_2}{k_2} + \frac{1}{h_o} \cdot \frac{r_i}{r_o}$$

saving in $\frac{kg_{oil}}{h \cdot m^2} = \frac{W}{m^2} \cdot \frac{3600}{\eta \cdot 41860 \cdot 10^3}$

η combustion assumed 0.85 (Net Heating Value as reference)

Net Heating Value of oil 41860 kJ/kg

ANNUAL ENERGY SAVING

50 mm insulation

5000 h/year

100m^2 bare surface

154.5 TOE/year

CHAPTER 9

COGENERATION PLANTS

9.1 INTRODUCTION

In conventional utility power plants, electric energy is generally produced with an overall efficiency, that is, the ratio between useful output and input power as fuel, of less than 34-38% because of the large quantity of heat discharged into the atmosphere through cooling towers, lakes or rivers, without recovery.
Differently, in cogeneration plants, either useful heat and mechanical or electric power are generated from fuel or power is produced by recovering low-level heat from processes. An overall efficiency ranging from 55% to 85% is obtainable, depending on the type of cogeneration plant.

Cogeneration is an effective method of primary energy conservation; from this point of view, cogeneration can be applied whenever it is economically justified. The correct use of this system implies a balance between electric-power requirements and process-heat requirements in quantity and in quality. If this balance does not exist, electric power must be exchanged with utility and additional heat must be produced by boiler plants. The discharge of unnecessary heat should be avoided; otherwise, the system works with lower efficiency like a conventional utility plant.

Before the widespread diffusion of utilities to produce and distribute electric energy to end users, cogeneration plants were very common in industry. Afterwards, because of local regulations and tariffications, these plants have been unevenly located and used, the choice depending on the cost of primary energy, mainly coal and oil, and electric energy tariffs.

9.2 FORMS OF COGENERATION

Cogeneration plants can be grouped basically into two types referred to as topping cycles or bottoming cycles.

The topping cycles produce power, mechanical or electric, before delivering thermal energy to the end users. Typical examples are the backpressure or non-condensing steam turbine cycle, the gas turbine and combined cycles and the diesel engine cycle where exhausts are utilized as heat for end-user needs.

The bottoming cycle recovers thermal energy, which would normally be discarded, to produce process steam and electricity. In this cycle, first thermal energy is used for the process, then the exhaust energy is used to produce mechanical or electric power at the bottom of the cycle. This cycle is most attractive where there is a large quantity of thermal energy at a temperature of 623 K (350°C; 662°F) or greater associated with exothermic reactions, as in many chemical processes and in rotary kilns and furnaces. Recovery of the steam to be used in a steam turbine is the commonest bottoming cycle; for lower temperatures other cycles, like the saturated Rankine cycle with organic fluids, are sometimes used.

The electric generator may be synchronous or asynchronous; the choice between them depends on the working mode: if the system is independent of the utility grid the synchronous generator must be used; if the system is interconnected with the utility grid, either type can conveniently be used. The generator voltage can be at low or medium level depending on the size and on the layout of the internal distribution network.

9.3 THE BACKPRESSURE OR NON-CONDENSING STEAM TURBINE CYCLE

Fig. 9.1 shows typical energy flows in a utility plant designed to generate electric power alone and in a cogeneration plant with backpressure steam turbine exhausting steam headers to the plant process.

The amount of power which can be produced by expanding steam in a prime mover is limited by the Available Energy (AE) between the inlet and outlet of the

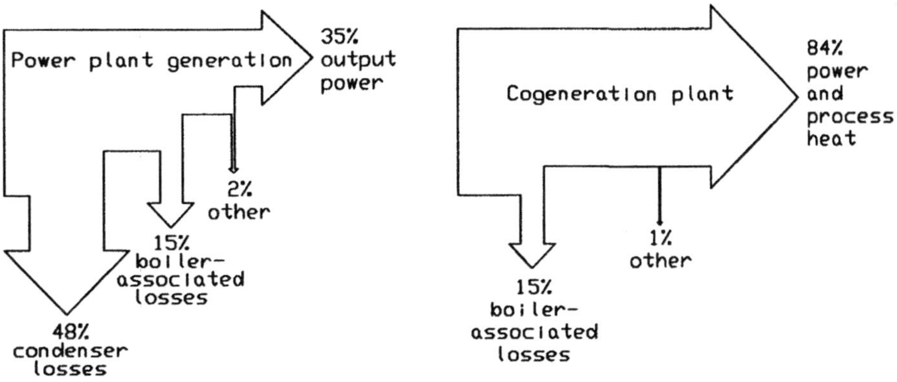

Fig. 9.1 Energy balance in traditional steam turbine power plants and in cogeneration plants

steam turbine. This energy is the enthalpy difference between the inlet superheated steam, at high pressure and temperature, and the outlet steam at lower pressure along an ideal isoentropic expansion. The Mollier diagram or equivalent steam tables (see Chapter 6.4) can conveniently be used for this purpose (see Fig. 9.2). Alternatively, Theoretical Steam Rate tables such as those published by ASME can be used; these report the Theoretical Steam flow Rate (TSR) required to generate 1 kWh in a 100% efficiency expansion process (see Table 9.1). TSR is the ratio between the energy content of 1 kWh (3600 kJ/kWh) and the Available Energy AE (kJ/kg); it represents the amount of steam theoretically needed to produce 1 kWh:

$$TSR(kg/kWh) = 3600\,(kJ/kWh)\,/\,AE\,(kJ/kg)$$

where

3600 kJ/kWh = conversion factor (see Table 2.4)

$AE = h_{in} - h_{out}$ along an isoentropic expansion

h_{out} = enthalpy of the steam at outlet condition

h_{in} = enthalpy of the steam at inlet condition

Typical values of Theoretical Steam flow Rate (TSR) are 7-8 kg of steam/kWh with a pressure drop from 4 MPa (40 bar; 590 psi) to 0.4 MPa (4 bar; 59 psi).

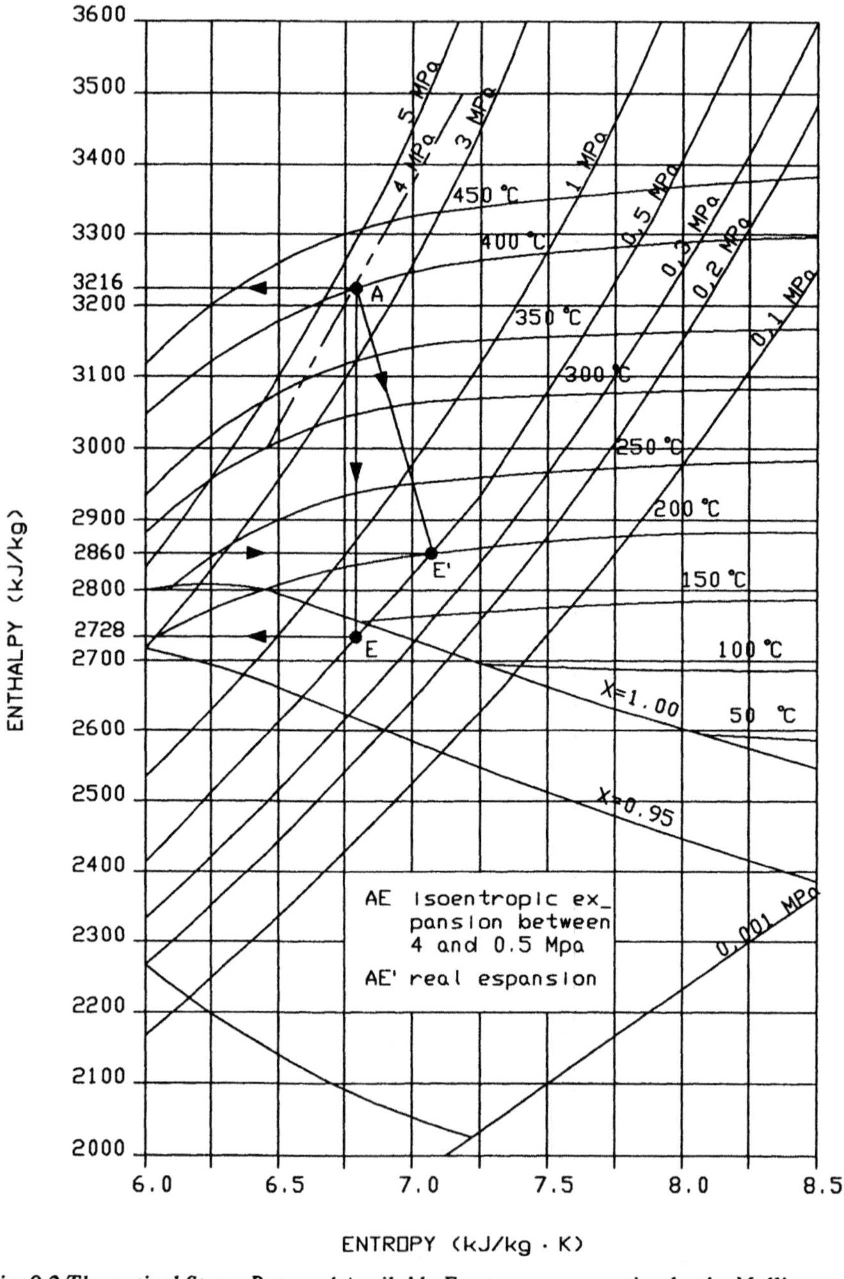

Fig. 9.2 Theoretical Steam Rate and Available Energy representation by the Mollier diagram

Table 9.1 Theoretical Steam Rate

EXHAUST STEAM PRESSURE	MPa psi bar	0.034 5 0.345	0.172 25 1.724	0.345 50 3.448	0.690 100 6.895	1.034 150 10.343	1.379 200 13.790	0.000333 0.048 0.0033
CASE 1								
(Input 1.034 MPa)								
TSR VALUES	lb/kWh	21.7	31.1	46.0				10.88
	kg/kWh	9.8	14.1	20.9				4.9
AE VALUES	kJ/kg	365.74	255.19	172.53				729.475
CASE 2								
(Input 1.724 MPa)								
TSR VALUES	lb/kWh	16.6	21.7	28.2	45.2	76.5		9.34
	kg/kWh	7.5	9.8	12.8	20.5	34.7		4.2
AE VALUES	kJ/kg	478.11	365.74	281.44	175.59	103.74		849.752
CASE 3								
(Input 2.758 MPa)								
TSR VALUES	lb/kWh	13.0	16.0	19.4	26.5	35.4	48.2	8.04
	kg/kWh	5.9	7.3	8.8	12.0	16.1	21.9	3.6
AE VALUES	kJ/kg	610.51	496.04	409.10	299.49	224.20	164.66	987.150
CASE 4								
(Input 4.137 MPa)								
TSR VALUES	lb/kWh	11.1	13.2	15.4	19.4	23.8	29	7.25
	kg/kWh	5.0	6.0	7.0	8.8	10.8	13.2	3.3
AE VALUES	kJ/kg	715.01	601.26	515.36	409.10	333.47	273.67	1094.715
CASE 5								
(Input 5.861 MPa)								
TSR VALUES	lb/kWh	9.8	11.5	13.1	15.9	18.6	21.5	6.72
	kg/kWh	4.4	5.2	5.9	7.2	8.4	9.8	3.0
AE VALUES	kJ/kg	809.86	690.14	605.85	499.16	426.70	369.14	1181.054
CASE 6								
(Input 8.619 MPa)								
TSR VALUES	lb/kWh	8.8	10.1	11.3	13.3	15.1	16.8	6.26
	kg/kWh	4.0	4.6	5.1	6.0	6.8	7.6	2.8
AE VALUES	kJ/kg	901.89	785.81	702.36	596.74	525.60	472.42	1267.840
CASE 7								
(Input 9.998 MPa)								
TSR VALUES	lb/kWh	8.4	9.5	10.5	12.2	13.8	15.2	6.01
	kg/kWh	3.8	4.3	4.8	5.5	6.3	6.9	2.7
AE VALUES	kJ/kg	944.84	835.44	755.87	650.54	575.12	522.15	1320.579

Initial steam condition							
	CASE 1	CASE 2	CASE 3	CASE 4	CASE 5	CASE 6	CASE 7
MPa	1.034	1.724	2.758	4.137	5.861	8.619	9.998
psi	150	250	400	600	850	1250	1450
bar	10.3	17.2	27.6	41.4	58.6	86.2	100.0
°C	186	260	343	399	441	482	510
°F	366	500	650	750	825	900	950
kJ/kg	2781	2935	3105	3209	3281	3346	3399
Btu/lb	1196	1262	1335	1380	1411	1438	1461

The previous TSR value can conveniently be converted into an Actual Steam Rate (ASR) by introducing the efficiency of the turbine which takes into account the shift from the isoentropic expansion and the efficiency of the electric generator. Then:

$$ASR(kg/kWh) = TSR/\eta = (3600/AE) \cdot 1/(\eta_T \cdot \eta_G)$$

where

η_T = turbine efficiency

η_G = electric generator efficiency from the shaft to the electric output section.

The electric output is then the ratio between the Actual Steam Flow (ASF) through the turbine as required by the process and the Actual Steam Rate (ASR):

$$P_e(kW) = ASF\ (kg/h)/ASR\ (kg/kWh)$$

Typical values are 10000 kg/h of steam to produce 1000 kW of electric power with a pressure drop of 4 MPa (40 bar; 590 psi).

If ASR is established, depending on the turbine cycle and operating characteristics, the relationship between the inlet and outlet steam enthalpy is as follows:

$$h_{out} = h_{in} - AE \cdot \eta_T = h_{in} - 3600/(\eta_G \cdot ASR) \qquad (kJ/kg)$$

Depending on the size of the turbine, on the quantity and quality of process steam demand and on other operating factors, several options are available (see Fig. 9.3): straight non-condensing turbine, single or multiple extraction noncondensing turbine.

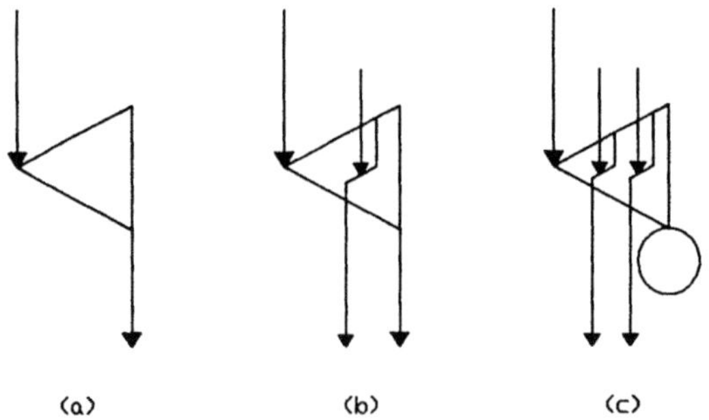

Fig. 9.3 Steam turbine types for cogeneration systems: (a) straight noncondensing (b) single extraction noncondensing (c) double extraction condensing

Fig. 9.4 illustrates a typical industrial steam turbine cycle. It consists of a high-pressure boiler, generally 4-10 MPa (40-100 bar; 590-1470 psi) generating superheated steam for admission to a backpressure or noncondensing steam turbine. The steam turbine drives either an electric generator or other equipment such as compressors, pumps, etc.. The majority of the steam energy content remains in the outlet steam which will be utilized in the process; the energy required for mechanical power and related losses is delivered between the inlet and outlet of the turbine.

The main factors governing the optimal exploitation of a steam turbine cycle based on a fixed exhaust pressure and a constant net heat to process can be summarized as follows:

— the overall efficiency of a steam turbine plant is influenced by the inlet volume flow, inlet/outlet pressure ratio, geometry associated with turbine staging, throttling losses, mechanical coupling and electric generator losses.

— Table 9.2 shows typical values of turbine and overall efficiency and ASR coefficient for a range of unit sizes of backpressure and condensing turbines. Small single-valve, single-stage units have low efficiency, less than 50%. The multivalve, multistage units may reach an efficiency of up to 80% in large power plants; in consequence, the greater the efficiency the greater the electric power generated with the same steam flow. A proper sizing of the prime mover to meet the process requirements is the first step in the optimization of a cogeneration plant;

— the overall efficiency of a steam turbine diminishes with the output power rate, so there is always a minimum useful value below which operation is not economically worthwhile. Technical characteristics given by manufacturers must be consulted;

— an increase in the initial steam pressure and/or steam temperature will increase the amount of electric energy generated because of the increase in the enthalpy inlet-outlet difference (see Table 9.1). Since an increase of the inlet steam temperatures generally results in an increase of the temperature of the steam supplied to the process, a top limit exists to prevent outlet steam being too highly superheated, as this would necessitate a desuperheating process before delivery to the end users. In such a case, desuperheating water must be added downstream of the turbine outlet; in consequence, the steam flow through the turbine must generally be reduced by the quantity of water added so that the electric power generated is not increased by using higher inlet steam conditions;

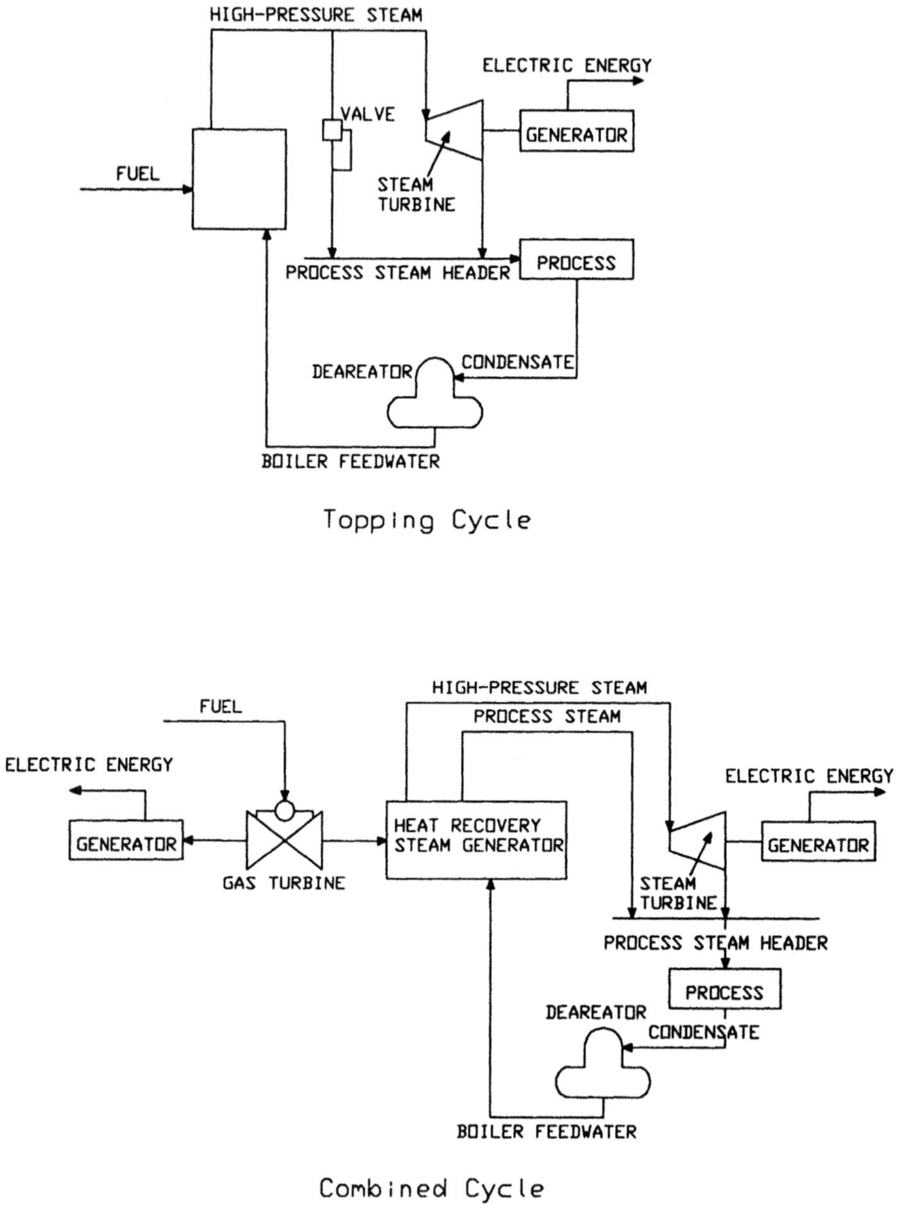

Fig. 9.4 Plant combined cycle cogeneration steam system

— an increase of the energy available for power generation is also possible if the outlet pressure is reduced with a given set of initial steam conditions. The outlet pressure must be the lowest value compatible with the end-user needs;
— an increase in the power generated, with the same fuel consumption in the boiler plant, is possible if feedwater is heated by using steam extracted from turbine stages or exhausted from the process. The level of the power increase depends on the number of heaters and on the temperatures;
— to avoid damage like the erosion of the turbine blades by liquid droplets, inlet steam is always superheated. The outlet steam has generally a quality index (see quality index in Chapter 6.4) of not less than 90%.

Table 9.2 Overall efficiency and ASR coefficient for backpressure and condensing turbines

Type of Unit	Size MW	AE kJ/kg	range η_T %		range $\eta_T \cdot \eta_G$ %		range ASR			
							lb/kWh		kg/kWh	
BACKPRESSURE Single- valve/ single-stage	0.1-1	515.36	40	50	38	48	40.5	32.4	18.4	14.7
Multivalve/ multistage	1-5	515.36	65	75	62	71	24.9	21.6	11.3	9.8
	5-25	515.36	75	80	71	76	21.6	20.3	9.8	9.2
CONDENSING Single- valve/ single-stage	0.1-1	1320.579	40	50	38	48	15.8	12.7	7.2	5.7
Multivalve/ multistage	3-20	1320.579	70	76	67	72	9.0	8.3	4.1	3.8
	2-50	1320.579	76	80	72	76	8.3	7.9	3.8	3.6

Notes
Typical operating conditions for medium power backpressure turbine
See CASE 4 in Table 9.1 with output pressure 0.34 MPa, 3.4 bar, 50 psi
Input steam 4.137 MPa, 41.37 bar, 600 psi

Typical operating conditions for condensing turbines
See CASE 7 in Table 9.1 with atmospheric output pressure

Typical values for industrial applications with an electric power ranging between 500 kW and 5000 kW are: inlet pressure 4-10 MPa, outlet pressure 0.3-1 MPa, Actual Steam Rate 10-15 kg of steam/kWh, oil consumption in the boiler plant 0.7-1 kg of oil/kWh.

An example of a steam turbine cycle is given in Chapter 9.7.

9.4 THE GAS TURBINE CYCLE

The gas turbine as prime mover associated with a heat recovery boiler or with direct use of the exhausts in the process is another highly efficient topping cycle. It is available for both mechanical and electric generator drive in a wide range of sizes from a few hundreds to hundreds of thousands of kW. One can distinguish between industrial turbines, in the range from 1 MW to 200 MW, and aero-derivative turbines in the range from 2 MW to 40 MW. The revolution speed is generally 3000-3600 r/min for power higher than 60 MW and 6000-12000 r/min for lower power.

Gas turbine power plants may operate on either an open or a closed cycle, generally the Brayton cycle, as shown in Fig. 9.5. In the open cycle, which is the commoner, atmospheric air is continuously drawn into the compressor; then, air at high pressure enters a combustion chamber where it is mixed with fuel and combustion occurs resulting in combustion exhaust at high temperature. The combustion products expand through the turbine and they are discharged into the environment. Part of the mechanical power (typically 60%) is generally used to drive the compressor and the remainder to generate electricity or to drive other loads.

Fig. 9.6 shows typical energy flows for a gas turbine plant without heat recovery and for a cogeneration plant with exhaust heat recovery (50%) for steam generation or for direct recovery in the process.
The maximum mechanical power available at the shaft does not generally exceed 20-25% of the turbine input power as fuel in small and medium-sized units (1-5 MW). The amount of the recoverable heat depends on the bottom temperature level required by the end user and in consequence on the stack-gas temperature.
Typical input power values, in the case of natural gas input, are 545 Sm^3/h (5228 kW) to produce 1000 kW of electric power and 2500-3700 kW of recoverable heat. Specific consumptions, Sm^3 of natural gas per kWh, are lower for bigger gas turbine units.

The overall efficiency of the system, that is, the ratio of total output to input power, is roughly 70-80%, depending on the bottom level of the thermal energy user which determines the temperature of the stack exhausts. Higher values are not generally possible, because of the great amount of the air flow and so of the exhausts. Gasoil instead of natural gas can also be used with appropriate burners.

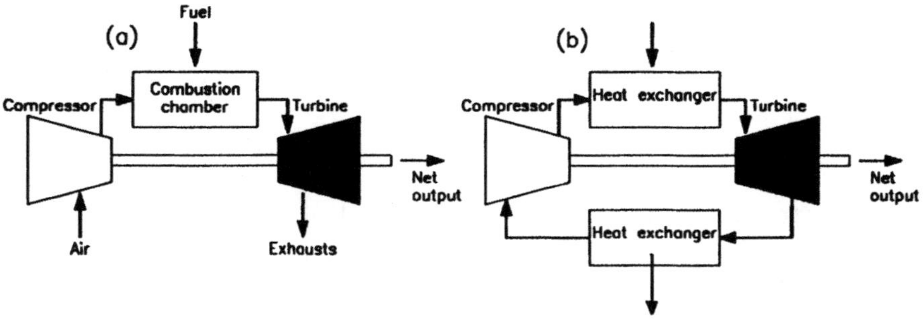

Fig. 9.5 Simple gas turbine: a) open to the atmosphere; b) closed

Combined cycles, where exhausts with or without supplementary burners produce steam and then electric energy in a steam turbine cycle, can also be used for sizes higher than 20 MW of total electric output. These cycles allow much higher power-producing capability per unit of steam than the backpressure system or the gas turbine by themselves.

Combined cycles with gas turbine and condensing or backpressure turbines are widely used with the ratio of output electric power to input power equal to 46-52%.

The principal data to take into account for selecting gas turbines are:

— unit fuel consumption/output power. Typical values range between 4.5 kW to 5.5 kW input (0.47-0.57 Sm^3 of natural gas) per 1 kW of output shaft power. Gasoil can be used instead of natural gas;

— exhaust flow and temperature values, on which the selection of the recovery system can be based;

— the bottom level of temperature required from the process end users which limits the temperature at which the exhaust can be cooled;

— ambient temperature on which depend both the density and so the mass of air flowing through the compressor, which is a nearly constant volume flow-rate machine, and the bottom temperature of the cycle;

— atmospheric pressure which varies according to the altitude. In consequence, the density and so the mass of air flowing through the compressor changes;

— pressure drops upstream of the compressor and turbine system and downstream of the turbine that may influence the gas turbine useful output;

— the inlet air compression ratio and temperature, to which efficiency is closely related. Air compression ratios of 5 to 7 are quite common; higher ratios can

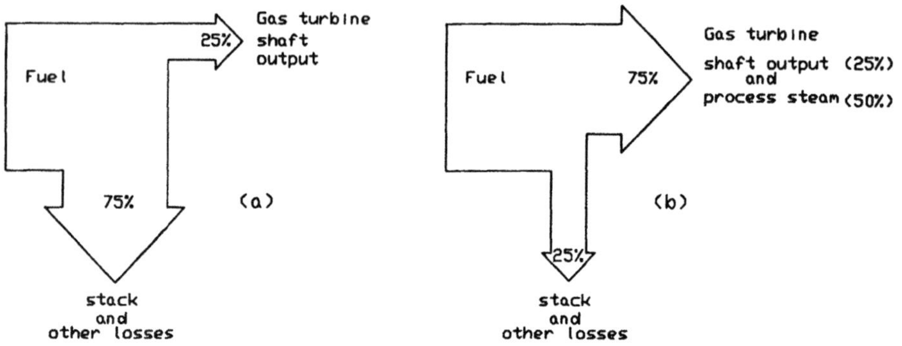

Fig. 9.6 Energy balance (a) in a gas turbine and (b) in a cogeneration plant with a gas turbine (exhausts are recovered to produce steam)

be used to improve efficiency. Note that the influence of air compressor inlet temperature and of altitude is considerable since they affect the weight of air (and oxygen) available for combustion. Standard conditions for land-based systems are generally 288 K (15.6°C; 60°F) intake at 0.1 MPa pressure (1 atm; 1 bar; 14.2 psi). A reduction in ambient air density, due to either altitude or high temperature, results in a derating of the turbine power output. Typical deratings are as follows: 10% per 1000 m of increase of altitude; 2% per 977 Pa (100 mm H20; 3.9 inch H20) due to pressure drops downstream and upstream of the turbine; 7-10% per 10 K or °C (18°F) of increases in ambient temperature;

— the mass of air is considerably higher than the theoretical value needed for combustion (excess air equal to 300-400%) since it is imposed by the operating constraints of the air compressor-turbine system;

— if natural gas is used, the pressure of the gas which is roughly 1.5 the air pressure at the inlet of the turbine, must be at least 1-1.2 MPa. In fact, typical values of air pressure are 0.6-1.5 MPa for industrial turbines and 1.5-3 MPa for aero-derivative turbines. A gas compressor may be required to ensure these conditions.

With steam production, heat-recovering steam generators are installed downstream of the turbine exhaust outlet. The exhaust temperature is about 723-823 K (450-550°C; 842-1022°F), the lower for aero-derivative turbines, with an excess air of 400%. The exhaust is 15-18% oxygen.

According to the classification made in Chapter 6, the generators may be of the unfired, supplementary fired or the fired type.

Unfired generators use the exhaust of the gas turbine unit and they are convective heat exchangers. The exhaust gas temperature is roughly 773-823 K (500-550°C; 1000°F).

Supplementary fired generators use a supplementary burner located downstream of the gas turbine duct to raise the temperature of the exhaust to a maximum of 1088-1143 K (815-870°C; 1500-1600°F). Superheated steam at high pressure, suitable for steam turbines in combined cycle plants, is produced.

Fired generators are similar to steam boilers. Gas turbine exhausts are used to preheat the combustion air. Outlet air can be added if necessary to meet the steam demand. Turbine exhausts are bypassed if they are in excess.

The energy recovery from the exhausts is closely related to the saturation temperature of the steam required by the process, which affects the minimum temperature value at which the exhaust can be cooled through the unfired boiler. Lower values can be obtained if an economizer is installed to pre-heat feedwater (see Chapter 6.9).

A preliminary evaluation of the amount of steam that can be generated is as follows:

$$m_{steam} = m_g \cdot c_p \cdot (t_g - t_s) \cdot \eta/(h_s - h_i)$$

where

m_{steam} = steam flow-rate produced by the recovery system in kg/s
m_g = exhaust gas flow-rate in kg/s
c_p = specific heat of the exhaust; typical value is 1 kJ/kg · K (0.24 Btu/lb · °F)
t_g = temperature of the exhaust gas
t_s = saturation temperature of the steam
h_s = enthalpy of the superheated steam or of the saturated steam
h_i = enthalpy of the saturated liquid in the steam drum
η = effectiveness (typical value 0.95)

In the case of supplementary burners, the same relationship can be used by varying the inlet gas temperature until the desired steam flow is reached.

Typical values of operating parameters for several gas turbines are shown in Table 9.3. Note that the output power can be regulated by varying the input fuel or by throttling the air input in the compressor.

Additional equipment to start the compressor, such as electric motor, reciprocating engine motor or compressed air tanks, are always required.

An example of a gas turbine cycle is given in Chapter 9.7.

Table 9.3 Technical parameters for standard gas turbines

Size Unit	Specific consumption	Input fuel as natural gas			Input air flow			Heat recovery	Output electric energy	Total efficiency	Electric efficiency
MW	kJ/kWh	10^6kJ/h (c)	10^6Btu/h	Sm^3/h	10^3Sm^3/h	10^3kg/h	10^3lb/h	10^6kJ/h (a)	10^6kJ/h (b)	% (a+b)/(c)	% (b)/(c)
0.61	18515	11.3	10.7	329	14.7	19.0	41.9	5.7	2.2	69.9	19.4
1	18916	18.9	17.9	551	24.6	31.8	70.1	9.5	3.6	69.5	19.0
5	14995	75.0	71.1	2184	97.7	126.0	277.8	37.8	18.0	74.4	24.0
10	12410	124.1	117.6	3615	161.7	208.6	459.9	62.6	36.0	79.4	29.0
25	12200	305.0	289.1	8886	397.4	512.7	1130.3	153.8	90.0	79.9	29.5
40	12000	480.0	455.0	13984	625.5	806.8	1778.8	242.1	144.0	80.4	30.0
100	12000	1200.0	1137.4	34960	1563.7	2017.1	4447.0	605.1	360.0	80.4	30.0

Notes
— air density in standard conditions 1.29 kg/Sm^3
— combustion air 45 Sm^3 per unity of Sm^3 of natural gas
— air pressure ratio in the range 5-7
— exhaust temperature 723 K, 450°C, 842°F; stack exhaust temperature 423 K, 150°C, 302°F
temperature drop available for recovery 573 K, 300°C, 540°F

9.5 THE DIESEL ENGINE CYCLE

Reciprocating engine types, principally the spark-ignited gas engine for natural gas or the Diesel engine for liquid fuel, are widely applied in cogeneration systems to drive electric generators and mechanical loads such as compressors and pumps. Alternating machines can be designed to run on either gas or gasoil.

Diesel engines are available in a wide range of power, from several to thousands kW at different operating speeds ranging from 100 r/min to 1800 r/min according to the size and technical characteristics of the system.

Fig. 9.7 shows a typical energy flow in a cogeneration plant with a diesel engine: shaft output (35%), steam from exhaust (20%), hot water from cooling (30%), losses (15%).

Exhaust gas, at a temperature in the range 623-723 K (350-450°C; 662-842°F), permits steam generation at a saturation pressure of 0.3-1 MPa, suitable for industrial applications.

Jacket and piston water cooling as well as lubricant cooling water can provide hot water at an average temperature of 343-353 K (70-80°C; 158-176°F) which can be used for space-heating or industrial low-temperature end users. Superheated water can also be produced with special design of the diesel engine and recovery system.

The maximum mechanical power available at the shaft generally ranges from 28 to 40% of the diesel engine input power as fuel. The higher values refer to turbocharged machines. In consequence, the quantity of heat rejection equals 72-60% of the fuel input. The amount of the recoverable heat depends on the bottom temperature level required by the end user in the form of hot water.

Cooling equipment must also be provided if users of hot water are not in operation. Exhausts will be discharged into the atmosphere.

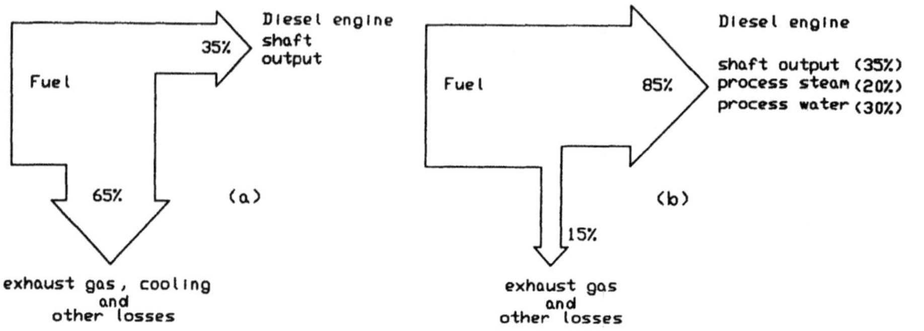

Fig. 9.7 Energy balance (a) in a diesel engine and (b) in a cogeneration plant with a diesel engine (exhausts are recovered to produce steam)

Typical input-power values, in the case of natural gas input, are 345 Sm^3/h (3300 kW) to produce 1000 kW of electric power and 1750 kW of recoverable heat. The overall efficiency of the system, that is, the ratio between output and input power, is roughly 83%. The ratio of output electric power to input power is roughly 30%.

Typical values of operating parameters for several diesel-engines fed by natural gas are shown in Table 9.4. As a general rule, this type of prime mover is attractive if the process requires large quantities of low-level heat recovered from the jacket water and lubricant oil cooling systems which may reach 50% of the heat rejection. An example of a diesel engine cycle is reported in Chapter 9.7.

9.6 DETERMINING THE FEASIBILITY OF COGENERATION

Cogeneration feasibility is based on economic and technical factors which have to be correlated to complete a valid evaluation. Major factors for consideration are:

— the ratio between the electricity demand and the fuel for the factory defined as daily, monthly and yearly ratio. This ratio must be consistent with the ratio between electric output power and heat recovery for the cogeneration system chosen;

Table 9.4 Technical parameters for diesel engines

Size Unit	Specific consumption	Input fuel as natural gas			Input air flow			Heat recovery	Output electric energy	Total efficiency	Electric efficiency
MW	kJ/kWh	10^6kJ/h (c)	10^6Btu/h	Sm^3/h	$10^3Sm^3/h$	10^3kg/h	10^3lb/h	10^6kJ/h (a)	10^6kJ/h (b)	% (a+b)/(c)	% (b)/(c)
0.25	12000	3.0	2.8	87	1.3	1.7	3.7	1.5	0.9	78.83	30.00
1	11500	11.5	10.9	335	5.0	6.5	14.3	5.6	3.6	80.14	31.30
5	11500	57.5	54.5	1675	25.1	32.4	71.5	28.1	18.0	80.14	31.30
10	11250	112.5	106.6	3277	49.2	63.4	139.8	54.9	36.0	80.83	32.00
25	11250	281.3	266.6	8194	122.9	158.5	349.5	137.3	90.0	80.83	32.00
40	11250	450.0	426.5	13110	196.6	253.7	559.3	219.7	144.0	80.83	32.00
100	11250	1125.0	1066.4	32775	491.6	634.2	1398.2	549.4	360.0	80.83	32.00

Notes
— air density in standard condition 1.29 kg/Sm^3
— combustion air 15 Sm^3 per unity of Sm^3 of natural gas

— the profile of thermal demand, including temperature levels of end user requirements and typical fluctuations of the demand (daily, monthly, yearly). Temperature levels must be consistent with the level of heat rejected from the cogeneration system;
— the profile of electric demand and typical fluctuations as for the thermal profile. Thermal and electric profiles must be correlated with each other;
— purchased fuel and electricity costs, present and projected future costs;
— working hours per year and per the total life of the plant;
— plant system sized for the present factory needs and for the future;
— capital cost of the cogeneration plant and operating cost during the life of the plant;
— environmental issues.

Many cogeneration approaches can be followed in order to make a choice among system types and sizes. In order to ensure, however, the highest efficiency of the system, the recovery of the rejected heat must be effective in any operating condition of the cogeneration plant. Additional boiler plants will satisfy the end-user requirements, if these are higher than the recovery heat. Depending on the industrial processes, this constraint can be more or less important in determining the size of the plant.

A first approach is to design a system which is capable of meeting thermal load requirements, regardless of the electric demand. It is connected to the utility grid and sells excess or buys additional electricity depending on the factory's thermal and electric profile and on the operating conditions.

A second approach is to design a system which is capable of meeting either peak or base electric load requirements, regardless of the thermal demand which nevertheless must be greater than the heat rejected. It is connected to the utility grid and sells excess or buys additional electricity depending on the sizing and on the operating conditions.

A third approach is to design a system independent of the utility grid. It requires overcapacity or redundant equipment to ensure reliability which is guaranteed by the utility in the first two approaches. These systems have traditionally been oversized to meet peak electric demand, with supplementary equipment to satisfy the thermal demand if necessary.

9.7 PRACTICAL EXAMPLES

Three examples of cogeneration plant are given below. The saving in primary energy in comparison with thermal utility plants is also shown.

Notice that cogeneration plants save primary energy, not energy entering the factory. They can effect an energy-cost saving only if the balance between electric energy and thermal energy costs is favorable for the factory.

Example 1
Cogeneration plant with steam turbine

Table 9.5 lists data and operating parameters of a steam turbine in two typical working conditions (input-output pressure drop equal to 3.793 MPa and 8.963 MPa). Specific consumptions attributed to the production of electric energy are calculated as kJ/kWh or kg_{oil}/kWh entering the boiler plant.

Primary energy saving in comparison with standard thermal utility plants and variation of the factory energy consumption (additional fuel consumption and reduction of kWh from utilities) are also shown.

Economic evaluations can also be made by introducing the cost of electric energy and fuel. For a preliminary evaluation, the average cost of electric energy purchased from utilities can be introduced (see Table 20.3); for a more detailed analysis, it is necessary to calculate the purchased energy and the consequent cost corresponding to the new demand profile of the plant for utilities-energy. As a general rule, the new cost per unit of purchased electric energy will be higher than the current cost because of a lower exploitation of the power supplied under the utility contract (as happens if the cogeneration plant covers the base of the electric demand and the utility supply the additional demand). Local regulations concerning the selling of energy to the utilities and its purchase from them in emergency must also be considered.

Maintenance costs, too, must be taken into account. They can be introduced as a fixed cost per unit of kWh produced.

Example 2
Cogeneration plant with gas turbine

Table 9.6 lists data and operating parameters of a gas turbine in typical working conditions. Specific consumptions attributed to the production of electric energy are calculated as kJ/kWh or Sm^3/kWh of natural gas entering the gas turbine.

Primary energy saving in comparison with standard thermal utility plants and variation of the factory energy consumption (additional fuel consumption and reduction of kWh from utilities) are also shown.

The recoverable heat varies according to the end-user requirements: hot air for drying, steam and hot water with or without additional burners.

Economic evaluations can also be made by introducing the cost of electric energy and fuel. For a preliminary evaluation, the average cost of electric energy purchased from utilities can be introduced (see Table 20.3); for a more detailed analysis, it is necessary to calculate the purchased energy and the consequent cost corresponding to the new demand profile of the plant for utilities-energy. As a general rule, the new cost per unit of purchased electric energy will be higher than the current cost because of a lower exploitation of the power supplied under the utility contract (as happens if the cogeneration plant covers the base of the electric demand and the utility supply the additional demand). Local regulations concerning the selling of energy to the utility and its purchase from them in emergency must also be considered.

Maintenance costs, too, must be taken into account. They can be introduced as a fixed cost per unit of kWh produced.

Example 3
Cogeneration plant with diesel engine

Table 9.7 lists data and operating parameters of a diesel engine in typical working conditions. Specific consumptions attributed to the production of electric energy are calculated as kJ/kWh or kg_{gasoil}/kWh or Sm^3/kWh of natural gas entering the engine.

Primary energy saving in comparison with standard thermal utility plants and variation of the factory energy consumption (additional fuel consumption and reduction of kWh from utilities) are also shown.

The recoverable heat varies according to the end-user requirements: hot air for drying, steam and hot water at different temperatures to which the possibility of a complete exploitation of the rejected heat is correlated.

Economic evaluations can also be made by introducing the cost of electric energy and fuel. For a preliminary evaluation, the average cost of electric energy purchased from utilities can be introduced (see Table 20.3); for a more detailed analysis, it is necessary to calculate the purchased energy and the consequent cost corresponding to the new demand profile of the plant for utilities-energy. As a general rule, the new cost per unit of purchased electric energy will be higher than the current cost because of a lower exploitation of the power supplied under the utility contract (as happens if the cogeneration plant covers the base of the electric demand and the utility supply the additional demand). Local regulations concerning the selling of energy to the utility and its purchase from them in emergency must also be considered.

Maintenance costs, too, must be taken into account. They can be introduced as a fixed cost per unit of kWh produced.

Table 9.5 Examples of steam turbine cycles (reference 1 t of steam as input)

	INPUT (1 t of steam)			
a	pressure	MPa	4.137	9.997
b	temperature	°C	399	510
b'		°F	750	950
c	enthalpy	MJ	3209	3399
	OUTPUT (1 t of steam)			
d	pressure	MPa	0.344	1.034
e	theoretical enthalpy	MJ	2693.64	2824
f = c − e	available energy AE	MJ	515.36	575
g	expansion efficiency	%	75	75
h = f · g/100	useful drop	MJ	386.52	431.25
i = c − h	actual output enthalpy	MJ	2822.48	2967.75
l	temperature	°C	180	220
l'		°F	356	428
m	average condensate enthalpy	MJ	440	440
n = i − m	useful heat for process	MJ	2382.48	2527.75
	BOILER			
o = c	output steam enthalpy	MJ	3209	3399
p = m	average condensate enthalpy	MJ	440	440
q = o − p	necessary heat	MJ	2769	2959
r	boiler efficiency (NHV as reference)	%	89	90
s = q · 100/r	fuel input	MJ	3111.235	3287.777
t = s/41.86	oil	kg_{oil}	74.431	78.654
t' = s/34.325	natural gas	Sm^3	90.219	95.339
	ELECTRIC GENERATOR			
u	efficiency	%	96	94
v = h · u/100	output electric energy	MJ	371.059	405.375
v' = v · 1000/3600		kWh	103.073	112.605
w = 1000/v'	**SPECIFIC CONSUMPTION**	kgvap/kWh	9.701	8.880

Table 9.5 (continued)

	PLANT			
A	useful electric power	kW	1000	1000
B = w · A/1000	steam flow	t/h	9.701	8.880
C = s · B	input fuel	MJ/h	30184.73	29197.33
D = t · B	oil	kg_{oil}/h	722.122	698.5
E = t' · B	natural gas	Sm^3/h	875.300	846.667
F = n · B	useful heat for process	MJ/h	23114.45	22447.85
G = A · 3600/1000	output electric power	MJ/h	3600	3600
H = F + G	total output	MJ/h	26714.41	26047.81
I = H/C	plant efficiency	%	88.5	89.21
L = (C – F · 100/r)/A	equiv.specific consumption	MJ/kWh Btu/kWh kcal/kWh	4.213 3933 1006.554	4.255 4033 1016.548
L' = L/41.86		kgoil/kWh	0.101	0.101
	PRIMARY ENERGY SAVING			
M = F/A	useful heat/electric power	MJ/kWh	23.114	22.447
N = 9.63 – L (*)	primary energy saving	MJ/kWh Btu/kWh kcal/kWh	5.414 5132 1293.445	5.372 5092 1283.451
O = N · A/41860		TOE/h	0.129	0.128
P	operating hours	h/year	6000	6000
Q = O · P	primary energy saving	TOE/year	776.067	770.07
	FACTORY ENERGY CONSUMPTION			
R = L' · P · A/1000	factory additional oil	TOE/year	603.932	609.929
S = A · P/1000	reduction kWh from utilities	MWh/year	6000	6000

Note
(*) reference specific consumption in utility power plant 9630 kJ/kWh

Table 9.6 Examples of gas turbine cycles (reference 100 MJ as input or $94.7 \cdot 10^3$ Btu)

	INPUT (100 MJ as input fuel)		
a	natural gas	MJ	100.00
$b = a \cdot 1000/34325$		Sm^3	2.91
$c = b \cdot 0.75$		kg	2.18
	OUTPUT		
d	shaft energy	MJ	21.00
$e = d \cdot 1000/3600$		kWh	5.83
f	exhaust gas mass	kg	141.6
$g = f/1.29$	exhaust gas volume	Sm^3	109.43
h	exhaust gas temperature	°C	500.00
h'		°F	932
i	exhaust gas specific heat	$kJ/kg \cdot K$	1.08
$l = f \cdot h \cdot i/1000$	exhaust gas enthalpy	MJ	76.23
m	exhaust circuit losses	MJ	3.81
n	typical recoverable exhaust	MJ	50.69
$o = l - m$	max recoverable from exhaust	MJ	72.42
	ELECTRIC GENERATOR		
p	efficiency	%	94.00
$q = d \cdot p/100$	output electric energy	MJ	19.74
$r = q \cdot 1000/3600$		kWh	5.48
	SPECIFIC CONSUMPTION		
$s = a/r$	Input fuel	MJ/kWh	18.24
$t = b/r$		Sm^3/kWh	0.53

Table 9.6 (continued)

	PLANT		
A	useful electric power	kW	1000.00
B = A · s	input fuel	MJ/h	18237.08
C = A · t	natural gas input	Sm^3/h	531.3
D = A · f/r	exhaust flow-rate	kg/h	25744.05
E = A · n/r	typical recoverable heat	MJ/h	9244.38
F = A · 3600/1000	useful electric power	MJ/h	3600.00
G = E+F	useful output	MJ/h	12844.37
H = G/B	plant efficiency	%	0.70
I = (B-E/0.85)/A	specific consumption	MJ/kWh	7.36
		Btu/kWh	6977.58
		kcal/kWh	1758.56
	PRIMARY ENERGY SAVING		
L = (E/0.85)/A	recoverable heat/electric power	MJ/kWh	10.86
M = 9.63 – I (*)	primary energy saving	MJ/kWh	2.27
N = M · A/41860		TOE/h	0.054
O	operating hours	h/year	6000.00
P = N · O	primary energy saving	TOE/year	325.17
	FACTORY ENERGY CONSUMPTION		
Q = (B – E/0.85) · · O/34.325	factory additional fuel	Sm^3/year	1286760
Q' = Q · 34.325/41860		TOE/year	1061.58
R = A · O/1000	reduction kWh from utilities	MWh/year	6000

Notes
equivalent boiler efficiency 85% in step I (Net Heating Value as reference)
(*) reference specific consumption in utility power plant 9630 kJ/kWh
c_p air at 500°C, 932°F equal to 1.08 kJ/kg · K
natural gas density 0.75 kg/Sm^3
air density at 500°C, 932°F equal to 0.45 kg/Sm^3

Table 9.7 Examples of diesel engine cogeneration systems (reference 100 MJ as input or 94.7 · 10^3 Btu)

	INPUT (100 MJ as input fuel)		
a	natural gas	MJ	100.00
b = a · 1000/34325		Sm^3	2.91
c = b · 0.75		kg	2.18
	OUTPUT		
d	shaft energy	MJ	35.10
e = d · 1000/3600		kWh	9.75
f	exhaust gas mass	kg	50.00
g = f/1.29	exhaust gas volume	Sm^3	38.76
h	exhaust gas temperature	°C	475.00
h'		°F	887
i	exhaust gas specific heat	kJ/kg · K	1.08
l = f · h · i/1000	exhaust gas enthalpy	MJ	25.65
m	exhaust circuit losses	MJ	1.28
n	typical recoverable exhaust	MJ	19.83
o = l − m	max recoverable from exhaust	MJ	24.37
p	heat recovery from cooling equipment	MJ	29.75
	ELECTRIC GENERATOR		
q	efficiency	%	94.00
r = d · q/100	output electric energy	MJ	32.99
s = r · 1000/3600		kWh	9.17
	SPECIFIC CONSUMPTION		
t = a/s	Input fuel	MJ/kWh	10.91
u = b/s		Sm^3/kWh	0.32

Table 9.7 (continued)

	PLANT		
A	useful electric power	kW	1000.00
B = A · t	input fuel	MJ/h	10911.07
C = A · u	natural gas input	Sm^3/h	317.88
D = A · f/s	exhaust flow-rate	kg/h	5455.54
E = A · (n + p)/s	typical recoverable heat	MJ/h	5410.26
F = A · 3600/1000	useful electric power	MJ/h	3600.00
G = E+F	useful output	MJ/h	9010.26
H = G/B	plant efficiency	%	0.83
I = (B – E/0.85)/A	specific consumption	MJ/kWh	4.55
		Btu/kWh	4309.06
		kcal/kWh	1086.02
	PRIMARY ENERGY SAVING		
L = (E/0.85)/ A	recoverable heat/electric power	MJ/kWh	6.37
M = 9.63 – I (*)	primary energy saving	MJ/kWh	5.08
N = M · A/41860		TOE/h	0.121
O	operating hours	h/year	6000.00
P = N · O	primary energy saving	TOE/year	728.28
	FACTORY ENERGY CONSUMPTION		
Q = (B – E/0.85) · · O/34.325	factory additional fuel	Sm^3/year	794763
Q' = Q · 34.325/41860		TOE/year	651.70
R = A · O	reduction kWh from utilities	MWh/year	6000

Notes
equivalent boiler efficiency 85% in step I (Net Heating Value as reference)
(*) reference specific consumption in utility power plant 9630 kJ/kWh
c_p air at 500°C, 932°F equal to 1.08 kJ/kg · K
natural gas density 0.75 kg/Sm^3
air density at 500°C, 932°F equal to 0.45 kg/Sm^3

CHAPTER 10

FACILITIES—PUMPS AND FANS

10.1 INTRODUCTION

In process and HVAC applications, pumps and fans are widely used to move fluids, liquid or air or gas, by using mechanical energy to overcome the resistance of the flow circuit. The prime mover at the shaft is generally an electrical drive, but also other types of drive can be used. The power required at the shaft depends on the volume and the density of the fluid moved, on the pressure difference across the fan or the pump and on the mechanical design and efficiency.

Pumps and fans can be defined as continuous rotary machines in which the rapidly rotating element accelerates the fluid passing through it, converting the velocity head into pressure, partially in the rotating element and partially in stationary diffusers or in blades.

Pumps are similar to compressors but work essentially with incompressible fluids.

Fans or blowers can be defined as compressors that develop a very low density increase from inlet to discharge (less than 5-7% density increase).

For pumps moving a liquid, density variations are generally negligible. For fans moving a gas, density changes become significant and must be taken into account. Variation of density, however, is strictly related to the compression ratio; in fans with a head of 500 Pa (a quite high value for most applications), the pressure ratio equals 1.005, so the density variation is negligible.

The most commonly used machines (see also Chapter 11 for air compressors) can be classified as centrifugal or axial machines. The basic types are shown in Fig. 10.1. Others types of machine can be used for special purposes.

Centrifugal machines are dynamic machines in which one or more rotating impellers accelerate the fluid. The main fluid flow is radial.

Axial machines are dynamic machines in which each stage consists of two rows of blades, one rotating and the other stationary. The main fluid flow is axial.

The bladed rotor accelerates the fluid; the stator converts kinetic energy into pressure. Note that in low-head machines (fans, in particular), the stator does not exist or it can be unbladed.

In industrial applications, centrifugal machines are widely used in pumps and fans; axial machines particularly in fans.

One of the major problems in operating fans and pumps is the control of the volume flow-rate to meet user needs which may require fluid flow inferior to the rated one, sometimes for long operating periods.

Common mechanical devices for the control of fans are: inlet louvres, inlet vanes, outlet dampers, different kinds of damper such as butterfly and guillotine and variable pitch for axial fans. With these controls and in particular, but quite common, operating conditions, an excessive power consumption may occur, together with high levels of sound and vibration and fan-volume shortfalls. Two-speed fan motors, especially if combined with either inlet-vane or inlet-louvre control, are a considerable improvement. Nevertheless, variable-speed electrical drives are the most efficient method of fluid flow control.

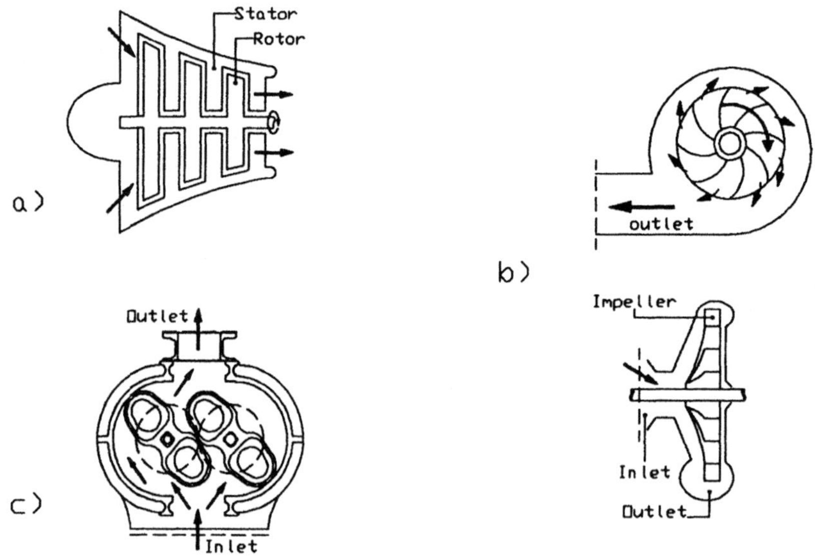

Fig. 10.1 Basic types of rotating pumps, fans and compressors: a) axial flow; b) centrifugal; c) lobe

The most usual mechanical devices for the control of the fluid flow of pumps are throttling valves which cause additional pressure losses in the circuit and so reduce the fluid flow. Because of these losses, no significant decrease of power consumption occurs. This control, together with recycling systems, is one of the so-called dissipative methods. As with fans, the most efficient control system employs variable-speed electrical drives to regulate the shaft speed.

10.2 BASIC PRINCIPLES OF PUMP AND FAN OPERATION

In the diagram of total pressure or total head versus volume flow-rate or capacity, the working point is defined as the intersection between the characteristic curves of the machine and of the load.

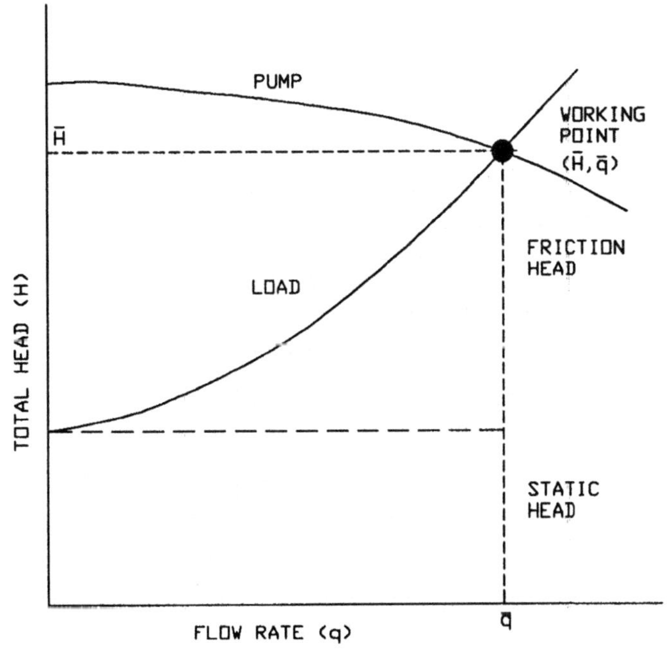

Fig. 10.2 Pump and load characteristics

In what follows, the characteristics of centrifugal pumps are discussed, but the same considerations can easily be extended to any centrifugal machines (fans, compressors) and to axial machines. The latter have a head-capacity curve much steeper than that of a centrifugal machine, hence the safe operating range is much smaller.

The basic curve for the pump at a set of given values for impeller diameter and shaft speed, and the basic curve for the load, that is total pressure or head, static and friction head, versus volume flow-rate or capacity, are shown in Fig. 10.2. The intersection between the two curves defines the working point ($\overline{H}$, $\overline{q}$) of the system, which cannot change without modification of either the load circuit or the pump.

The pump manufacturer provides different sets of performance curves: total head versus flow-rate or capacity for different impeller diameters, shaft power versus flow-rate or capacity and sets of points with the same efficiency. Any set of curves is given for a fixed speed of the impeller (see Fig. 10.3).

From fluid mechanics laws and practice, the following relationships for centrifugal pumps can be derived to take into account the impeller speed variation.

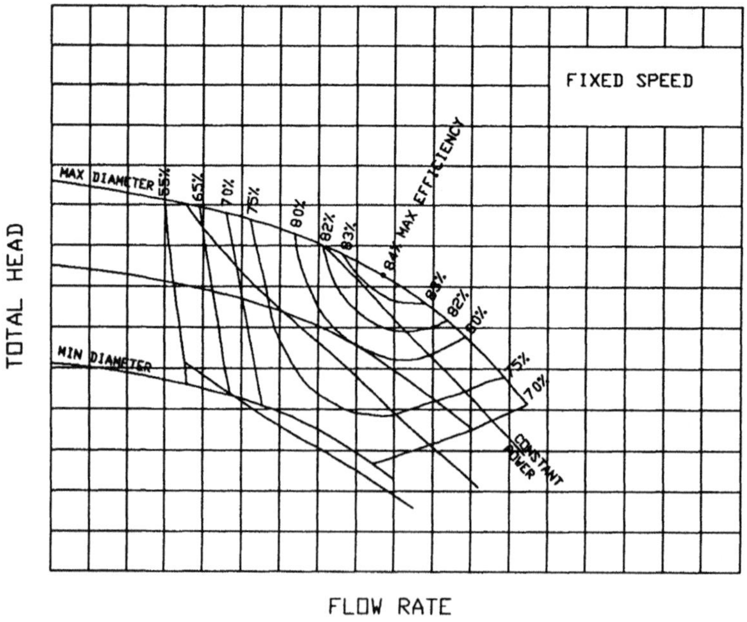

Fig. 10.3 Set of curves for a given pump

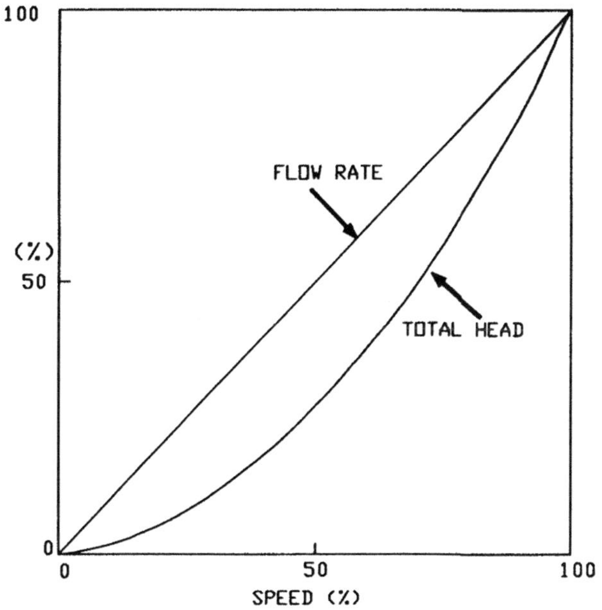

Fig. 10.4 Pump flow-rate and total head versus speed

If n_1 and n_2 are two impeller speeds corresponding to two working points (1, 2), and if efficiency is assumed to be constant (in practice it varies with the speed following an empirical law), it follows that:

Flow capacity	$q_2 = q_1 \cdot (n_2/n_1)$	**varies directly as the speed**
Total Head	$H_2 = H_1 \cdot (n_2/n_1)^2$	**varies as the square of the speed**
Shaft Power	$P_2 = P_1 \cdot (n_2/n_1)^3$	**varies as the cube of the speed**

Curves representing flow-rate and total head versus impeller speed are shown in Fig. 10.4.

These relationships can be written as a function of the impeller diameter instead of the speed by simply substituting impeller diameter values D_1 and D_2 for n_1 and n_2.

If other types of pumps are involved, consult the manufacturer for instructions.

The volume flow-rate can be calculated by multiplying the fluid velocity (typical values for water are 2-3 m/s) by the pipe section (m^2). The mass flow-rate is equal to the volume flow-rate multiplied by the fluid density (kg/m^3).

From the above-mentioned relationships (see also Chapter 4.5), it follows that the power P required at the shaft is:

$$\text{Power Shaft } P = \frac{q \cdot H \cdot \rho \cdot (g/g_c)}{\eta} \qquad \left(W, \frac{ft \cdot lbf}{s}\right)$$

where

q = volume flow-rate (m^3/s, ft^3/s)

H = total head = H static + H losses (m, ft)

ρ = density of the fluid (kg/m^3, lb/ft^3)

η = pump efficiency

g = gravitational acceleration 9.81 m/s^2 = 32.17 ft/s^2

$$g_c = \text{conversion factor} = \frac{1\ kg \cdot (m/s^2)}{N} = 32.17 \frac{lb \cdot (ft/s^2)}{lbf}$$

Notice that, in practice, power varies with the impeller speed more than with the cube because of the reduction of efficiency with speed.

Fig. 10.5 shows two typical operating situations:

(a) friction or dynamic lower than static losses.
Pump inlet and outlet are connected with two tanks by short pipelines without significant friction losses. If the difference in level between the tanks is constant, the static head is also constant. Friction losses, due to the movement of the fluid, are negligible as already assumed;

(b) friction or dynamic greater than static losses.
The pipeline between the two tanks is quite long and friction losses due to the movement of the fluid must be added to the static losses.

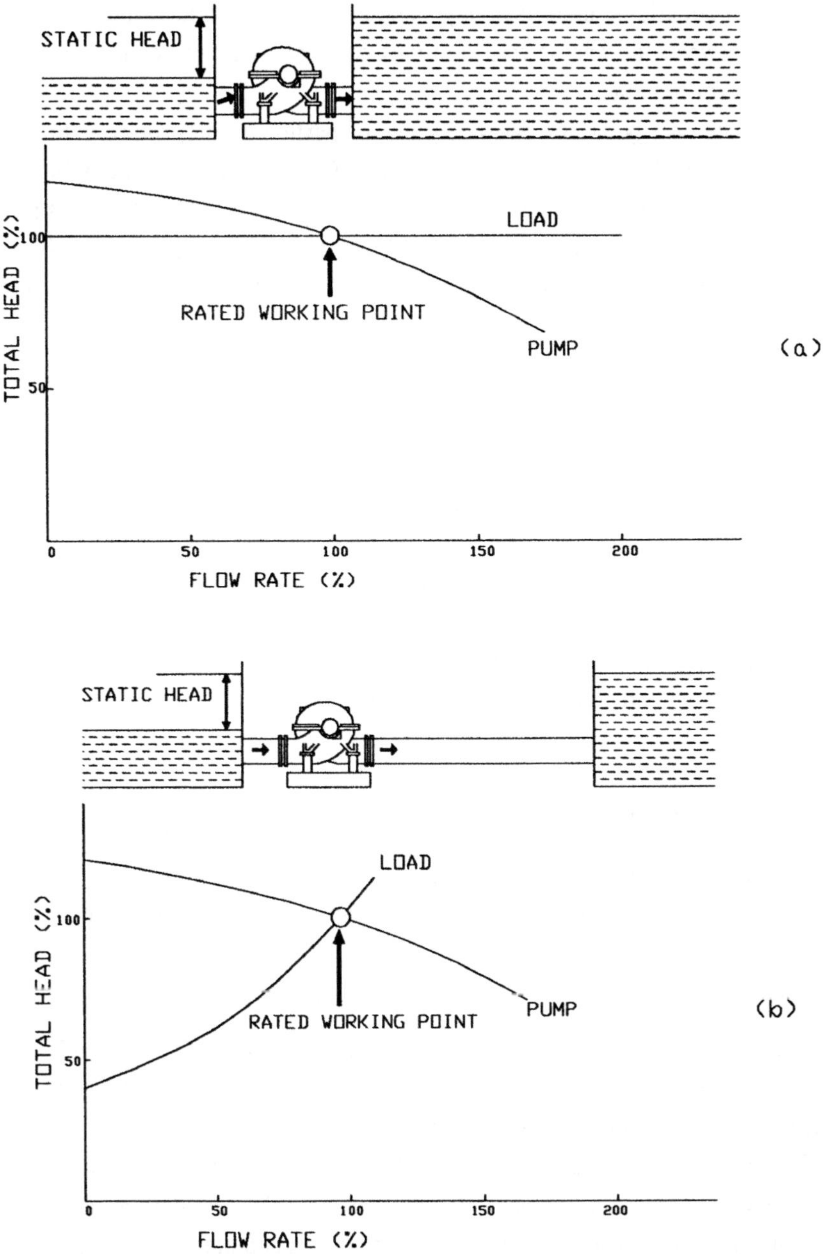

Fig. 10.5 Typical load circuits: (a) dynamic losses lower than static losses; (b) dynamic losses greater than static losses

In both situations, there is always a single working point, which is the intersection between the characteristic curves of flow capacity versus head for the pump and for the load.

From the energy-saving point of view, of particular importance is the operation at reduced flow-rate in the same circuit. To reach this working point, one of the two curves, either the pump or the load curve, must be modified. As already mentioned, the pump curve varies with the shaft speed and the impeller diameter. The load curve can be modified by introducing additional equipment, such as throttling valves, between the pump and the load.

10.3 METHODS OF REGULATING THE WORKING POINT

Methods of regulating the working point, that is, regulating head and flow, can be classed as dissipative and non-dissipative. The first category is based on the introduction of an additional head loss inside the circuit without changing the pump characteristics and by shifting the end-user load curve; the second is based on the introduction of different pump characteristics such as new impellers or new shaft speed and consequently new pump curves.

Dissipative methods

Dissipative methods are of two types:

- — introduction of head friction loss by means of a throttling valve or similar devices;
- — recycling part of the flow inside a closed ring circuit.

Fig. 10.6 shows curves corresponding to the operating conditions in Fig. 10.5, and indicates the new working point corresponding to a 50% reduction of the rated flow by means of a throttling valve.

If the static head is greater than the dynamic one, dissipative regulation does not provoke significant losses; in the opposite case, this regulation increases the dynamic losses to a level which would be unacceptable for a long operating period.

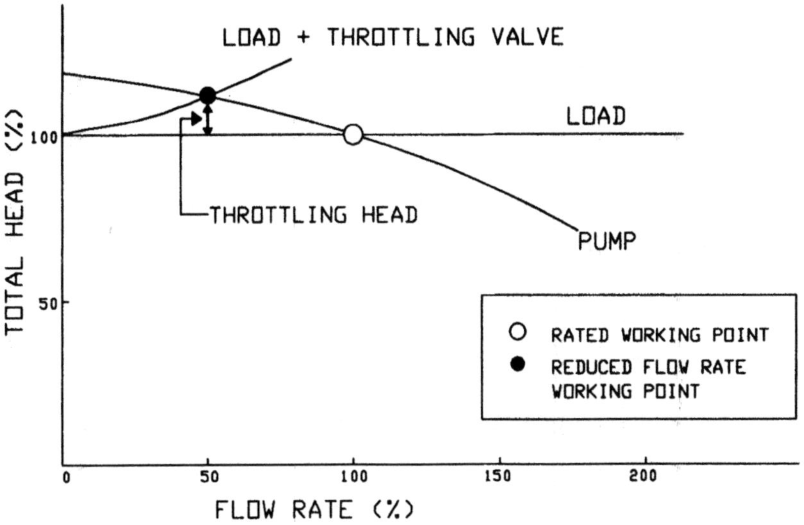

(a)

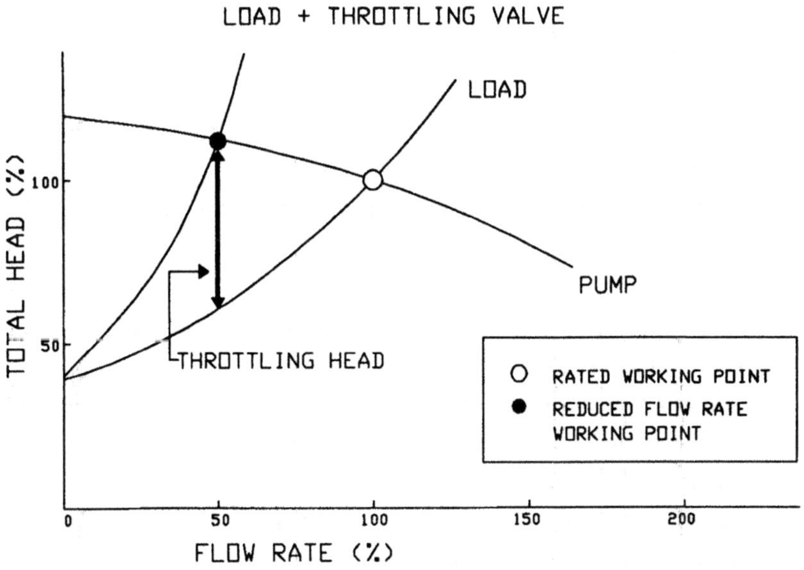

(b)

Fig. 10.6 Pump and load curves corresponding to the circuits in Fig. 10.5

The power lost in the throttling valve is calculated with the same relationship used before for shaft power:

$$P_{losses} = \frac{q \cdot \Delta H \cdot \rho \cdot (g/g_c)}{\eta} \qquad (W, \frac{ft \cdot lbf}{s})$$

where

q = reduced capacity value flowing through the throttling valve

ΔH = additional throttling valve head losses.

Non-dissipative methods

These are based on modifying the pump curves without changing the load curve.

This means modifying the design of the pump (e.g. changing the geometry of the impeller) or modifying the shaft speed by means of variable-speed drives upstream of the pump shaft.

We can distinguish:

— use of centrifugal machines with variable geometry to obtain different curves, at constant shaft speed;

— use of hydraulic or electromechanical coupling to regulate the pump shaft speed at constant electric motor speed;

— use of electrical variable-speed drives to regulate the motor speed and so the pump shaft speed.

In large pump systems, drivers like steam engines or turbines, gas turbines and diesel engines can also be applied (see Chapter 9).

Fig. 10.7 shows a typical example of flow regulation obtained by varying the pump speed until the curve of the pump crosses the curve of the load at the desired working point.

Great precision is not generally necessary; note that the bottom speed limit does not usually go below 40-50% of the design speed.

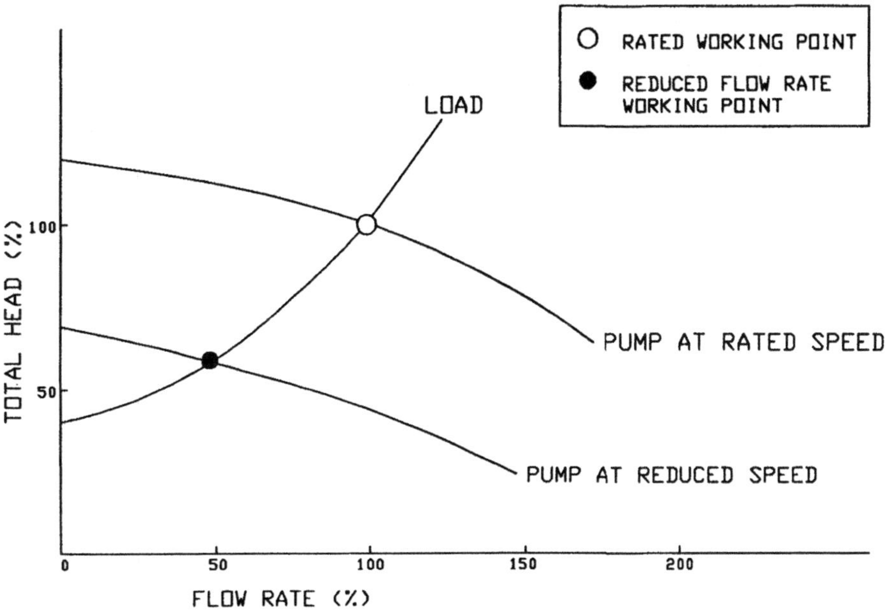

Fig. 10.7 Pump curves at different shaft speeds

Comparison of dissipative with non-dissipative methods underlines the following energy-saving advantages of the latter:

- **reduction of energy consumption at constant energy delivered to the load;**
- **reduction of the power demanded from the electric line due to the reduction of the power needed for driving the pump.**

The commonest electrical drives are d.c motors with a.c./d.c.converters and a.c. motors with inverters; the latter can also be applied to existing motors and are suggested for hazardous environmental conditions (see Chapter 7.4).

10.4 COMPARISON BETWEEN DISSIPATIVE AND NON-DISSIPATIVE REGULATION METHODS

An example of procedures for comparing some regulation methods is given below. The details of the calculations are set out in Chapter 10.6.

Fig. 10.8 shows the situations to compare. The load curve is given together with characteristic pump curves (see Fig. 10.9), corresponding to a fixed speed (in this figure it is 3550 r/min). The design working point is shown as A_n (q_n; H_n; η_n).

A reduction of the design flow rate to 50% is required. The working point should move from point A_n to A_1 ($q_1 = 0.5\ q_n$; H_1; η_1) always along the load curve to reach the new working condition.

Two alternative procedures are considered:

— use of a dissipative regulation that modifies the load curve at constant speed by introducing a throttling valve. The new working point is now point A_2 ($q_2 = 0.5 \cdot q_n$; H_2; η_2) with the same flow as point A_1 but with increased head. This head increase corresponds to a required shaft power:

$$PA_2 - PA_1 = \frac{(0.5 \cdot q_n) \cdot (H_2 - H_1) \cdot \rho \cdot (g/g_c)}{\eta_2} \qquad \text{(W)}$$

where the efficiency value in the point A_1 is provided by the pump manufacturer's performance curves (see Fig. 10.9).

The power now required in the new working point A_2 is:

$$PA_2 = \frac{(0.5 \cdot q_n) \cdot H_2 \cdot \rho \cdot (g/g_c)}{\eta_2} \qquad \text{(W)}$$

— use of a non-dissipative regulation that varies the shaft speed and so the pump curve until the new curve crosses the load curve at the desired working point A_1 ($q_1 = 0.5\ q_n$; H_1; η_1). Curves at different speeds are given by the pump manufacturer or can be constructed point by point following the relationships described in Chapter 10.2, where efficiency is assumed not to change appreciably for reasonable speed variations. The new speed equals 2750 r/min. The total power at the pump shaft is:

$$PA_1 = \frac{(0.5 \cdot q_n) \cdot H_1 \cdot \rho \cdot (g/g_c)}{\eta_1} \qquad \text{(W)}$$

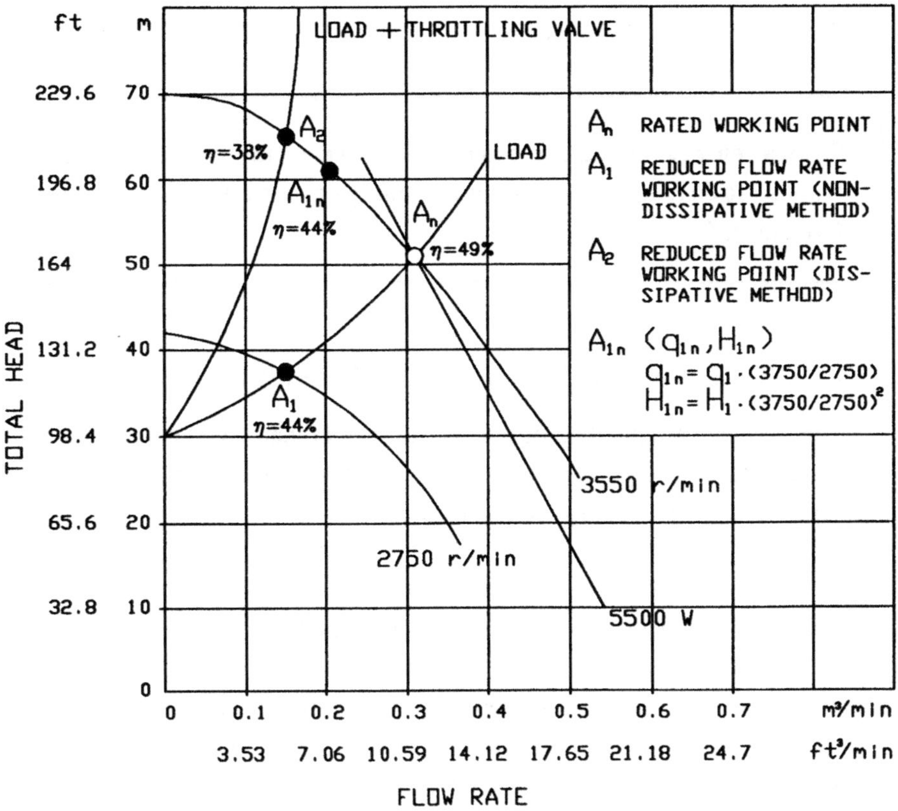

Fig. 10.8 Working points with dissipative and non-dissipative regulation (see Example Chapter 10.6)

where the efficiency must be calculated from data supplied by the pump manufacturer. The efficiency corresponding to point A_1 will be greater than the efficiency at point A_2 (the value of the efficiency at A_1 can be assumed to be equal to that at the working point A_{1n} on the curve of nominal speed with q and H which are related to q_1 and H_1 as shown in Chapter 10.2);

—the head difference between A_2 and A_1 represents the saving in power and so in energy due to the use of a non-dissipative method. This saving depends both on the head increase and on the difference of efficiency between the two points;

— the energy saved depends on the working hours

$$ES = (PA_2 - PA_1) \cdot \text{working hours} \qquad \text{(Wh)}$$

Typical energy saving for a 50% reduction of the capacity ranges from 25 to 50% of the corresponding energy needed with non-dissipative methods.

10.5 CENTRIFUGAL MACHINES REGULATION METHODS

Non-dissipative regulation of pumps and fans in industrial applications is an effective area of energy saving in many processes like paper, chemical and refinery and in many water-pumping systems.

In the low- and medium-power range, from 1 to 200 kW, electrical variable-speed drives can conveniently be applied. Synchronous and asynchronous motors, fed at variable frequency and voltage, are particularly suitable for operating in hazardous locations with a proper choice of the motor and related equipment. Higher-power ranges, too, can be covered with the same drives, but also gas turbines, steam motors and turbines and diesel engines can be employed economically if the waste heat is recovered as already shown in cogeneration systems (see Chapter 9).

The amount of energy saved depends on the need for low-rate capacity, on the working hours of the system and on the electric energy cost, but it is always significant.

Other advantages, in addition to the energy saving which is the main factor, are: greater simplicity due to the elimination of throttling valves and similar devices and simpler design of the mechanical and electric circuits made possible by softer starting.

10.6 PRACTICAL EXAMPLES

Example

Compare the energy consumption of a centrifugal pump if flow is reduced to 50% of the rated value with (i) a dissipative method and (ii) a non-dissipative method

An example is given below of technical and economic comparison between a dissipative system (throttling valve) and a non-dissipative one (electrical variable-speed drive). This example refers to a low-power pump which works at 50% of rated capacity for all the year, but it can be assumed as a basic model for other situations.

Table 10.1 (a) and Table 10.1 (b) show the technical characteristics of the pump and the data elaboration procedures with both SI and non-SI units. The characteristic curves of the pump and of the circuit are those reported in Fig. 10.8 and in Fig. 10.9, where the comparison between dissipative and non-dissipative methods is illustrated. The comprehension of correlations between the figures and the table content is facilitated by introducing working point indexes (A_n, A_1, A_2).

The energy saving is shown in Table 10.1 (a) and in Table 10.1 (b), and the economic evaluation in Table 20.3.

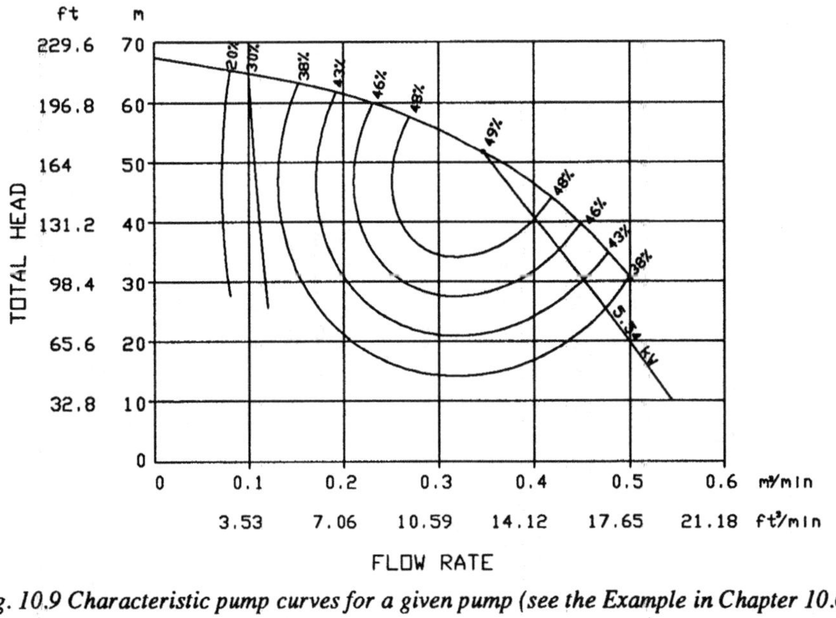

Fig. 10.9 Characteristic pump curves for a given pump (see the Example in Chapter 10.6)

Table 10.1 (a) Comparison between dissipative and non-dissipative methods (see Fig. 10.8 and Fig. 10.9)

SI units			AT RATED CAPACITY (100%)		AT REDUCED CAPACITY (50%) DISSIPATIVE METHOD (I)		AT REDUCED CAPACITY (50%) NON-DISSIPATIVE METHOD (II)	
			Working point		Working point		Working point	
a	CAPACITY	m^3/min	An	0.312	A1/A2	0.16	A1	0.16
a' = a/60		m^3/s		0.005		0.0027		0.0027
b	HEAD (basic load)	m	An	53.20	A1	37.50	A1	37.50
c		m			A2	67.00		
d	PUMP EFFICIENCY	%	An	49.00	A2	38.00	A1 (A1n)	44.00
	POWER DELIVERED							
$e = \frac{a' \cdot b \cdot 9.81 \cdot r}{1000}$	to the circuit	kW	An	2.72	A1	0.98	A1	0.98
	REGULATING LOSSES							
$f = \frac{a' \cdot (c - b) \cdot 9.81 \cdot r}{1000}$	valve losses	kW		0	A2-A1	0.77		0
$g = f \cdot 100/d$	valve losses at pump shaft	kW		0	A2-A1	2.03		0
$h = g + e \cdot 100/d$	TOTAL SHAFT PUMP POWER	kW	An	5.54	A2	4.61	A1	2.23
i	ELECTRIC MOTOR EFFICIENCY	%		90.00		90.00		90.00
$l = h \cdot (100/i - 1)$	ELECTRIC MOTOR LOSSES	kW	An	0.62	A2	0.51	A1	0.25
m	DRIVE CONTROL LOSSES	kW		0.00		0.00	A1	0.28
	ELECTRIC POWER							
$n = m + l + h$	supplied by the line	kW	An	6.16	A2	5.12	A1	2.75
o	WORKING HOURS	h/year	An	8640	A2	8640	A1	8640
	ELECTRIC ENERGY							
$p = n \cdot o$	supplied by the line	kWh/year	An	53222	A2	44237	A1	23760
	ENERGY SAVING AT REDUCED CAPACITY							
$q = p(I) - p(II)$	METHOD(I)–METHOD(II)	kWh/year						20477
r	density of fluid	kg/m^3		1000		1000		1000

Table 10.1 (b) Comparison between dissipative and non-dissipative methods (see Fig. 10.8 and Fig. 10.9)

Non SI units			AT RATED CAPACITY (100%)		AT REDUCED CAPACITY (50%)			
					DISSIPATIVE METHOD (I)		NON DISSIPATIVE METHOD (II)	
			Working point		Working point		Working point	
a	CAPACITY	ft^3/min	An	11.02	A1/A2	5.651	A1	5.651
b	HEAD (basic load)	ft	An	174.54	A1	123.26	A1	123.26
c		ft			A2	219.82		
d	PUMP EFFICIENCY	%	An	49.00	A2	38.00	A1 (A1n)	44.00
$e = a \cdot b \cdot r/60$	POWER DELIVERED to the circuit	ft · lb/s	An	2001	A1	724.75	A1	724.75
$e' = e \cdot 1.356/1000$		kW		2.7		1.0		1.0
	REGULATING LOSSES							
$f = \frac{a \cdot (c - b) \cdot r}{60}$	valve losses	ft · lb/s		0	A2-A1	567.76		0
$g = f \cdot 100/d$	valve losses at pump shaft	ft · lb/s		0	A2-A1	1494.1		0
$g' = g \cdot 1.356/1000$		kW				2.02		
$h = g + e \cdot 100/d$	TOTAL SHAFT PUMP POWER	ft · lb/s	An	4084	A2	3401.35	A1	1647.16
$h' = h \cdot 1.356/1000$		kW		5.54		4.61		2.23
i	ELECTRIC MOTOR EFFICIENCY	%		90.00		90.00		90.00
$l = h' \cdot (100/i - 1)$	ELECTRIC MOTOR LOSSES	kW	An	0.61	A2	0.51	A1	0.25
m	DRIVE CONTROL LOSSES	kW		0.00		0.00	A1	0.28
$n = m + l + h'$	ELECTRIC POWER supplied by the line	kW	An	6.16	A2	5.12	A1	2.75
o	WORKING HOURS	h/year	An	8640	A2	8640	A1	8640
$p = n \cdot o$	ELECTRIC ENERGY supplied by the line	kWh/year	An	53222	A2	44237	A1	23760
$q = p(I) - p(II)$	ENERGY SAVING AT REDUCED CAPACITY METHOD(I)–METHOD(II)	kWh/year						20477
r	density of fluid	lb/ft^3		62.43		62.43		62.43

CHAPTER 11

FACILITIES—AIR COMPRESSORS

11.1 INTRODUCTION

Air compressor applications may be broadly classified as follows: (i) providing power, as in tools or other machines and in control systems for HVAC and for industrial equipment; (ii) handling process gas, usually in refineries and chemical plants, and transporting and distributing gas in pipelines; (iii) providing air for combustion as in gas turbine cycles. An additional use is for cleaning or drying material more efficiently than thermal energy.

Compressors in general, not only those for air compression, can be classified as (1) positive-displacement or intermittent-flow, and (2) continuous-flow.

In class (1), successive volumes of gas are confined within a closed space and elevated to a higher pressure. Examples are: reciprocating compressors and rotary compressors such as sliding vane, straight lobe, liquid piston, helical lobe or screw compressors.

Class (2) comprises dynamic compressors such as centrifugal and axial compressors (see also Chapter 10) where the velocity of the rotating elements is converted into pressure partially in the rotating element and partially in stationary diffusers or blades; and also mixed-flow compressors or ejectors. Note that ejector types use a high-velocity steam or gas to transport the inflowing gas, then convert the velocity of the mixture to pressure in a diffuser (see Chapter 7.5).

All major types of compressor (reciprocating, vane, screw, dynamic) are used for air compression. If the compression rate, that is the ratio between the absolute discharge pressure and the absolute intake pressure, is too high, it may become necessary to combine groups of elements in series to form a multistage system

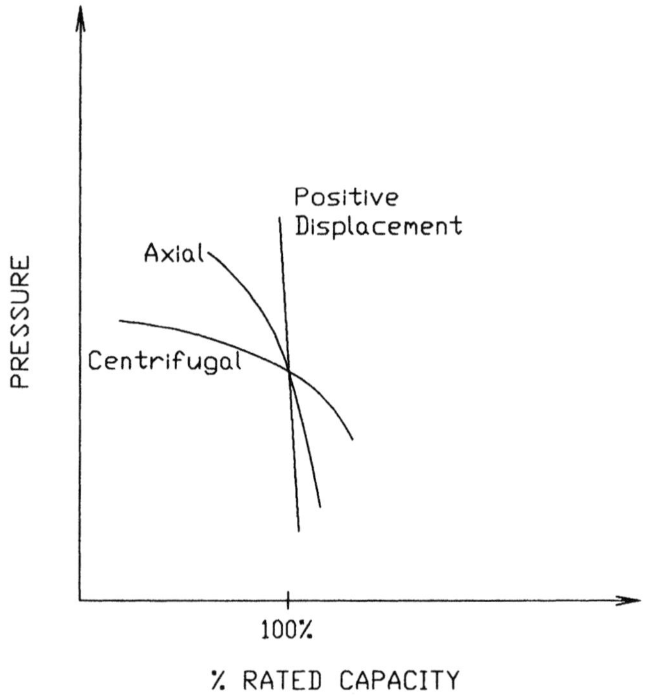

Fig. 11.1 Characteristic curves for different compressor types

with two or more steps of compression. The gas is often cooled between stages to reduce the temperature and thus the volume entering the following stage. Theoretical minimum power with perfect intercooling and no pressure losses between stages is obtained if the ratio of compression is the same in all stages.

Fig. 11.1 shows characteristic curves for the different compressor types: positive-displacement compressors show a constant volume and a variable pressure ratio; centrifugal compressors (low-speed type) work with almost constant pressure ratio and variable volume; axial and high-speed centrifugal compressors have characteristics in between those of these two types.

11.2 BASIC PRINCIPLES OF COMPRESSED-AIR SYSTEMS

Two basic theoretical compression processes, the isothermal and the adiabatic, are applicable to both positive-displacement and continuous-flow compressors. Neither of the two processes is actually obtainable, but they can be used for basic comparison.

Isothermal compression

The temperature is kept constant as the pressure increases. This requires heat to be removed continuously and expelled from the compressor.

The compression from state 1 to state 2 follows the formula:

$$p_1 v_1 = p_2 v_2 = \text{constant}$$

Adiabatic compression

No heat is added to or removed from the compressor during the compression process which is isoentropic (or reversible).

The compression from state 1 to state 2 follows the formula:

$$p_1 \cdot v_1^k = p_2 \cdot v_2^k = \text{constant}$$

where k is the ratio of the specific heat parameters c_p/c_v (k = 1.4 for air).

Fig. 11.2 shows the relationship between p pressure and v specific volume for the foregoing processes; note that area ADEF represents the work required during isothermal compression; the area ABEF the work for adiabatic compression. Compressors are often designed to remove as much heat as possible but the heat of compression can never be removed as rapidly as it is generated and the temperature increases during compression. It is evident that the isothermal area

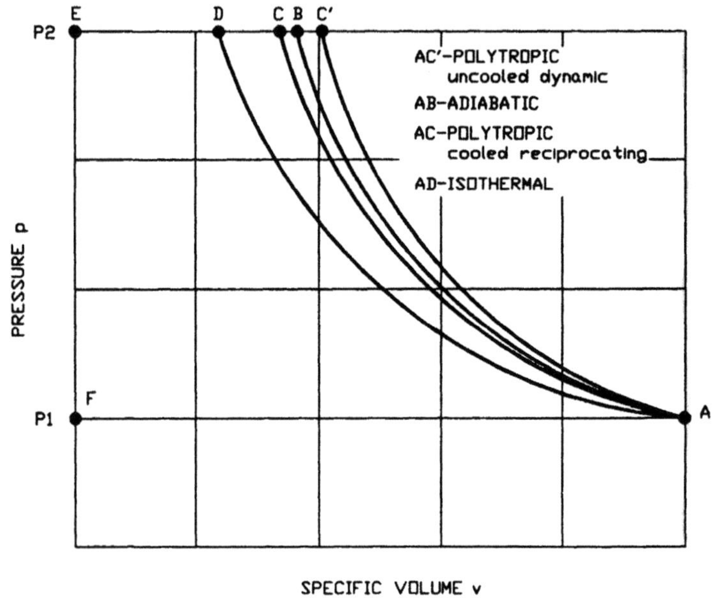

Fig. 11.2 Pressure-specific volume curves for various gas compression processes.

is much smaller than the adiabatic; in practice, neither the isothermal nor the adiabatic compression is obtainable.

Most positive-displacement units approach an adiabatic process quite closely; dynamic units generally follow a polytropic process with a formula similar to the adiabatic one, but the exponent is different. Of course, the exponent in all cases depends on the machine and on its operating conditions.

Polytropic compression

The compression from state 1 to state 2, which is irreversible, follows the formula:

$$p_1 \cdot v_1^n = p_2 \cdot v_2^n = \text{constant}$$

The exponent n is determined experimentally and may be lower or higher than the adiabatic coefficient k, depending on the level of cooling. Since it changes during the compression of real gases, an average value is generally used (n = 1.3 or higher than 1.4 are typical polytropic values).

Fig. 11.2 also shows polytropic curves for an uncooled dynamic unit and for a reciprocating water-cooled unit; the areas ACEF and AC'EF represent the work for the compression.

The power requirement of a theoretical process with isothermal compression ($T_2 = T_1$), shown as the area ADEF in Fig. 11.2 for mass unit, is as follows (notice that the relationship is transformed from the mass unit reference to the mass flowing by multiplying the specific volume by the mass flow-rate):

$$P_{is} = p_1 \cdot q_1 \cdot \ln \frac{p_2}{p_1} \cdot \frac{1}{1000} = 101325 \cdot q_1 \cdot \frac{1}{1000} \ln \frac{p_2}{p_1} =$$

$$= 101.325 \cdot q_1 \ln \frac{p_2}{p_1} \qquad \text{(kW)}$$

where

p_2 = outlet absolute pressure (Pa)

p_1 = inlet pressure (usually the atmospheric pressure = 101325 Pa)

q_1 = inlet air flow-rate (m^3/s) referred to inlet condition = specific volume · inlet air mass flow-rate (usually inlet conditions are standard atmospheric conditions: 0.1013 MPa; 15.6°C, 60°F. See note to Table 2.5).

The power requirement of the theoretical adiabatic compression process, shown as the area ABEF in Fig. 11.2 for mass unit, is as follows:

$$P_p = p_1 \cdot \frac{k}{k-1} \cdot \left(\left(\frac{p_2}{p_1} \right)^{\left(\frac{k-1}{k} \right)} - 1 \right) \cdot q_1 \cdot \frac{1}{1000} =$$

$$= 101.325 \cdot \frac{k}{k-1} \cdot \left(\left(\frac{p_2}{p_1} \right)^{\left(\frac{k-1}{k} \right)} - 1 \right) \cdot q_1 \qquad \text{(kW)}$$

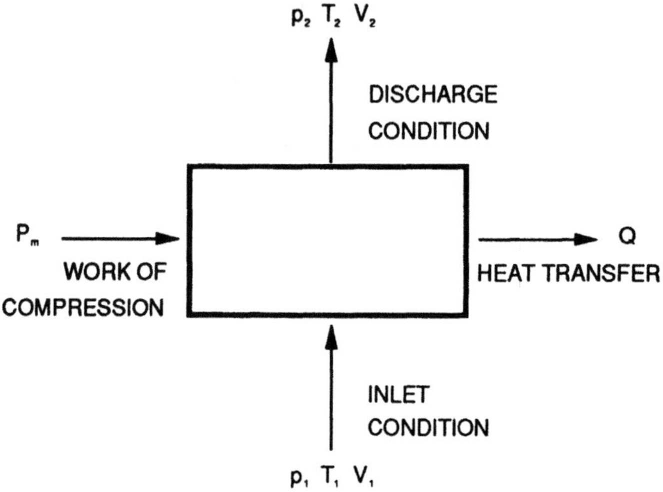

Fig. 11.3 Energy balance in a controlled-volume system

The relationship between the inlet (T_1) and outlet (T_2) temperatures is:

$$T_2/T_1 = (p_2/p_1)^{((k-1)/k)}$$

Then:

$$T_2 = T_1 \cdot (p_2/p_1)^{((k-1)/k)}$$

Note that all pressures and temperatures are absolute. If the compression ratio is defined, the outlet temperature depends only on the inlet temperature.

All the foregoing relationships can be used for polytropic compression processes by substituting n for the parameter k.

Depending on the type of compression cycle, different heat and energy balances can be made.

Energy balance can be written with reference to a controlled-volume system as represented in Fig. 11.3:

$$\begin{aligned} P_m = & +Q - m \cdot \left(u_1 + p_1 v_1 + V_1^2/2 + H_1 g/g_c\right) + \\ & + m \cdot \left(u_2 + p_2 v_2 + V_2^2/2 + H_2 g/g_c\right) = \\ & = +Q - m\left(h_1 + V_1^2/2 + H_1 g/g_c\right) + m\left(h_2 + V_2^2/2 + H_2 g/g_c\right) \end{aligned}$$

where

P_m = mechanical power positive if entering the system
Q = heat transfer-rate positive if leaving the system
m = mass flow-rate
h_1, h_2 = inlet, outlet enthalpy (which is defined as the sum of the internal energy, generally designated u, and the product of pressure and specific volume = u + pv)
V_1, V_2 = inlet, outlet velocity
H_1, H_2 = inlet, outlet static head
g = gravitational acceleration 9.81 m/s^2 = 32.17 ft/s^2
g_c = conversion factor $= \frac{1\ \mathrm{kg} \cdot (\mathrm{m/s^2})}{\mathrm{N}} = 32.17 \frac{\mathrm{lb} \cdot (\mathrm{ft/s^2})}{\mathrm{lbf}}$

If the change in potential energy is neglected, as typically happens, then:

$$P_m = Q + m \cdot \left(h_2 - h_1\right) + m \cdot \left(V_2^2 - V_1^2\right)/2$$

The contribution of kinetic energy is negligible in most cases, so:

$$P_m = Q + m \cdot \left(h_2 - h_1\right)$$

This means that the mechanical work equals the heat transferred (which is zero in the adiabatic process) plus the air enthalpy difference (which is zero in the isothermal process) between outlet and inlet. Depending on the known parameters, air enthalpy can be calculated from charts or by multiplying the specific heat at constant pressure by the temperature, as for an ideal gas (see Chapter 13.2).

Notice that the difference between the work required in an actual process and in an adiabatic process represents the amount of losses inside the machine.

The sum of input energy to the compressor in the form of mechanical power and any other energy as heat added or removed, is transformed into the following terms: heat radiated from the frame and other parts to the atmosphere; heat conducted from the frame to the foundation; heat carried from inside by lubricating and cooling oil and radiated to the atmosphere or transferred to cooling fluid for disposal; heat carried away by the cooling medium in cylinder jackets (for reciprocating compressors); heat carried away by the cooling medium in intercoolers; heat carried out of the compressor in the compressed air leaving the system.

The efficiency of the compression process can be defined as the ratio between the power requirement in adiabatic or isoentropic compression P_{ad} and the actual power P_{actual} at the shaft of the compressor:

$$\eta_{ad} = P_{ad}/P_{actual} = (h_{2ad} - h_1)/(h_2 - h_1) = (T_{2ad} - T_1)/(T_2 - T_1)$$

The efficiency of the compression process can also be defined as the ratio between the power requirement in isothermal compression P_{isoth} already defined and the actual power P_{actual} at the shaft of the compressor. Notice that the compressor isothermal efficiency takes into account losses inside the whole system. It is:

$$\eta_{isoth} = P_{isoth}/P_{actual}$$

In addition, a so-called polytropic efficiency, which can be considered as the true aerodynamic efficiency of the compressor itself with ideal gas, is introduced by developing the classic relationships. Then:

$$\eta_p = (n/(n-1))/(k/(k-1))$$

The polytropic efficiency of a compressor is always greater than the adiabatic efficiency already defined.

Table 11.1 shows the theoretical values of kW needed for an inlet air flow of 1 Sm^3/min in standard atmospheric conditions (60°F, 15.6°C; 0.1 MPa. See note to Table 2.5) and the values of final temperatures with different compression processes (isothermal, adiabatic, polytropic).
If the efficiency of the drive (80-90%) and other losses are considered, typical values are 0.12-0.15 kWh per 1 Sm^3 of inlet air in standard atmospheric conditions to compress at 0.8 MPa (8 bar, 116 psi). It follows that a typical power range between 6 and 8 kW is needed per 1 Sm^3/min.

Table 11.1 Theoretical values of power requirements for different air compression processes (1 Sm^3/min = input air in standard atmospheric conditions, 15.6 °C, 60 °F and 0.1013 MPa; see note to Table 2.5)

OUTLET PRESSURE absolute			COMPRESSION RATIO	ISOTHERMAL 15.6°C = 60°F	POLYTROPIC n = 1.3			ADIABATIC k = 1.4		
				Power	Power	Outlet temperature		Power	Outlet temperature	
MPa	bar	psi		kW	kW	°C	°F	kW	°C	°F
0.2	2	29	1.97	1.15	1.24	65	148	1.27	78	172
0.3	3	44	2.96	1.83	2.08	98	208	2.15	121	249
0.4	4	58	3.95	2.32	2.73	123	254	2.84	154	310
0.5	5	73	4.94	2.70	3.26	144	292	3.42	182	360
0.6	6	87	5.92	3.00	3.71	162	324	3.91	207	404
0.7	7	102	6.91	3.26	4.11	178	352	4.36	228	443
0.8	8	116	7.90	3.49	4.47	192	378	4.76	248	478
0.9	9	131	8.88	3.69	4.80	205	401	5.12	266	510
1	10	145	9.87	3.87	5.09	217	422	5.46	282	540
1.1	11	160	10.86	4.03	5.37	228	442	5.77	298	568
1.2	12	174	11.85	4.17	5.63	238	460	6.07	312	594
1.3	13	189	12.83	4.31	5.87	247	477	6.34	326	618
1.4	14	203	13.82	4.43	6.10	256	493	6.61	338	641
1.5	15	218	14.81	4.55	6.31	265	508	6.86	351	663

Notes

all pressures are absolute; relationships for temperature calculations use absolute temperatures (see Chapter 11.2)

1 $m^3/min = 35.3\ ft^3/min = 0.0167\ m^3/s = 0.588\ ft^3/s$

to convert from the actual volume to the standard volume see note to Table 2.5

practical values of driver input power (usually electric) are 30-60% higher than values of theoretical processes

11.3 AUXILIARY EQUIPMENT

In practice, a compressor system includes auxiliary equipment which is essential for good operation.

Standard equipment for reciprocating compressors includes intake filter, aftercooler and very often air receiver.

Dynamic compressors require an inlet filter, no receiver and sometimes an aftercooler.

Every compressor must be equipped with an air cleaner to supply adequate cool, clean and acid-free air at all times. If the air cleaner has to be placed at a distance from the compressor, care must be taken to provide a suitable suction line to the compressor.

Coolers can be classified as precooler, intercooler and aftercooler depending on their functions. They are basically heat exchangers designed to reduce the gas temperature before, during or after compression.

Aftercoolers condense and remove moisture from compressed air before it enters the distribution system. They include condensate traps; they can be air-cooled or watercooled. The latter are the better. Cooling water must be clean, cool and soft. Cooling water generally removes 15-40% of the total heat of compression, depending on the machine size and operating characteristics. Values up to 80-90% can be reached in intercooled centrifugal machines.

The cooling of air both reduces the air temperature and volume, and condenses and removes water vapor and other condensable fluids. The shaft power requirement is reduced if the cooling is effected during the compression.

After the compression, cooling may improve safety by reducing the condensable liable to cause various problems: washing away of lubricants, rusting of some parts, freezing in exposed lines during cold weather, condensation and freezing of moisture in the outlet when the expansion is considerable.

Notice that air entering the compressor at the first stage is a mixture of dry air and water vapor. The humidity ratio (mass of water vapor/mass of dry air) varies with the temperature; if the mixture temperature is reduced at constant mixture pressure, the water vapor state moves to the dew point at constant partial pressure and at constant humidity ratio. Then, if cooling goes on, condensation occurs and the final temperature is lower than the initial temperature. The system will consist of a saturated mixture of dry air and water vapor at partial pressure less than the initial one because the mole fraction of water vapor in the final state is less than the value at the initial temperature. A fuller explanation of the basic principles of air treatment and the definition of dew-point temperature are given in Chapter 13.2.

In application at 0.7-0.8 MPa, which is a typical operating pressure in industrial plants, a dew point at 10-1.5°C (50-35°F) is generally adequate, depending on the humidity ratio of the outlet air.

If compressed air at this pressure never goes below the dew point after leaving the dryer, there will be no further condensation. When pressure is reduced before final use, the dew point corresponding to the reduced pressure is lower than the previous one.

Chemical drying and absorbing systems are dried-air or dehumidification systems commonly used instead of cooling equipment. Notice that all these systems are intended to remove water vapor and not lubricant oil in vapor state. If absolutely oil-free air is required, non-lubricated systems are the best.

Cooling systems use refrigerating compressors (see Chapter 12) and may use direct expansion (cold refrigerant gas expands in the air/refrigerant exchanger) or indirect expansion by using chilled water as a medium to refrigerate air. In the latter case, regenerative systems can conveniently be used: the inlet air is partially cooled by the discharge air, thus reducing the power of the refrigerating compressor and reheating the outlet air to the distribution line.

Advantages of reheating the discharge air are: the air volume is increased and thus less free air is required to move equipment and the possibility of condensation down the dew point along the distribution lines is further reduced.

Air receivers are used with reciprocating compressors. Dynamic compressors do not require receivers because they have capacity control which should equalize line output with demand.

Air receivers are very important additional equipment which ensure correct operation of the compressor. An air receiver absorbs pulsations in the discharge line from the compressor and smooths the flow of air to the user lines. It works as a reservoir for storing compressed air to cope with sudden peak demands exceeding the capacity of the compressor. Compressors work between a lower and an upper limit of discharge pressure; the dimension of the storage must be selected with regard to technical and economic criteria, taking into account the load requirement, the compressor rated air flow and the demand pressure. Air receivers also allow moisture to precipitate and condense at the bottom, and prevent it from being carried along the air distribution pipeline.

If excess condensate is produced, a water-cooled aftercooler and separator should be added between the compressor and the receiver. This equipment dries and cools air, which means higher efficiency and safe operation for the compressor plant. When the water supply is scarce or too expensive, air-cooled aftercoolers are technically and economically justified.

11.4 TECHNICAL DATA ON INDUSTRIAL COMPRESSORS AND CONTROL SYSTEMS

Table 11.2 lists the main properties of typical industrial compressors (shaft power, shaft speed, volume of compressed air and pressure). Efficient operation is always influenced by the following factors: intake air temperature (as cool as possible), minimum system pressure, absence of leaks along the pipeline, heat recovery, selection of compressors appropriate in type and size for each application.

Table 11.2 Characteristics of typical medium-power air compressors

TYPE OF COMPRESSOR	POWER kW	SPEED r/min	VOLUME FLOW Sm^3/min (*)	ABSOLUTE PRESSURE MPa	psi	COMPRESSOR RATIO	ISOTHERMAL POWER(15.6°C) kW	ISOTHERMAL EFFICIENCY %
Rotary, air-cooled	20	2950-3550	2.77	0.79	115	7.8	9.6	48%
	56	2950-3550	9.76	0.45	65	4.4	24.4	44%
	112	1450-1775	20	0.80	116	7.9	69.7	62%
	146	1450-1775	23	0.85	123	8.4	82.5	57%
	162	1450-1775	23	1.00	145	9.9	88.8	55%
Rotary water-cooled	20	2950-3550	2.77	0.79	115	7.8	9.6	48%
	56	2950-3550	9.91	0.79	115	7.8	34.3	61%
	90	1450-1775	17.8	0.79	115	7.8	61.6	68%
	298	1450-1775	58.6	0.79	115	7.8	202.8	68%
Reciprocating	56	400-450	9.9	0.79	115	7.8	34.3	61%
	165	550-600	23	0.90	131	8.9	84.7	51%
	300	550-600	64	0.79	115	7.8	221.5	74%

Note
(*) standard atmospheric conditions 15.6°C, 60°F, 0.1013 MPa

The intake air temperature must be kept as low as possible to reduce the work of the compressor which varies with the volume of air linearly dependent on the temperature. Table 11.3 shows how the power requirement varies with air temperature.

Table 11.3 Power saving versus air-inlet temperatures at constant inlet pressure and compression ratio (consumptions are compared with power requirements in atmospheric conditions)

Temperature of air inlet		Inlet volume required to deliver 1 Sm^3 (*)	Power saving for compression
°C	°F	m^3	%
0	32	0.946	5.40
5	41	0.963	3.67
10	50	0.981	1.94
15	59	0.998	0.21
15.6	60	1.000 (*)	0.00
20	68	1.015	-1.52
25	77	1.033	-3.26
30	86	1.050	-4.99
35	95	1.067	-6.72
40	104	1.085	-8.45
45	113	1.102	-10.18
50	122	1.119	-11.91
a	b	$c = (a+273.15)/(15.6+273.15)$	$d = (1-c)\cdot 100$

Notes
for every 5°C (9°F) rise in inlet temperature, a 1.7% rise in energy consumption will be needed to achieve equivalent output
(*) standard atmospheric conditions 15.6°C, 60°F, 0.1013 MPa

The end users may require different pressures depending on the characteristics of the equipment; as a general rule, the lower the system pressure, the lower the compressor shaft power which varies more than linearly. Significant energy saving can thus be obtained by reducing the system pressure to the right value (see Example 2). A typical approach, when various end-user machines have different power requirements, is to provide compressors which can meet the highest pressure requirement and to reduce the pressure near the low-pressure users. An alternative approach is to divide the load into different groups, on the basis of pressure needs, and to install different compressors.

Compressors are used with air- or water-cooling equipment depending on the availability of refrigerating media and on their temperatures all the year round. Air cooling is suggested for low-and medium-power sizes (less than 100 kW). Heat recovery can conveniently be applied, particularly with large water-cooled units (see Chapter 12.4).

Positive-displacement compressors can be cooled by water which can reach a maximum temperature of 35-40°C (95-104°F); the use of this water for space heating is linked to the water/air exchangers which must be properly designed and equipped with forced ventilation systems. During hot seasons, heat must be discharged by cooling systems.

Screw compressors are oil-flooded compressors where oil is injected to seal clearances, absorb the heat of compression, prevent metal contact between rotors and lubricate the bearings. Oil separation is performed immediately after compression. The oil and air mixture passes through the discharge valve to an oil reservoir, or a combination of an oil reservoir and an air receiver. The air then passes through a separator before flowing to the discharge line where most of the remaining oil is removed. Oil temperature controls are installed in all units to keep the injected oil within a range of 54°C to 71°C (130°F to 160°F).

In screw compressors, water- or air-cooled heat exchangers are usually installed; recoverable heat temperature can be 60-80°C.

The control of air compressors

The output of compressors must be controlled to match the system demand. The parameter to control is usually the discharge pressure. A range of variation of about 10% between the discharge pressure and the end-user requirement is generally accepted.

i) Reciprocating compressors. Automatic start-and-stop control in which compressors run at full load for a period and then are stopped can be used if starting unloaders are applied and if it is not started too often.

Constant-speed control is effected by different methods (intake throttling, external bypassing, inlet valve loading).

Variable-speed control, which is the ideal control for positive-displacement compressors, can conveniently be achieved by installing variable-speed drives, particularly a.c. drives for safety reasons. Variable-speed drives also provide a soft-start feature with minimum starting current, thus avoiding damage to the motor through frequent starting. This allows the compressor to stop completely and start again more frequently than in installations where electric motors are fed directly from the line;

ii) screw compressors. Automatic start-and-stop control can be used if an automatic start unloading system is installed. To unload the compressor and thus reduce the power required, systems which include closing the inlet valve, releasing oil reservoir pressure and reducing oil flow to the compressor are convenient.

Variable-speed drives can also be used;

iii) centrifugal compressors. There are three main methods of controlling centrifugal compressors: variable speed, blowoff of unwanted output, intake throttling.

Speed is varied by means of variable-speed drives, either electrical or steam- or gas-turbine depending on the size of the system. Power is less following the speed reduction (see also the section on pumps in Chapter 10). Blowoff and recirculation control can be used to release any excess air into the atmosphere or to throttle the excess air to intake pressure and put it through the compressor again.

Intake throttling reduces discharge pressure for a given intake volume or reduces volume for a given discharge pressure. This method is widely used with constant-speed machines.

For large units, adjustable inlet guide vanes are installed ahead of the first impeller. The power saving is less than with variable-speed drives, but it is still significant;

iv) axial compressors. Adjustable vanes are typically used on the first several stages of axial compressors, since they are very sensitive to the gas angle attack and have a very short operating range.

11.5 COMPRESSED-AIR SYSTEMS: DISTRIBUTION LINES AND END USERS

The distribution lines

The aim of the distribution system is to carry compressed air from the compressor to the end users at the required pressure, generally in the range 0.5-0.8 MPa (5-8 bar, 73-116 psi). The speed generally ranges from 5 to 10 m/s; the shorter the line, the higher the speed.

Most systems consist of a main line from which secondary lines go to different production areas where individual feeder lines carry the compressed air to the end users.

For heavy intermittent demands, pressure losses can be reduced by installing air receivers as near as possible to the demand point.

Pressure losses must be considered in selecting the pipe sizes by using tables and charts from specialized literature.

Pressure losses range between 400 Pa/m and 40 Pa/m with an air-flow speed from 5 to 10 m/s. Higher speeds will provoke higher losses.
Leaks along the pipeline may cause significant losses. Notice that a hole with a diameter of 5 mm will lose at least 5 kW of electric input power (with a leak of roughly 1 Sm^3/min).
Table 11.4 lists values of losses in typical operating conditions.

A useful test is to shut down all the equipment needing compressed air and to measure the compressor power requirement: the resulting value represents the system leaks because no useful work is delivered to the end users.

Table 11.4 Pressure and leakage losses in air mains and related power in typical operating conditions

Nominal bore	Air flow in standard conditions	Air flow at 0.8 MPa (absolute)	Air speed at 0.8 MPa (absolute)	Pressure drop per 100 m			Equivalent power pressure losses (100 m)	Equivalent power leakage losses	Pipe efficiency
mm	Sm^3/min	m^3/min	m/s	MPa	bar	psi	kW	kW	%
40	11.9	1.5	19.9	0.07	0.7	10.15	1.7	4.1	86
50	9.2	1.2	9.9	0.015	0.15	2.18	0.3	3.2	89
65	15.8	2.0	10.1	0.01	0.1	1.45	0.3	5.5	89
80	21.1	2.7	8.8	0.006	0.06	0.87	0.3	7.3	90
100	29.	3.7	7.8	0.005	0.05	0.73	0.3	10.1	90
125	46.1	5.8	7.9	0.004	0.04	0.58	0.4	16.1	90

Notes
standard atmospheric condition 15.6°C, 60°F, 0.1013 MPa
compression ratio equal to 7.9; m^3 air flow at 0.8 MPa = Sm^3/7.9
percentage of flow leaks assumed equal to 10% of the pipe air flow
equivalent power is evaluated by the isothermal compression formula. Practical values are higher (see Note Table 11.1)

Flow leaks through an orifice can be calculated by using the Bernoulli equation for the flow of compressible gas and by considering the process as adiabatic:

$$\frac{k}{k-1}\cdot\frac{p}{\rho}+\frac{1}{2}\cdot V^2=\text{constant.}$$

The air speed in the pipe is negligible in comparison with that immediately outside the orifice (typical ratio between the two speeds is roughly 30-40).
It follows:

$$m(\text{kg/s}) = C_d\cdot A\cdot\rho_i\cdot\sqrt{2\cdot\left(\frac{k}{k-1}\right)\cdot\frac{p_i}{\rho_i}\cdot r_p^{2/k}\cdot\left(1-r_p^{(k-1)/k}\right)}$$

where
C_d = discharge coefficient = 0.96 (typical value)
A = orifice area (m^2)
ρ_i, p_i = air density (kg/m^3) and pressure (Pa) in the pipeline
k = adiabatic coefficient = 1.4
r_p = ratio between the external and the internal pressure (p_o/p_i = atmospheric pressure/p_i)

Data for calculation (example: hole diameter 0.005 m)
p_o = 0.1013 MPa
ρ_o = 1.23 kg/Sm^3
p_i = 0.8 MPa

$$\rho_i = 1.23\cdot\frac{0.8}{0.101}\ \text{kg/m}^3 \text{ (see also Note Table 2.5)}$$

$$A=\frac{\pi}{4}\cdot\left(5\cdot10^{-3}\right)^2\ \text{m}^2$$

r_p = 0.1013/0.8
m = 0.0212 kg/s

$$Q\,(\text{Sm}^3/\text{min}) = m\,(\text{kg/s})\cdot\frac{60}{1.23} = 1.034\ \text{Sm}^3/\text{min}$$ (See Table 11.1 for related power)

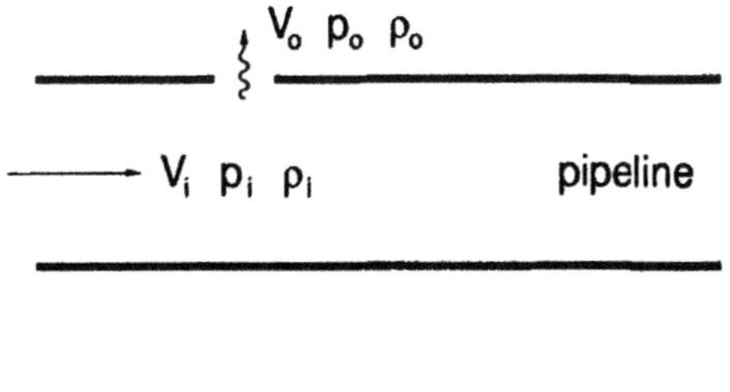

The efficiency of the distribution line between two sections A and B (see Fig. 11.4) can be calculated as follows:

$$\eta = (\ln (p_B/p_1) / \ln (p_A/p_1)) \cdot (100 - \text{leak}\%)$$

where

p_1 =	**compressor inlet pressure (usually the atmospheric pressure)**
p_A =	**line upstream pressure or discharge pressure from the compressor (section A)**
p_B =	**line downstream pressure (section B, $p_B < p_A$)**
$p_A - p_B$ =	**pressure drop between sections A and B**
leak% =	**percentage of flow leaks between sections A and B**

Typical values of the flow leaks range from 10% to 30% depending on the operating conditions and on the maintenance of the pipeline; the total efficiency of the distribution line ranges from 80% to 60%.

Fig. 11.4 shows the energy balance of a complete system: electric input energy, output energy as heat (losses and recovered heat), compressed air, pipe distribution losses (leaks and pressure losses), end users.

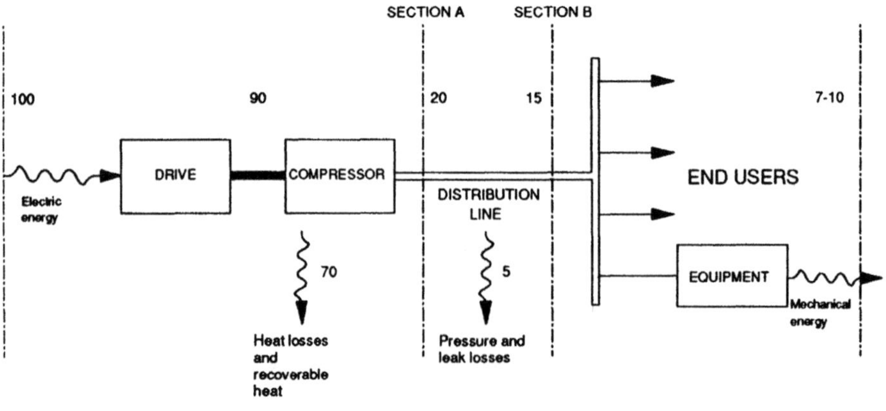

Fig. 11.4 Energy flow from input section to mechanical energy end-user section in a compressed-air system

The end users

The end users of compressed air for power purposes can be classified as positive-displacement machines, where a piston is given a reciprocating motion within a cylinder by alternate feeding of the air supply from one side of the piston to the other, and rotating machines where pistons or turbines or blades transform the energy accumulated in the compressed air into a rotating motion.

Table 11.5 lists typical values of air consumption for different tools and other machines together with a coefficient which takes into account the discontinuous operating mode. In practice, the capacity of the air compressors is equal to the sum of the single end-user air consumptions with a security factor of at least 20%.

The power of an end-user reciprocating motor depends mainly on the mean pressure differential across the air inlet and outlet, the displacement of the piston through the cylinder at maximum stroke and the number of strokes per unit of time. The degree to which air can be expanded is determined by air moisture freezing which may block the exhaust passages with ice, and not by the efficiency of the system which would impose high expansion ratios. Freezing can happen with expansion ratios as small as 2 to 1.

The work diagram of a piston motor includes two phases: a constant-pressure phase (air at the distribution-line pressure enters the cylinder through valves and moves the piston, thus performing work) followed by a limited expansion (inlet valves are closed and the piston cylinder performs work by using the energy delivered during the expansion). The first phase is generally the major one; air consumption can be reduced for the same output work by cutting off the air entering as early as possible (first phase), and then allowing expansion (second phase). This, the opposite of the compression phase, is always a small part of the total work in the end-user equipment.

The energy balance already introduced for the compression process can be used for the end-user cycle.

End-user requirements are generally expressed in m^3/min; power (W) can be calculated approximately by multiplying the air flow in m^3/s by the pressure (Pa), if only the constant-pressure work is considered. Higher values can be obtained if the expansion work (the opposite of the compression work) is added.

Table 11.5 Air requirements of compressed-air end-user machines (0.7 MPa absolute, 100% use factor)

Type of machine	Air flow at 0.7 MPa		Air flow referred to standard conditions
	m^3/min	m^3/s	Sm^3/min
Screw driver	0.15-0.3	0.0025-0.005	1-2
Concrete vibrator	1-2	0.016-0.032	7-14
Paint spray	0.2-0.5	0.003-0.0086	1.4-3.5
Steel drills	0.5-2	0.008-0.032	3.5-14
Chipping hammers	0.35-1	0.006-0.016	2-7

11.6 ENERGY-SAVING INVESTMENTS

As discussed in Chapter 11.2, energy can be saved by improving the overall efficiency of the system.

Typical suggestions are:

— the discharge pressure must be kept as low as possible, depending on the real end-user needs. Reduction of pressure by throttling valves must be avoided;

— the compressed-air temperature must be lowered in order to reduce the volume flowing through the compressor;

— leaks along the pipeline must be carefully avoided by proper maintenance;

— heat recovery from cooling systems, particularly for high-power systems, is always advisable;

— proper sizing of the compressors and related equipment allows operation at high efficiency.

11.7 PRACTICAL EXAMPLES

Example 1
Reduction of the compressor discharge pressure to the real end-user needs

A factory has an air-compressor plant which works at a discharge pressure of 0.9 MPa. Most equipment may correctly operate at 0.6-0.65 MPa, except one which needs 0.8-0.85 MPa with a small air flow requirement (roughly 10% of the total air flow). Throttling valves are used to reduce the pressure to the end-user requirements.

Evaluate the energy saved if the air-compressor plant operates at lower pressure, 0.7 MPa, to supply the main equipment. The user at 0.8-0.85 MPa will be supplied with a separate air compressor.

Table 11.6 shows the energy saving associated with a reduction of the discharge pressure from 0.9 MPa to 0.7 MPa for the main equipment (energy saving roughly 20%) and the additional energy required to supply the higher-pressure end user with a separate compressor.

For 5000 h/year, the energy saved is 95492 kWh/year.

Economic evaluation is shown in Table 20.3.

Example 2
Heat recovery from a screw compressor

A screw compressor works between inlet air at atmospheric pressure and outlet air at discharge pressure equal to 0.7 MPa. The air is cooled by oil injected in the rotor of the compressor. The outlet air temperature is relatively low (55-65°C, 131-149°F) and no additional cooling is needed.

The basic operating data are as follows:

SCREW COMPRESSOR

- inlet air in standard atmospheric conditions (0.1 MPa and 60°F, 15.6°C)
- air discharge pressure 0.7 MPa
- air volume flow-rate equal to 9.3 Sm^3/min
- mechanical input power 55 kW
- outlet air average temperature 55°C (131°F)

RECOVERY SYSTEM

— heat exchanger oil/water
— inlet oil temperature 95°C (203°F)
— water volume flow-rate 16 m^3/h
— water inlet temperature 60°C (140°F)
— water outlet temperature 80°C (176°F)
— working hours 5000 h/year.

Table 11.7 lists basic data for evaluation of the energy saved when water is used for space heating and other services inside the factory.

The energy saved is estimated equal to 18.1 TOE/year.

Economic evaluation is shown in Table 20.3.

Notice that the results of this example can easily be extended to other similar compressors with oil injection or to compressors cooled by means of air/water or air/air heat exchangers.

Table 11.6 Energy saving associated with a reduction in the delivery pressure from 0.9 MPa to lower values; inlet air flow rate 30 Sm^3/min in standard atmospheric conditions

			Reference plant	New plant (1)+(2)	
				(1)	(2)
a	Discharge pressure	MPa	0.9	0.7	0.9
b=a/0.006895		psi	130.5	101.5	130.5
c	Inlet air flow	Sm^3/min	30	27	3
d=c/60		Sm^3/s	0.5	0.45	0.05
e=a/0.1013	Compressor ratio		8.89	6.9	8.89
f(°)	Electric input power	kW	184	147	18.44
	POWER SAVING (1)				
g=f(REF)-f(1)	PS	kW		37.54	
g' = g · 100/f(REF)		%		20.35%	
h=f(2)	ADDITIONAL POWER (2)				
	PA	kW			18.44
i	Working hours	h/year	5000	5000	5000
l=(g-h)·i	ENERGY SAVING	kWh/year	95492		

Notes
(°)Electric input power = theoretical isothermal power/0.6 (see Chapter 11.2)
theoretical isothermal power = 101.325·d·ln(a/0.1033)

Table 11.7 Energy saving associated with heat recovery from a screw compressor

Inlet air flow in standard atmospheric conditions 9.3 Sm^3/min = 0.15 Sm^3/s
Discharge pressure 0.7 MPa

a	Mechanical input power	kW	55
b = a/0.9	Electric input power (°)	kW	61.1
c = 0.6·a	Heat recovery (°°)	kW	35.75
d = c·860		kcal/h	30745
e = c·3600/1.055		Btu/h	121990.5
f = c/0.85	Equivalent fuel (°°°)	kW	42.1
g = f·860		kcal/h	36170.58
h = f·3600/1.055		Btu/h	143518.2
i = f·3600/41860		kgoil/h	3.6
l	Working hours	h/year	5000
m = i·l/1000	ENERGY SAVING	TOE/year	18.1

Notes
(°) electric motor efficiency equal to 0.9
(°°) heat recovery assumed equal to 65% of mechanical input power
(°°°) boiler efficiency 0.85 (Net Heating Value as reference)
oil net heating value 41860 kJ/kg

CHAPTER 12

FACILITIES—INDUSTRIAL COOLING SYSTEMS

12.1 INTRODUCTION

Refrigeration systems for process application and air conditioning convert work into a flow of heat from a process source or from a refrigerated space to an environmental sink. The main types can be described as vapor compression, absorption, and Brayton cycles.

Vapor compression systems are the commonest; they use, for heat transfer, the condensation and evaporation of a volatile medium such as ammonia, light hydrocarbon or fluorocarbon. The input power is given by the compressor driver. The basic principles of compressor operation have been discussed in Chapter 11.

Absorption systems need a high-temperature source as main power input and mechanical power only as additional power to drive a pump. Ammonia-water and lithium bromide-water systems are generally used.

Brayton cycles are used for gas refrigeration systems, where a change in phase does not occur. They are chosen to achieve very low temperatures for the liquefaction of air and other gases and for specialized applications such as aircraft-cabin cooling.

12.2 BASIC PRINCIPLES OF VAPOR COMPRESSION SYSTEMS

Basic cycles

Fig. 12.1 shows the basic scheme of a vapor-compression refrigeration system, where the flow of power and heat transfers are also shown (they are positive in the direction of thr arrows).

The main components are:

- **evaporator where the heat transferred from a process source or from a refrigerated space (cold region) results in vaporization of the refrigerant;**
- **compressor where the gas refrigerant is compressed to a relatively high pressure and temperature;**
- **condenser where the refrigerant condenses and the resulting heat is transferred from the refrigerant to the cooler surroundings (warm region);**
- **expansion valve where the liquid refrigerant expands to the evaporator pressure and leaves the valve as a two-phase liquid-vapor mixture.**

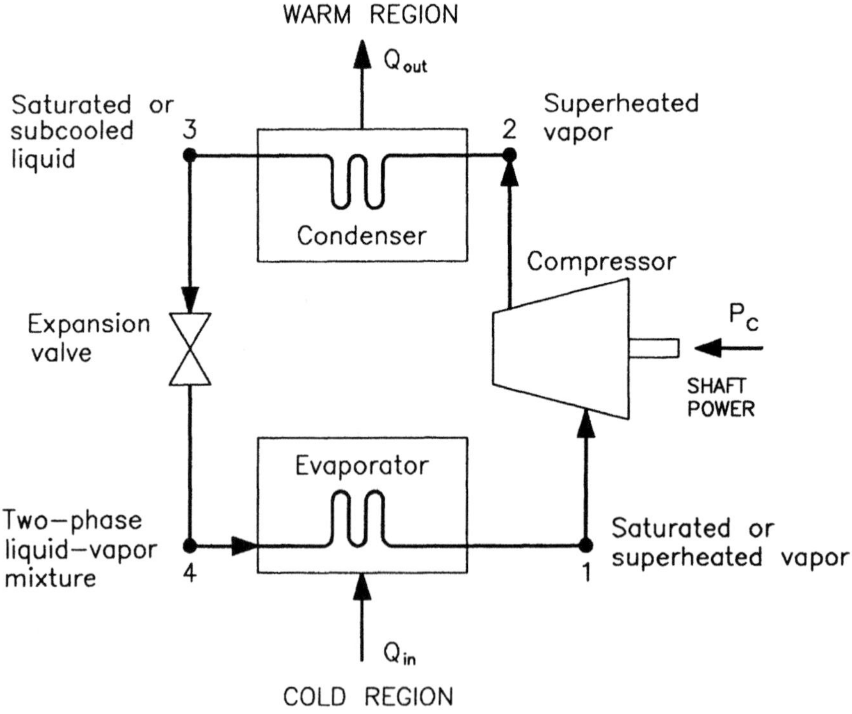

Fig. 12.1 A vapor - compression refrigeration system

As a first step, let us consider an idealized cycle with no irreversibilities which represents an upper limit to the performance of the vapor-compression refrigeration cycle. This means that there are no frictional pressure losses, and that the refrigerant flows at constant pressure through the condenser and the evaporator. In addition, the compression is assumed to be an isoentropic process and the expansion through the throttling valve an adiabatic process.

The main difference between the idealized and the actual cycle is in the compression process. In the latter case it is only adiabatic and the entropy increases. The pressure drops, which always occur in the cycle, are generally ignored.

A temperature-entropy or pressure-enthalpy diagram can be used to represent the cycle. Both are shown in Fig. 12.2.a and Fig. 12.2.b.

The four phases of the cycle are described below.

Evaporator (process from 4 to 1' or 1)

When the refrigerant, as a two-phase liquid-vapor mixture, passes through the evaporator, it absorbs heat from the cold space (cold region) and it changes into saturated vapor at point 1'. This happens at constant pressure and temperature. The evaporator temperature must be lower than the cold space temperature (cold-region temperature). In practice, the final state is a superheated vapor at point 1.

The rate of heat-transfer per unit of refrigerant mass flow-rate is:

$$\frac{Q_{in}}{m} = h_1' - h_4 \text{ or } h_1 - h_4$$

where

Q_{in} = heat-transfer rate or refrigeration capacity (W)

m = mass flow-rate (kg/s)

h'_1, h_1, h_4 = enthalpy per unit mass (J/kg)

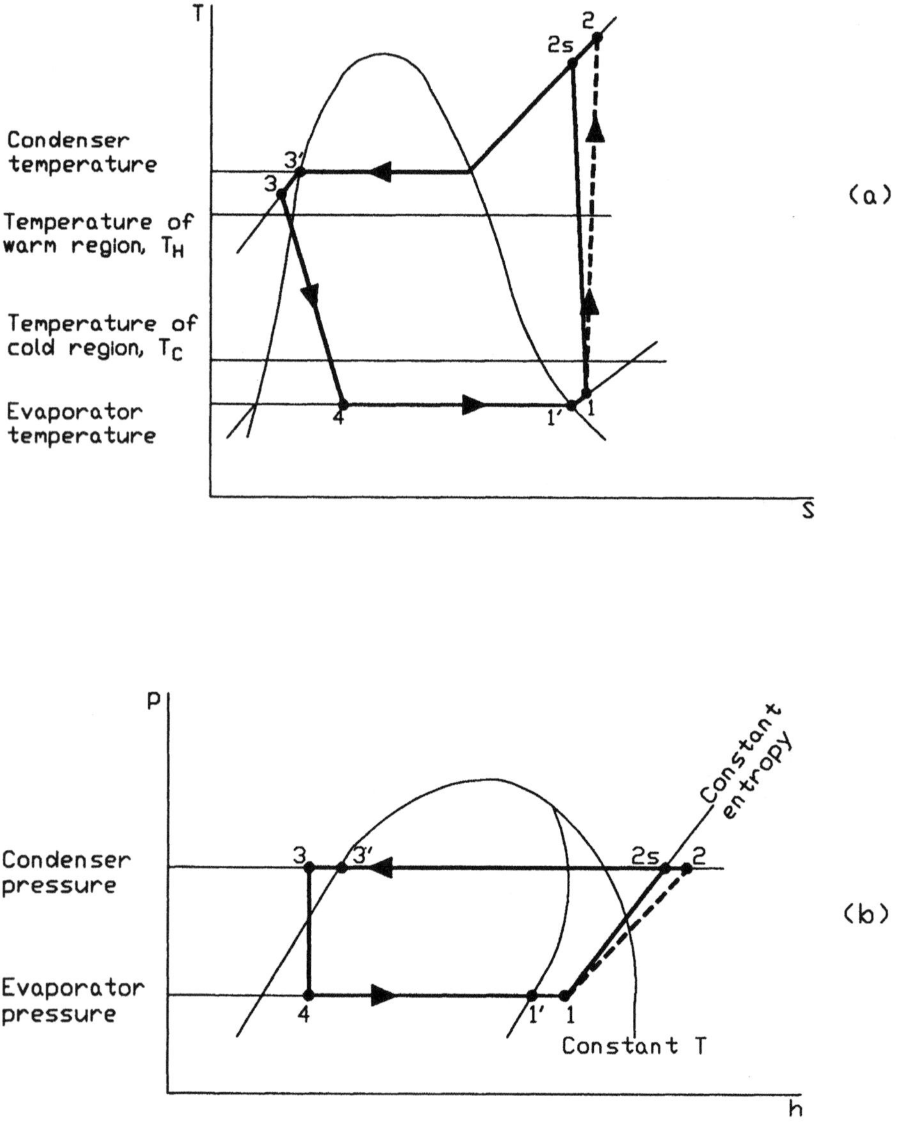

Fig. 12.2 Diagrams of vapor-compression refrigeration cycles:
a) Temperature-entropy diagram
b) Pressure-enthalpy diagram

The SI unit of refrigeration capacity is W.
Capacity, however, is frequently expressed in tons, where one ton of refrigeration capacity is defined as the transfer of heat at the rate of 3.52 kW (12000 Btu/h) per ton of refrigerant, which is roughly the rate of cooling obtained by melting ice at the rate of one ton per day. The heat-transfer rate, in practice, is also expressed as kfrig/h which is the heat, in kcal/h, transferred from the cold region to the refrigerant.

Compressor (process from 1 to 2s or to 2)

The refrigerant, in saturated or superheated vapor state, is compressed to higher pressure and temperature. The latter value must be higher than that in the warm region. Note that the process in an idealized cycle is isoentropic, without heat transfer to or from the compressor. The vapor at point 2s is generally superheated.

In an actual cycle, instead of 2s the final point is 2 to which corresponds a higher value of entropy.

The power input as mechanical power per unit of refrigerant mass flow-rate is:

$$\frac{P_c}{m} = h_{2s} - h_1' \text{ or } h_2 - h_1$$

where
P_c = input power as mechanical power at the shaft of the compressor (W)
h_2, h_{2s}, h_1', h_1 = enthalpy per unit mass (J/kg)

The effect of the irreversible compression in the actual cycle can be taken into account by introducing isoentropic compressor efficiency between point 1 and point 2:

$$\eta_c = \frac{h_{2s} - h_1}{h_2 - h_1}$$

In addition, mechanical efficiency must be introduced (friction losses, auxiliary power such as lubricant oil pump, etc.). Notice that in hermetic compressors (in which the electric motor is placed in the same crank as the compressor), widely used in small-medium sized equipment, the heat related to the motor losses must also be introduced in the thermodynamic cycle.

Because of the above-mentioned efficiencies, the shaft power is then greater than in ideal cycles by a factor 1.3-1.4.

Condenser (process from 2 to 3' or 3)

The refrigerant passes through the condenser where it changes from the superheated state to the saturated vapor state, and then it condenses by transferring heat to the surrounding medium (warm region) which must be at a lower temperature than the refrigerant. The final state is saturated liquid (3') or subcooled liquid (3).

The rate of heat transfer from the refrigerant per unit of refrigerant mass flowrate is:

$$\frac{Q_{out}}{m} = h_2 - h_3' \text{ or } h_2 - h_3$$

where

Q_{out} = heat-transfer rate to the warm region (W)

h_2, h_{2s}, h_3', h_3 = enthalpy per unit mass (J/kg)

Expansion valve (process from 3' or 3 to 4)

The refrigerant, in the saturated or subcooled liquid state, expands through the expansion valve to lower pressure and temperature in a two-phase liquid-vapor state. During this process, enthalpy remains constant because vaporization occurs by subtraction of heat from the liquid so that the temperature decreases.

Point 4 represents the end of the cycle, after which it starts again.

In the complete cycle, the net input power is the mechanical-compressor input power (or mechanical and electric in hermetic compressors), since the expansion valve does not involve any power.

The coefficient of performance COP of the vapor-compression refrigeration system shown in Fig. 12.1 is:

$$COP = \frac{(Q_{in}/m)}{(P_c/m)} = \frac{h_1' - h_4}{h_{2s} - h_1'} \text{ for idealized cycle}$$

$$COP = \frac{(Q_{in}/m)}{(P_c/m)} = \frac{h_1 - h_4}{h_2 - h_1} \text{ for actual cycle}$$

Note that the operating temperatures of the vapor-compression refrigeration cycle are mainly conditioned by the temperature T_C to be maintained in the cold region and by the temperature T_H of the warm region where heat must be discarded. As shown in Fig. 12.2a, the evaporation temperature must be lower than the cold-region temperature T_C and the condensation temperature must be higher than the warm-region temperature T_H.

As the evaporating temperature falls further below that of the cold region, and the condensation temperature rises further above that of the warm region, the coefficient of performance decreases. The warm- and cold-region temperatures represent the top and bottom limits of the Carnot cycle, as shown in Fig. 12.3. Of course, if the T_H or T_C temperature varies, the coefficient of performance also varies.

The maximum theoretical coefficient of performance for a refrigeration cycle operating between a cold region at a temperature T_C (K) and a warm region at higher temperature T_H (K) (the Carnot cycle as shown in Fig. 12.3) is:

$$COP_{max} = \frac{T_C}{T_H - T_C}$$

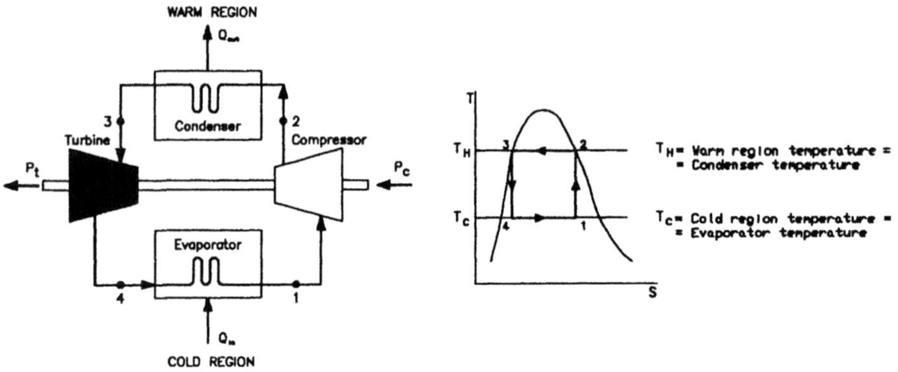

Fig. 12.3 Carnot vapor refrigeration cycle

The main differences between the Carnot cycle and any other cycle may be summarized as follows:

— to achieve a significant rate of heat transfer, the difference between the refrigerant temperature in the evaporator and the cold-region temperature T_C must be equal to 5-10°C (9-18°F). Also, the temperature of the refrigerant in the condenser must be several degrees above the T_H temperature. Consequently, the coefficient of performance, in practice, is lower than the Carnot coefficient;

— the Carnot cycle has a compression process with a two-phase liquid-vapor mixture or wet compression. This is avoided in practice because liquid droplets can damage the compressor, so dry compression, with only vapor, saturated or superheated, is required. In consequence, the coefficient of performance is reduced still further below the Carnot coefficient;

— the expansion process, which in a Carnot cycle occurs in a turbine which performs mechanical work, takes place in a throttling valve with a saving in capital and maintenance costs but with a further reduction of the coefficient of performance.

Practical vapor compressor cycles

In practice, the vapor enters the compressor as superheated instead of as saturated vapor. When superheating occurs in the evaporator or in the relative pipeline inside the space to be cooled, the refrigeration capacity increases. Therefore, the weight of the refrigerant in the compressor diminishes because of the increase in the specific volume of the vapor. In addition, the temperature increases in the final compression state (point 2 instead of 2s). Attention must be paid to this value in order to maintain the efficiency of the lubricant oil.

It is recommended that the refrigerant enter the throttling valve at a temperature lower than that of saturated liquid (point 3 instead of point 3' in Fig. 12.2). In this case, the evaporation temperature downstream of the throttling valve is reached with the vaporization of less liquid.

The COP for a practical cycle is roughly 40-50% of the corresponding Carnot cycle COP in HVAC applications and higher when the cold-region temperature is well below the ice point (273.15 K; 0°C; 32°F). Practical values of COP range between 0.75 and 3.5 depending on the operating conditions, temperatures and pressures.
The lower values occur with cold-region temperatures below ice point.

High-capacity systems with screw and centrifugal compressors usually have high COP values.

12.3 REFRIGERANT PROPERTIES AND APPLICATIONS

The main conditions (see typical parameter values in Table 12.1) which make a fluid a good medium for refrigerating cycles are:

— the evaporation enthalpy must be high and the specific heat of the fluid must be low. Consequently, during expansion, less vapor will be produced to cool the liquid to the evaporation temperature and the refrigeration capacity will be higher;

— the temperature corresponding to the critical point must be very high and greater than the condensation temperature which depends on the warm-region temperature;

— the freezing point must be lower than the minimum evaporation temperature, which depends on the temperature of the cold region;

— the specific volume of the vapor must be as low as possible in order to minimize the size of the reciprocating compressors. On the contrary, for a proper sizing of centrifugal compressors, a significant volume flow is required to avoid diameters being too small and revolution speeds too high;

— the operating pressures, which are linked to the operating temperatures, should avoid excessively low values in the evaporation process and excessively high values in the condensation process (typical values are 0.1-0.2 MPa for evaporator pressure and 0.8-2 MPa for condenser pressure). The evaporator pressure should be higher than the atmospheric one to avoid air input in the pipeline. Notice that a moderate pressure ratio should be guaranteed for the compressor to achieve better performance;

— other properties such as chemical stability, corrosiveness, toxicity, solubility in water, compatibility with lubricant and other materials, and cost have also to be considered when choosing refrigerants.

As a general rule, centrifugal compressors are best suited for low evaporator pressures and refrigerants with relatively high specific volume at low pressure. Reciprocating compressors work in higher pressure ranges and with low specific volume refrigerants.

The commonest refrigerants are ammonia and halogenated hydrocarbons. The latter include Refrigerant 12 (dichlorodifluoromethane CCl_2F_2), Refrigerant 22 (Monochlorodifluoromethane $CClHF_2$), Refrigerant 11 (trichlorofluoromethane CCl_3F), Refrigerant 502 (azeotrope mixture of refrigerants 22 and 115).

The main properties of refrigerants are reported in Table 12.1. Typical fields of application are: R12 (domestic refrigeration, small plants with T_C > -20°C, automotive air conditioning), R22 (food conservation, small and medium-size space air conditioning), R502 (low temperature food conservation), ammonia (low temperature food conservation and freezing), R11 (large air conditioning plants with centrifugal compressors). Some of these fluids have been considered harmful to the ozone layer in the upper atmosphere, so their use in the refrigeration industry is better avoided. Other fluids having similar thermodynamic behaviour are suggested (R22 instead of 502, etc.).

Table 12.1 Physical properties of refrigerants based on evaporation at 258 K and condensation at 305 K

Physical properties		R717	R22	R12	R11
Boiling point	K	240	232	243	297
	°F	–28	41.5	–21.6	74.5
	°C	–33.3	–40.8	–29.8	23.7
Freezing point	K	195.5	113	115	162
	°F	–107.9	–256	–252	–168
	°C	–77.7	–160	–158	–111
Critical temperature	K	406	369	385	471
	°F	271.5	204.8	233.6	388.4
	°C	133	96	112	198
Critical pressure	MPa	11.4	4.97	4.1	4.4
Specific volume at suction	m^3/kg	0.51	0.07	0.09	0.7
Latent heat at boiling point	kJ/kg	1367	234	234	182
Evaporator pressure	MPa	0.24	0.29	0.18	0.02
Condenser pressure	MPa	1.16	1.19	0.74	0.12
Compression ratio		4.94	4.03	4.07	6.25
Specific heat ratio c_p/c_v		1.187	1.19	1.14	1.12
Net refrigeration effect	kJ/kg	1100	163	117	156

Average evaporation temperature 258 K, 5°F, –15°C
Average condensation temperature 303 K, 86°F, 30°C

R717 Ammonia
R22 Monochlorodifluoromethane
R12 Dichlorodifluoromethane
R11 Trichlorofluoromethane

12.4 THE MAIN COMPONENTS OF VAPOR COMPRESSION SYSTEMS

As shown in Fig. 12.1, the main components of a vapor compression system are: evaporator, compressor, condenser and expansion valve.

Evaporator

The evaporator is basically an exchanger where the refrigerant, as a liquid or liquid-vapor mixture at low temperature and pressure, evaporates by subtraction of heat from the process or from the space to be cooled.

A liquid-vapor mixture exists when the evaporator is fed immediately downstream of the expansion valve.

Two basic types of evaporator are used in the refrigeration industry: (i) the direct expansion type, in which the fluid evaporates inside tubes, suitable for direct air cooling; (ii) the flooded type (pool-boiling) where the fluid boils in contact with tubes which contain the warmer liquid (as for the production of chilled water).

Attention must be paid to the formation of ice around evaporators working in air at a temperature below 0°C (32°F). The defrosting can be done by recirculating hot gas or by inverting the cycle or by electric resistors. The energy consumption in this phase must be taken into account when calculating the COP of the system.

Compressor

The compressor has the task of increasing the pressure of the refrigerant from the evaporating value to the condensing value, to which corresponds the condensation temperature. This temperature must be compatible with the surrounding or warm-region temperature, as already explained.

Compressors for refrigeration can be classified mainly as reciprocating compressors, screw compressors and centrifugal compressors. The choice among them depends on the power range, the working temperatures, the range of demand from end users, etc.. Detailed information on compressors can be found in specialized textbooks.

As a general rule, reciprocating compressors are used up to a maximum refrigerating capacity of 500-600 kW; centrifugal compressors have practically no upper limits and their application starts from a lower limit of a few hundred kW; screw compressors usually cover a range between 200 and 4000 kW. Reciprocating and screw compressors can be classified as volumetric machines and can easily be adapted to different working temperatures and pressures. The centrifugal compressor adapts poorly to variations in pressure.

Centrifugal and screw compressors can regulate the useful power continuously; attention must be paid to the low operating limit, which always exists in centrifugal compressors, established by the surge region where instability may occur. If this limit is reached, either additional load must be placed on the machine by recycling vapor from discharge to suction or one compressor must be taken out of service. Screw compressors can regulate the capacity in a wide range, between 10% and 100%, by regulating the quantity of gas at suction.

Reciprocating compressors can regulate the useful power step by step by operating the suction valves sequentially.

Control of the speed by a variable-speed drive is always possible.

Condenser

The condenser is basically a heat exchanger. Its main function is to condense the refrigerant from a vapor state to a liquid state (saturated or subcooled) by means of a surrounding medium (warm region) at lower temperature. The rate at which heat is subtracted from the refrigerant, with reference to Fig. 12.1, is:

$$Q_{out} = Q_{in} + P_c \qquad \text{(W)}$$

where

Q_{in} = **refrigerating capacity (W)**
P_c = **mechanical power at the compressor shaft or electric power in hermetic compressors (W).**

Three main modes of operation occur in the condenser depending on the different states through which the refrigerant passes: from superheated to saturated vapor, from saturated vapor to saturated liquid, from saturated liquid to subcooled liquid. The sum of the three terms mentioned above is the Q_{out} already introduced.

i) The heat to be discarded in the first step is:

$$c_p \cdot m \cdot (t_2 - t_k)$$

where
c_p = specific heat at constant pressure of the superheated vapor
t_2 = temperature of vapor entering the condenser
t_k = condensation temperature corresponding to the working pressure
m = mass flow-rate of the refrigerant

ii) The heat to be discarded during the condensation phase is:

$$m \cdot h_l$$

where
h_l = latent heat at the working conditions (temperature and pressure)

iii) The heat to be discarded during the subcooling phase is:

$$c \cdot m \cdot (t_k - t_3)$$

where
c = specific heat of the liquid
t_3 = outlet temperature of the subcooled liquid.

Depending on local conditions and on technical considerations, water or air can be used as cooling media.

If water is used, condensers are refrigerant fluid/water exchangers. Water can be discarded, but it is better to cool and recirculate it by appropriate cooling equipment such as cooling towers. These systems are based on the evaporation of a small quantity of the circulating water by subtraction of heat from the remaining water which is then cooled. Note that the temperature of the cooled water is always higher than the wet-bulb temperature which indicates the lowest value to which the circulating water can be cooled. In practice, the design temperature at the evaporating tower outlet is assumed to be at least 5°C (9°F) higher than the wet-bulb temperature (see also Chapter 13.2).

When discarded water is used for cooling, practical values of water consumption, in the case of outlet-inlet temperature changes of 15-20°C (27-36°F), range between 0.1 and 0.2 m^3/kWh of input energy, generally electric, consumed by the compressor prime mover. Higher values occur if the water outlet-inlet temperature change is smaller. When evaporating towers are used for cooling, water consumption values range between 0.01 and 0.02 m^3/kWh. Electric energy consumption of auxiliary equipment (pumps, etc.) ranges between 0.05 and 0.1 kWh/kWh of input energy consumed by the compressor prime mover.

Evaporating condensers are also used, in which condensation is due to the evaporation of external water. The condensation temperature is generally 10-12°C (18-22°F) higher than the external air temperature.

If air is used, in industrial applications forced circulation of air must be employed to facilitate heat transfer in refrigerant fluid/air exchangers. This system, which requires electric power for fans, has the disadvantage of different working temperatures if the air temperature varies. If the temperatures are too high, excessively high condensation temperatures may be required. As a result, the COP is reduced.

Practical values of energy consumption in air-condenser fans are 0.05-0.1 kWh/kWh of input energy, generally electric, consumed by the compressor prime mover.

Expansion valve

The expansion valve is a typical reduction valve which lowers the pressure to a value suitable for the downstream evaporator operation. The pressure is controlled by the liquid fluid level in flooded evaporators and by the superheated vapor temperature in direct expansion evaporators (thus ensuring roughly 5-8°C; 9-14.4°F of superheating temperature increase at the suction line).

12.5 CASCADE AND MULTISTAGE SYSTEMS

In order to obtain a good coefficient of performance with very low evaporating pressure and relatively high condenser pressure, cascade and multistage cycles are generally used.

A cascade cycle, in which two vapor-compression cycles are located in series is illustrated in Fig. 12.4. A counterflow heat exchanger is situated between the

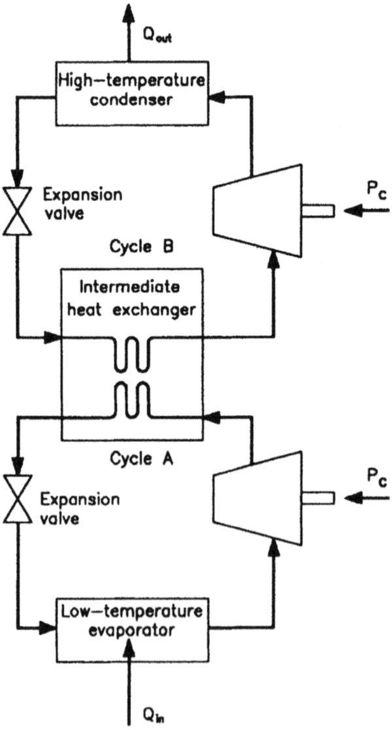

Fig. 12.4 Example of a cascade vapor-compression refrigeration cycle

two cycles: the heat rejected during the condensation of the low-temperature cycle (cycle A) is used to evaporate the refrigerant in the second cycle which works in a higher temperature range. If necessary, two refrigerants with different saturation pressure-temperature relationships can be selected to ensure the best temperature and pressure for the two temperature ranges: low evaporation temperature without excessively low pressure for the first cycle, relatively high condensation temperature without excessively high condensation pressure for the second cycle.

Multistage compression with intercooling between the stages, accomplished by means of the refrigerant itself, can also be used if big temperature differences between cold and warm regions must be obtained.

12.6 ABSORPTION REFRIGERATING SYSTEMS

Absorption refrigeration cycles which require thermal energy instead of mechanical energy as the main input power can be used in particular applications. The basic scheme is shown in Fig. 12.5.

They differ from the vapor-compression cycles in two important respects:

— downstream of the evaporator (point 1), the refrigerant in the vapor state is absorbed by a secondary substance, called an absorbent, to form a liquid solution. This solution is then pumped to a higher pressure (the condenser pressure) by means of a pump which requires much less work than vapor compression does, because of the low specific volume of the solution. The mechanical input power is consequently very low.

  Note that the formation of the liquid solution is an exothermic reaction. Since the amount of refrigerant that can be dissolved into the absorbent liquid increases as the solution temperature decreases, a cooling system must be installed to keep the temperature of the absorber as low as possible and to allow a strong refrigerant-absorbent solution;

— upstream of the condenser (point 2), the refrigerant vapor must be retrieved from the liquid solution (as a result, the strong refrigerant-absorbent solution becomes a weak solution) before entering the condenser. This is an endothermic process and consequently a thermal input power, which represents the main input of the system, is needed. Steam or waste heat are generally used; natural gas or other fuels can be burned to provide the heat source. The choice of heat source is determined mainly by cost.

The refrigerant vapor passes to the condenser and the remaining solution, which is a refrigerant weak solution, flows back into the absorber tank through a valve without entering the condenser.

The commonest cycles use lithium bromide-water solution in which water is the refrigerant and lithium bromide the absorbent; pressures are always lower than the atmospheric value. Another type of absorption system uses ammonia-water solution in which ammonia is the refrigerant and water the absorbent.

As a general indication, 1.5-2 kW of thermal power at temperatures between 100°C and 120°C (212-250°F) are required to produce 1 kW of refrigeration capacity. The coefficient of performance varies between 0.7 and 0.5, which is much lower than the corresponding coefficient for vapor-compression cycles. The heat to be discarded by means of refrigerating equipment (cooling tower, air condenser, etc.) for a refrigeration capacity of 1 kW is roughly 2.5 kW.

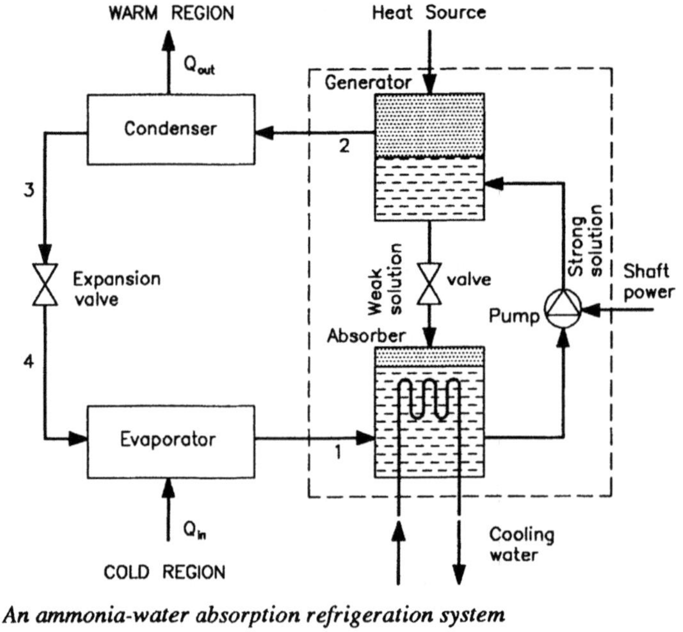

Fig. 12.5 An ammonia-water absorption refrigeration system

Higher COP up to 1 can be obtained in cycles using a direct fired system run on combustibles, with a two-stage absorption system.

Additional equipment, such as heat exchangers, which allow the strong solution entering the generator to be preheated by the weak solution leaving it, and rectifiers to remove any traces of water from the refrigerant before it enters the condenser, are needed to improve the coefficient of performance.

12.7 BRAYTON REFRIGERATION CYCLE

The Brayton refrigeration cycle is one of the gas refrigerating systems in which the process involves no change in phase. The basic scheme of this cycle is represented in Fig. 12.6. A modified scheme incorporating an intermediate heat exchanger between the fluids leaving the two basic exchangers is the form most often used.

12.8 ENERGY SAVING AND COMPUTER CONTROL

The coefficient of performance is the index commonly used to compare refrigeration systems.

As explained in Chapter 12.2, the upper limit is related to the Carnot cycle in which the COP index depends only on the cold and warm region temperatures.

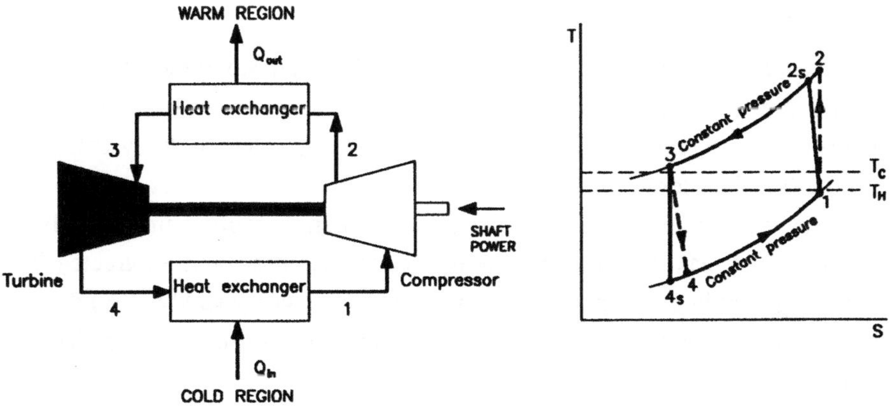

Fig. 12.6 Basic scheme of a Brayton refrigeration cycle

The need for a difference between the evaporation and condensation temperatures and those of the cold and warm regions (which are the T_C and T_H limiting temperatures of the Carnot cycle) together with the losses inside the components of the systems make the real cycle less efficient than the ideal one.

In order to obtain operating conditions with the best COP, and so to ensure significant energy saving, the following suggestions can be made for both plant management and capital investment:

Plant Management

— the evaporation temperature must be the highest permitted by the needs of the users. This can be achieved by setting the temperature control of the fluid to be cooled or the suction pressure of the valve;

— the condensation temperature must be the lowest permitted by the cooling system and by the warm region temperature;

— the load rate of the single compressors must be kept as high as possible by taking out the single machines in sequence in order to avoid low efficiency which occurs at low rate.

Capital investment

— to install a floating point for the control of the evaporating temperature according to external conditions and user needs;

— to improve the efficiency of the cooling equipment to allow lower condensation temperatures;

— to install variable-speed drive to regulate the refrigerant flow without dissipation;

— to install centralized computer control.

Costs and energy can also be saved by installing cold storage systems, such as ice or cold-water or eutectic storage, to shift the energy consumption from peak hours to low-rate hours and to smoothe the load demand and so the system power profile (see Chapter 4.7).

Heat recovery from the hot water of the condenser cooling circuit is always possible. The maximum recoverable heat equals the heat to be subtracted from the refrigerant in the condenser; actual values are lower (roughly 50%) depending on the temperatures of potential end users. Typical uses are space heating, sanitary hot water and cleaning.

Heat recovery from the lubricant oil cooling system allows higher temperatures than recovery from the condensation phase.

For high-power systems, drivers such as steam or gas turbines can be used in a cogeneration cycle.

12.9 HEAT PUMP OPERATION

A heat pump is a system which operates cyclically to transform low-temperature energy into higher-temperature energy by the application of external work. Both vapor-compression and absorption systems can be used as heat pumps with small modifications.

The coefficient of performance has the same meaning as in the refrigeration systems shown in Fig. 12.1 and in Fig. 12.2, but input and output energy play different roles. The useful output or desired effect is the heat directed to the mass to be heated (Q_{out}) so that:

$$COP = \frac{Q_{out}/m}{P_c/m} = \frac{h_{2s} - h_{3'}}{h_{2s} - h_{1'}} \quad \text{for the Carnot cycle}$$

$$COP = \frac{Q_{out}/m}{P_c/m} = \frac{(h_2 - h_3)}{h_2 - h_1} \quad \text{for the actual cycle}$$

Heat pumps can be used for space heating or for industrial processes to raise the temperature of available heat, generally waste heat at low temperature, or of ambient air or water sink, to a level acceptable to the end users. The prime mover is either an electric motor or a fuel engine.

Only if the COP is greater than 3 can heat pumps be regarded as an energy source economically more attractive than boilers. This depends on the overall efficiency of the prime mover, which may be a thermal or an electric motor, and on the end-user energy costs.

Primary energy saving, more than cost saving, depends on the efficiency of all the energy transformations. In the case of a fuel engine, the efficiency is the ratio between the mechanical power at the shaft and the input power as fuel in the engine itself. In the case of an electric motor, the efficiency is the ratio between the mechanical power at the shaft and the thermal input power needed in the utility power plant to produce electric energy for the motor drive; it is generally less than 33% (see Chapter 4.2).

As in refrigerating systems, the maximum theoretical COP obtainable is that of the Carnot cycle:

$$COP_{max} = \frac{T_H}{T_H - T_C}$$

One of the main disadvantages of heat pumps is the operating temperature range: evaporation or cold-region (heat-source) temperatures which are too low yield a low COP if relatively high temperatures are required as useful output. Contrariwise, in industrial applications, an additional constraint is that high-temperature waste above 95°C (200°F) cannot be supplied directly to the heat pump because of the limits imposed by the refrigerant characteristics.

Fig. 12.7 shows mean heat-pump COP values versus load (warm-region) temperatures for a number of source temperatures.

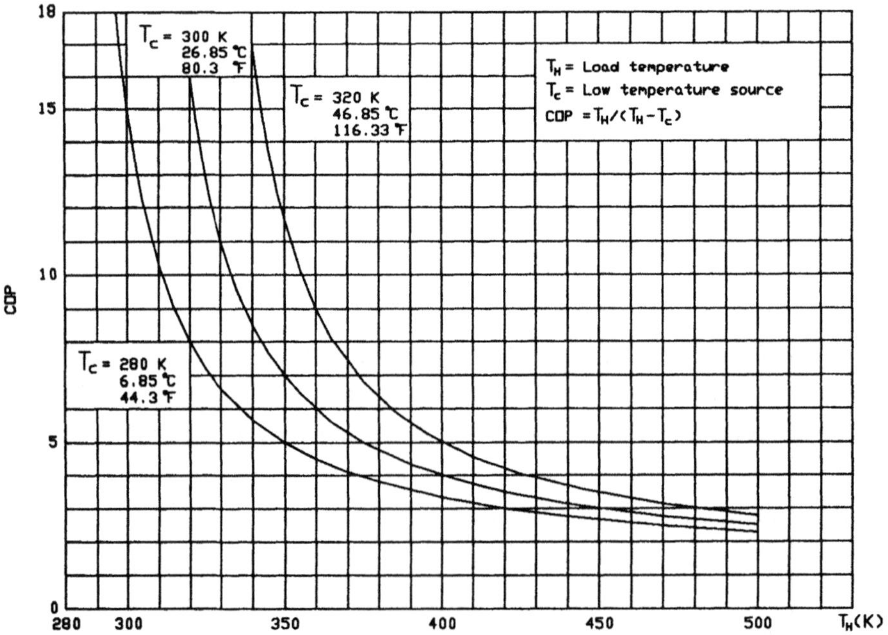

Fig. 12.7 Theoretical COP versus load temperatures for heat pumps

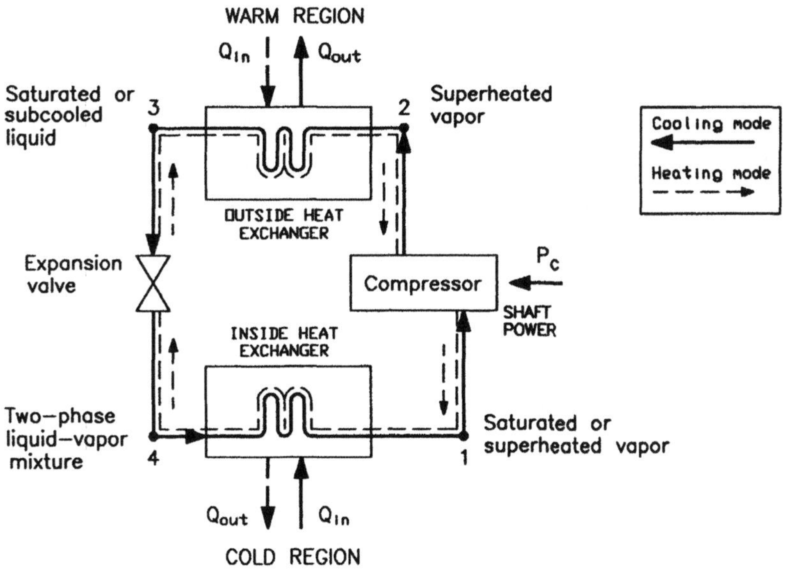

Fig. 12.8 Example of air-to-air reversing heat pump (operating point symbols and descriptions refer to the cooling mode)

In HVAC applications, heat pumps have evaporators in direct contact with the outside air which serves as energy source during the cold season. In the hot season, the same air source can provide cooling through a reversing valve. Fig. 12.8 shows how the refrigerant flows inside the same components in heating and in cooling modes. Note that in the heating mode the outside heat exchanger works as evaporator and the inside exchanger works as condenser. In the cooling mode, the outside heat exchanger becomes the condenser and the inside exchanger becomes the evaporator.

12.10 PRACTICAL EXAMPLES

Two examples of energy saving are given.

The first refers to the replacement of an absorption system by a rotating compressor system in a factory where waste heat streams are not available, so that all the heat needed by the absorption system must be supplied by means of fuels.

The second concerns heat recovery from condensation and lubricant oil cooling which is a possibility always present, but exploitable only when end users really need heat at low temperature.

Example 1
Replacement of an absorption system by a rotating compressor system

An absorption system with a rated capacity of 1744 kW (496 tons; 1500000 kfrig/h) is installed in a food-processing factory and it is fed by steam at 0.7 MPa. The actual refrigeration capacity is 995 KW with a steam flow-rate equal to 2299 kg/h. The replacement of this system by a rotating compressor is here examined from the energy-saving point of view, on the assumption that end-user needs are satisfied in both cases. The operating characteristics are detailed in Table 12.2 together with the technical comparison.

Economic evaluation is in Table 20.3.

Example 2
Heat recovery from condensation phase and lubricant oil cooling

A refrigerating plant with screw compressors has a nominal capacity of 1400 kW (1204000 kfrig/h; 397.7 tons). The input shaft power is 500 kW; the refrigerating plant works 5000 h/year at 70% of the rated capacity for food conservation at low temperature.

Heat recovery can be obtained in two ways: (i) from the condensation circuit by means of gas/water exchangers; (ii) from lubricant oil cooling equipment.

In case (i), hot water at 40-45°C (104-113°F) can be obtained if the condensation temperature is kept at 45-50°C (113-122°F), which is quite a high value but still acceptable for screw compressors. Hot water can be used for heating purposes in the HVAC plant of the factory; because of the relatively low temperature, it is advisable to feed the water/air exchanger used for cooling during the hot season (which has exchange surfaces bigger than those used for heating) and to use the boiler-fed hot water circuit for integration, when needed.

The recoverable heat is estimated to be equal to 60% of the sum of the shaft power and the refrigeration capacity at the average load, that is, 798 kW. Additional consumption of the compressor because of the higher condensation temperature is estimated as equal to 30 kW.

In case (ii), hot water at 45-50°C (113-122°F) can be obtained by means of a plate exchanger regardless of the condensation temperature; the discharge temperature of the lubricant oil ranges between 50 and 55°C (122 and 131°F), so this can be injected into the compressor at 45-50°C (113-122°F) with a temperature drop of 5°C (9°F). The recoverable heat is estimated as equal to 10% of the condensation heat, that is, 133 kW, much less than in case (i). Hot water can be used as in case (i) or for other users.

Energy saving in both cases has been calculated in Table 12.3, by introducing the efficiency of an equivalent boiler.

Economic evaluation is shown in Table 20.3.

Table 12.2 Comparison between an absorption and a rotating compressor system

	END-USER REQUIREMENTS			
a	cold-water flow	m^3/h	372	372
b	inlet temperature	°C	9	9
c	outlet temperature	°C	6.7	6.7
d = a · 4.186 · (b–c)/3.6	cooling load	kW	995	995
e	working hours	h/year	6000	6000
			ABSORPTION SYSTEM (1)	COMPRESSOR SYSTEM (2)
	REFRIGERATION PLANT			
f	rated capacity	kW	1744	1744
f' = f/3.52		tons	496	496
f'' = f · 860		kfrig/h	1500000	1500000
g = d · 100/f	% rated capacity		57	57
h = d	actual capacity	kW	995	995
i	input power (steam)	kgsteam/h	2299	0
i' = i · 2190/3600		kW	1399	0
l	boiler efficiency		0.8	
m = (i · 2190/l)/41860	input fuel	kgoil/h	150	
n	input power (electricity)	kW	0	245
o = h/i'	COP		0.711	
o = h/n	COP			4.067
	COOLING TOWERS AND AUXILIARIES			
p = d	end-user load	kW	995	995
q = i'	refrigeration system	kW	1399	
q = n				245
r = p+q	total to dissipate	kW	2394	1240
s	input electric power	kW	113	58
t	water consumption	m^3/h	7.5	3.9
	ENERGY CONSUMPTION			
A = m · e/1000	thermal energy	TOE/year	902	0
B = (s+n) · e/1000	electric energy	MWh/year	678	1818
C = t · e	water	$m^3/year$	45000	23400
	ENERGY SAVING			
A(1)–A(2)	thermal energy	TOE/year		902
B(1)–B(2)	electric energy	MWh/year		–1140
C(1)–C(2)	water	$m^3/year$		21600

Notes
useful enthalpy = steam enthalpy – condensate enthalpy = 2190 kJ/kg steam
41860 kJ/kgoil as net heating value
all input power figures of refrigeration plant and auxiliaries are average values referred to 6000 h/year

Table 12.3 Energy saving associated with heat recovery from a refrigerating plant

Case (i)	RECOVERY FROM CONDENSATION PHASE		
a	Rated refrigeration capacity	kW	1400
b	Rated shaft power	kW	500
c	Average load factor	%	70
$d = c \cdot (a+b)/100$	Condensation heat at average load	kW	1330
$e = 0.6 \cdot d$	Recovered heat	kW	798
$f = e/0.85$	Equivalent input power as fuel(*)	kW	938.8
$g = f \cdot 860$		kcal/h	807388.2
$h = f \cdot 3600/1.055$		Btu/h	3203568
$i = f \cdot 3600/41860$		kgoil/h	80.7
l	Working hours of recovered-heat users (**)	h/year	2400
$m = i \cdot l/1000$	ENERGY SAVING	TOE/year	193.8
Case (ii)	RECOVERY FROM LUBRICANT OIL COOLING		
a	Rated refrigeration capacity	kW	1400
b	Rated shaft power	kW	500
c	Average load factor	%	70
$d = c \cdot (a+b)/100$	Condensation heat at average load	kW	1330
$e = 0.1 \cdot d$	Recovered heat from lubricant	kW	133
$f = e/0.85$	Equivalent input power as fuel(*)	kW	156.5
$g = f \cdot 860$		kcal/h	134564.7
$h = f \cdot 3600/1.055$		Btu/h	533928
$i = f \cdot 3600/41860$		kgoil/h	13.5
l	Working hours of recovered-heat users (***)	h/year	6000
$m = i \cdot l/1000$	ENERGY SAVING	TOE/year	80.7

Notes
(*) boiler efficiency 0.85 (Net Heating Value as reference)
(**) working hours for heating
(***) working hours for heating and other uses
41860 kJ/kg as net heating value

CHAPTER 13

FACILITIES—HVAC SYSTEMS

13.1 INTRODUCTION

Heating, Ventilating, and Air Conditioning systems may be among the greatest users of energy, depending on the activities performed in the buildings where they are installed. In commercial buildings and hospitals, HVAC and lighting are the main energy users; in factories, on the other hand, process users are generally the most important.

In typical HVAC plants, the main electric energy users are the refrigeration plant and the air distribution fans with auxiliaries such as cooling towers and circulating pumps. Generally, the only user of fuel is the boiler plant (hot water or steam) for space-heating purposes in cold seasons and for re-heating after a dehumidification section in hot seasons. The relative importance of these energy consumptions depends mainly on the climate of the region and thus on for how long heating or cooling modes are needed.

HVAC plants are designed to provide a range of comfortable conditions most of the time for most people working in the building. It is not easy to formulate a universally applicable definition of comfortable conditions; an ambient temperature of between 23 and 25°C (73-77°F), a relative humidity of 40 to 60%, a specific humidity of 0.008-0.012 kg of water per kg of dry air, and maximum air velocity of less than 0.15-0.20 m/s (9-12 m/min, 30-40 ft/min) are generally assumed as optimum conditions. These values are different for industry, where temperatures can be lower and air velocity higher, up to 0.5 m/s.

The need for heating, cooling and ventilation depends on many factors such as external temperature and humidity, the number of people working in the room and their activities and clothing, the power installed (machinery and lighting), the type of processes performed (e.g. textile, food processing, chemical, pharmaceutical, etc.) and the functions of the buildings (commercial center, hospital, bank, industry, etc.).

Heating and cooling requirements usually range between 50 and 100 W/m^2 and higher for particular applications; considering the ceiling height, typical values are 20-30 W/m^3 for residential buildings and less than 10 W/m^3 for industries without particular requirements. Of course, climatic conditions, type of activity performed and indoor air requirements strongly influence these values. Typically, cooling requirements are lower than those of heating, except with heavy internal loads.
In HVAC systems, hot water for heating is generally supplied at 60-80°C (140-176°F) and chilled water at 5-10°C (41-50°F); hot air at 30°C (86°F) and cold air at 13-15°C (55-59°F). The difference between inlet and outlet water temperatures is 10-20°C (18-36°F) for heating and 5-10°C (9-18°F) for cooling.

13.2 BASIC PRINCIPLES OF HVAC SYSTEMS

The basic task of each HVAC system is to control the temperature and humidity of the air inside buildings and thus to ensure comfort; economics requires that this result be reached with the minimum energy consumption.

In small plants, air temperature control is the simplest technique, but humidity control plays a fundamental role in ensuring comfort for occupants and the right environment for many industrial processes which require control of the vapor content in the air.

Design and operation of HVAC are based on knowledge of the behavior of mixtures of dry air and water vapor, commonly called moist air, which can be treated as a mixture of ideal gases.

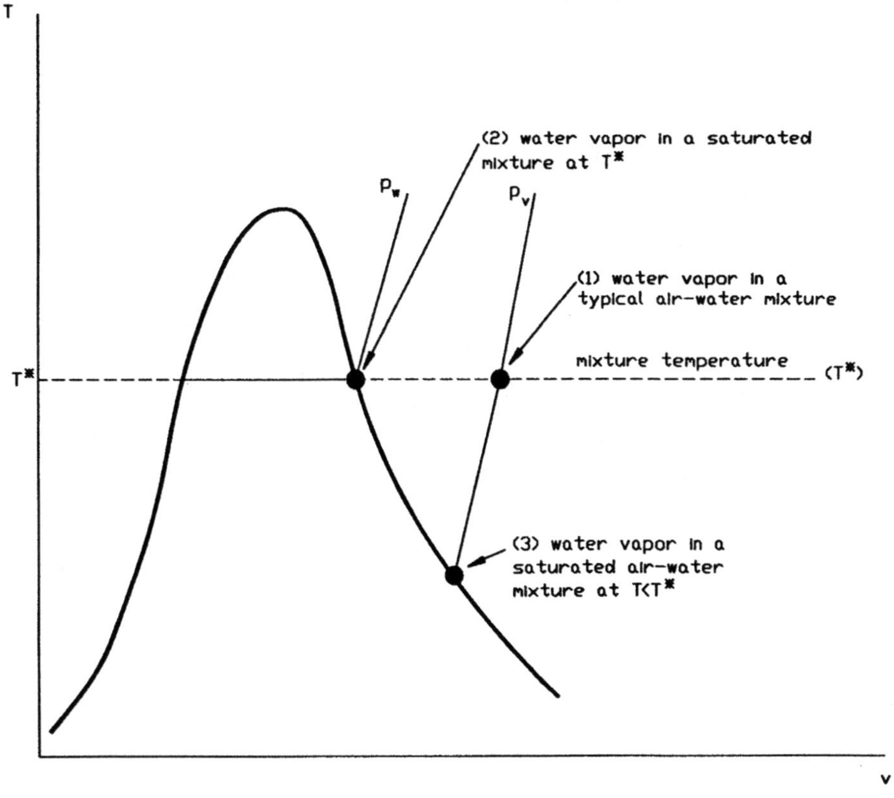

Fig. 13.1 Typical water vapor states in an air-water mixture

Typical states of water vapor in moist air are shown in the T,v diagram of Fig. 13.1: the dry air-water vapor mixture has a pressure p and a temperature T* and the components have a common temperature and pressures equal to the mixture pressure p multiplied by the single component mole fraction (p_v for vapor, p_a for air; $p_a > p_v$; $p = p_a + p_v$). At typical mixture temperatures and pressures, the vapor is superheated (p_v; typical situation as at point 1).

The mixture is said to be saturated (saturated air) when the partial pressure of vapor p_v has the same value as the saturation pressure of water p_w at the mixture temperature (p_w as at point 2; typically, p_w greater than p_v).

The quantity of water vapor in moist air varies from zero in dry air to a maximum value which depends on the pressure and temperature at which the mixture is saturated. Two different indexes of the quantity of water vapor in moist air are used: humidity ratio or specific humidity (ω) and relative humidity (Φ).

Humidity ratio or specific humidity (ω) is the ratio between the mass of water vapor (m_v) and the mass of dry air (m_a). This ratio reaches its maximum value in saturated air (practical values range from zero for dry air to a maximum value of less than 0.03 kg water per kg dry air). By using the ideal gas laws and the molecular weights of water and dry air, the humidity ratio can be expressed as the ratio between the molecular vapor weight multiplied by its partial pressure p_v and the molecular dry air weight multiplied by its partial pressure $(p - p_v)$, that is $0.622 \cdot p_v/(p - p_v)$.

Relative humidity (Φ) is the ratio between the mole fraction of water vapor in a given moist air sample and the mole fraction of water vapor in a saturated moist air sample at the same mixture temperature and pressure. This ratio equals 1 (or 100 if expressed as a percentage) for a saturated mixture so its values indicate how the the moist air should be treated.

Notice that air-water mixture in normal conditions has the adiabatic-saturation temperature close to the wet-bulb temperature which can be measured by a psychrometer.

The thermodynamic wet-bulb temperature is the temperature at which water, by evaporating into moist air at a given dry-bulb temperature (see Chapter 2.4) and humidity ratio, can bring air to an adiabatic saturation while the mixture pressure is constant. The psychrometer consists of two thermometers; one bulb is covered by a wick thoroughly wetted with water. When the wet bulb is placed in an airstream, water evaporates and an equilibrium temperature, called the wet-bulb temperature, is reached. This temperature can be assumed as the adiabatic-saturation temperature.

For practical purposes, the enthalpy of the mixture is calculated by summing the contributions of the dry air and of the water vapor. This value is usually referred to the mass of dry air (M_a) and the humidity ratio is then introduced:

$$H_{mix} = M_a \cdot h_a + M_v \cdot h_v$$

$$(H_{mix}/M_a) = h_{mix} = h_a + (M_v/M_a) \cdot h_v = h_a + \omega \cdot h_v$$

The enthalpy of dry air can be obtained from ideal gas tables or calculated by using specific heat (1.005 kJ/kg · K which varies slightly over the interval 10-30°C (50-86°F). The enthalpy of water vapor, which is superheated at operating points such as point 1 in Fig. 13.1, is assumed to be equal to the saturated vapor value corresponding to the given temperature (as in the steam table or the Mollier diagram; see Fig. 6.1 and Fig. 6.2). It can also be expressed as the sum of the enthalpies of the saturated vapor and of the superheated vapor (the latter enthalpy is equal to vapor specific heat, when the air-water mixture is in atmospheric conditions assumed as equal to 1.8 kJ/kg · K, multiplied by the temperature).

Approximately, the enthalpy of the mixture of dry air and water vapor is:

$$h_{mix} = 1 \cdot t + \omega \cdot (2500 + 1.8 \cdot t) \quad (kJ/kg)$$

where
1 kJ/kg · K is the specific heat of the dry air;
t is the difference between the mixture temperature in K and the standard temperature (273.15 K), that is, the temperature in °C;
2500 kJ/kg is the enthalpy of the saturated steam and 1.8 kJ/kg · K is the specific heat of the superheated steam when the air-water mixture is in atmospheric conditions ($p_v < p_a$; see Table 2.5).

The humidity ratio of a moist air mixture varies with the temperature; if the mixture temperature T* is reduced at constant mixture pressure, the water vapor state moves from point 1 to the dew-point temperature (point 3 in Fig. 13.1) at constant partial pressure p_v and at constant humidity ratio. Then, if cooling continues, condensation occurs and the final temperature is less than T*. The system will then consist of a saturated mixture of dry air and water vapor at a pressure less than p_v because the mole fraction of water vapor at the final state is less than the value at the initial temperature.

The dew-point temperature of a given air sample is the temperature of saturated moist air at the same pressure and with the same humidity ratio as the air sample.

By adding water in an adiabatic process to a moist air stream, the humidity ratio can be incremented as shown by the T,v diagram of Fig. 13.2. When a given mixture at known pressure and temperature (pressure p and temperature T*) comes into contact with water at temperature T_w (the example has T* greater than T_w), some of the water evaporates by using energy from the moist air and so the mixture temperature decreases. As the water is added continuously at the same rate at which water evaporates, the process can be assumed as an adiabatic one and the final mixture temperature, generally in a saturated condition, is equal to the make-up water temperature T_w.

A general relationship based on an energy balance (see Fig. 13.2 but it is valid for any similar situation) is:

$$(M_a \cdot h_a + M_v \cdot h_v) + [(M'_v - M_v) \cdot h_w] = (M_a \cdot h_a + M'_v \cdot h'_v)$$

where

M_a, h_a	=	mass and enthalpy of the dry air entering the system
M_v, h_v	=	mass and enthalpy of the water vapor in the moist air entering the system
M'_v	=	mass of the water vapor leaving the system
$(M'_v - M_v)$, h_w	=	mass and enthalpy of the make-up water equal to the saturated liquid enthalpy at the temperature T_W
h'_v	=	enthalpy of the water vapor in the saturated mixture leaving the system at the temperature T_w (assumed to be equal to the saturated water vapor enthalpy at the same temperature).

Graphic representations of the main properties of moist air mixtures are provided by psychrometric charts, where sets of curves are given for a mixture pressure generally equal to the atmospheric pressure (0.1 MPa). An example of a psychrometric chart, as suggested by ASHRAE (American Society of Heating, Refrigerating and Air Conditioning Engineers - USA), is illustrated in Fig. 13.3, where the coordinates are humidity ratio and enthalpy. Other charts, with a different choice of coordinates, can be used.

With reference to Fig. 13.3, notice that:

— constant humidity ratio lines are horizontal;

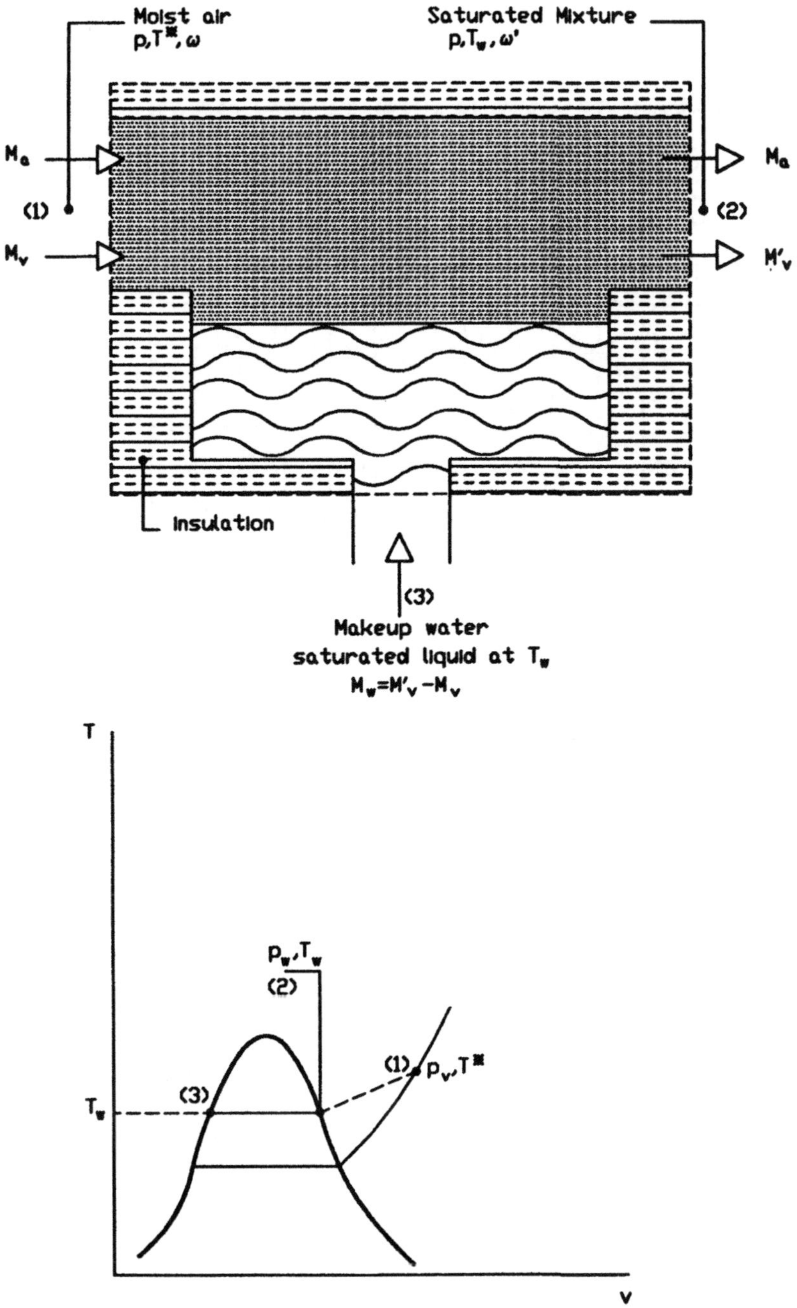

Fig. 13.2 Adiabatic saturator: basic equipment and T-v diagram representation

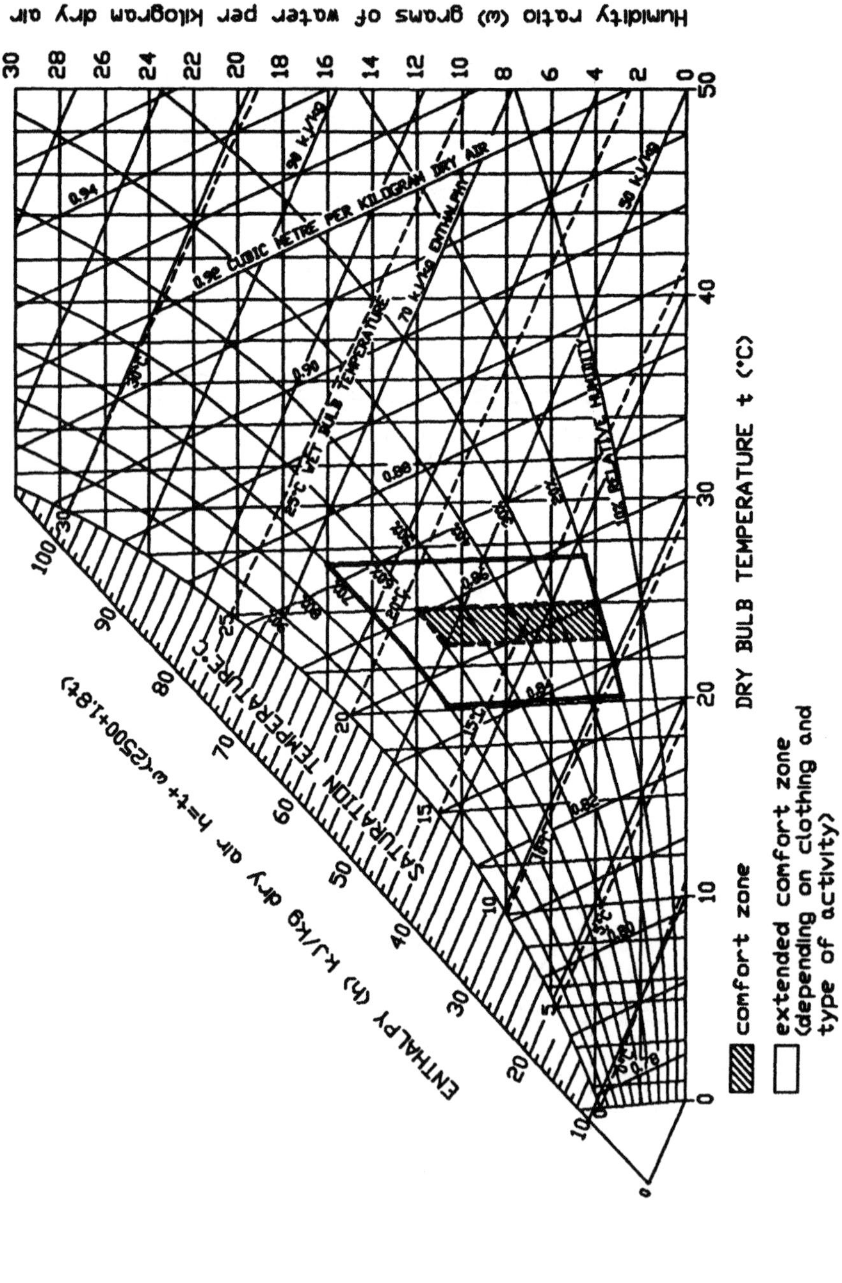

Fig. 13.3 An example of a psychrometric chart as suggested by ASHRAE

— constant enthalpy lines are oblique lines parallel to each other. The enthalpy values correspond to the relationships given above;
— constant thermodynamic wet-bulb temperature lines are oblique lines not parallel to each other and their direction differs slightly from that of the enthalpy lines;
— constant dry-bulb temperature lines are inclined slightly from the vertical direction and are not parallel to each other;
— constant relative humidity lines are shown at intervals of 10%. The line for the dry air (0%) is the horizontal line ($\omega = 0$). The saturation curve is the line corresponding to 100%; the dew-point temperature can be found at the intersection of constant ω line with the saturation curve;
— constant specific volume lines are straight but not parallel to each other.

Typical air conditioning processes

Most of the typical air conditioning processes (dehumidification, humidification, adiabatic mixing, etc.) can be described using the basic principles discussed above and psychrometric charts.

Dehumidification, when necessary to lower the relative humidity, is generally effected by a cooling section followed by a re-heating section. There are also other methods of dehumidification, such as chemical absorption by salts.

In a dehumidification process by a cooling section (see Fig. 13.4), the moist air stream (point 1) is cooled below its dew-point temperature, so that the water vapor, initially present, condenses. A saturated moist air mixture (with relative humidity 100% in ideal operating conditions and lower humidity ratio) leaves the dehumidification section (point 2) at the saturation temperature below the dew point (the final partial pressure of the water vapor in point 2 is lower than the initial partial pressure of the water vapor in the mixture because of the elimination of condensate). This moist air stream must generally be re-heated in a fol-

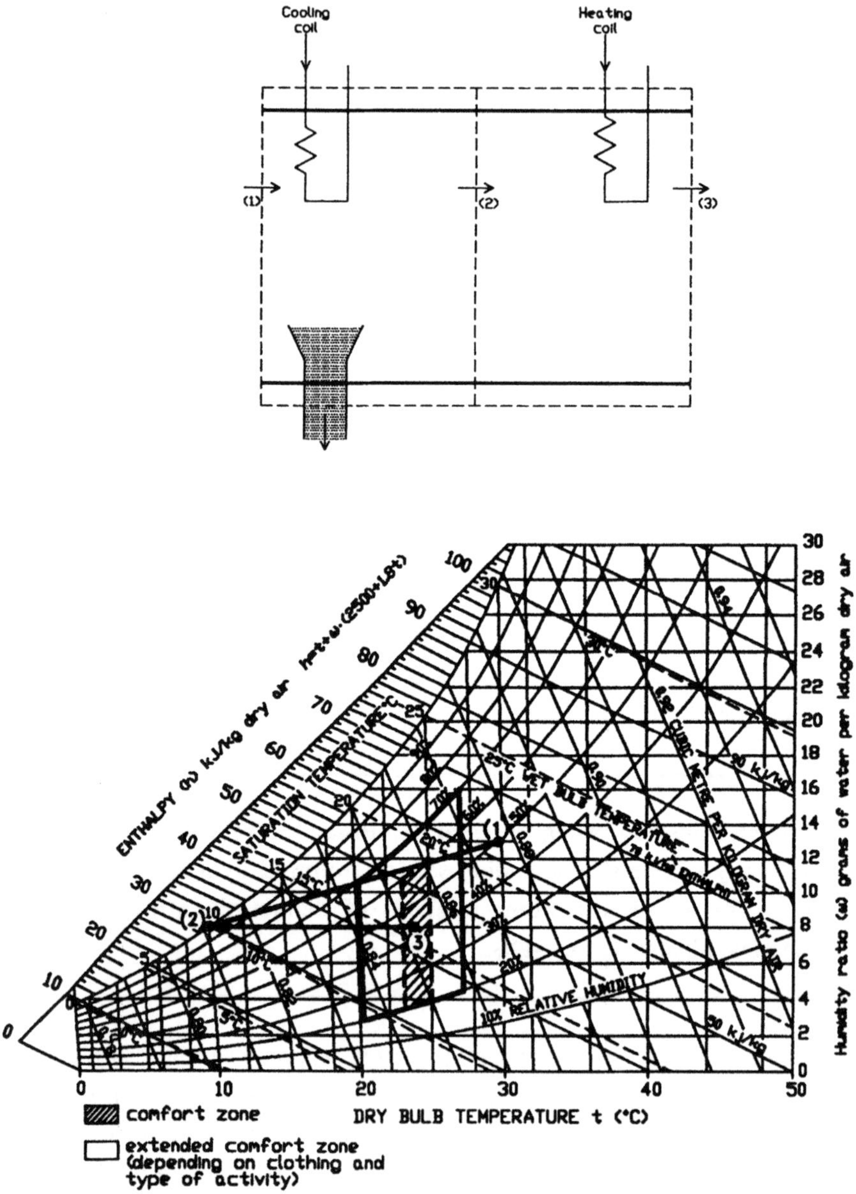

Fig. 13.4 Dehumidification by a cooling section followed by a re-heating section: basic equipment and psychrometric chart representation

lowing section in order to bring it to a condition most occupants would consider comfortable. This last phase occurs at constant partial pressure and constant humidity ratio so that a new temperature (point 3) is reached with a relative humidity less than 100%.

In order to avoid re-heating, which is an energy-consuming operation, dehumidification systems can bring the moist air mixture near the comfort zone instead of reaching saturation.

Humidification is the opposite of the foregoing and it is effected when it is necessary to increase the moisture content of the air.
It can be achieved by steam or by water injection. In the first case, the humidity ratio and the dry bulb temperature will increase. In the latter the temperature at the outlet will be lower than at the inlet.

Cooling by evaporation can be effected in hot dry climates. This is accomplished by spraying water into the air or by forcing contact with water. Because of the low humidity of the inlet air, part of the water evaporates, thus lowering the outlet temperature and increasing the relative humidity. If the inlet relative humidity value is small, the final humidity can be considered comfortable. No energy is added, so this phenomenon can be considered an adiabatic process (like humidification by water injection). As the constant enthalpy curves of the mixture are close to those of constant wet-bulb temperature, the process occurs at roughly constant wet-bulb temperature, as is shown by psychrometric charts.

Adiabatic mixing of two moist air streams may occur when hot and cold streams are mixed or when exhaust air or external air is added.

13.3 HEATING AND COOLING LOADS CALCULATION

The design of HVAC systems requires a detailed analysis of cooling and heating loads. The choice among different systems depends on many factors such as the building layout and occupancy, installation and maintenance requirements, capital and operating cost, energy cost, production requirements, etc..

The first step in designing a HVAC system is the calculation of heating and cooling loads.

To calculate the heating load, heat transmission losses through walls, glass, ceilings, floors and other surfaces must be considered together with infiltration losses. Required indoor and outdoor temperatures must obviously be taken into account. In addition, the energy required to warm the outdoor air to the space temperatures, must be considered. The sum of all these terms represents the total heating load.

The internal heat sources (sensible and latent heat produced by people, motors and machinery) may affect the size of the heating system and its operating modes. In industrial plants, heat sources can supply a significant portion of the heating requirements.

In cooling calculation, the heat (sensible or latent heat) which enters or is generated within a space must be calculated. It can be due to transmission (as in heating mode), to solar radiation, to outdoor infiltration and ventilation (sensible and latent heat), to people (100-400 W/person depending on the activity performed of which the latent heat represents 30-60%; the greater the physical effort, the higher the value), to other sources of sensible heat such as lighting, motors and other equipment (with values depending on their power and efficiency).

Details of these calculations can be found in specialized handbooks and in ASHRAE publications. Table 13.1 lists typical loads from occupants.

Table 13.1 Typical loads from occupants in conditioned spaces

Type of activity	Total heat W	Sensible heat W	Latent heat W
Sleeping	70	50	20
Sitting quietly	110	75	35
Sitting light work	140	80	60
Standing light work	150	80	70
Light work in factory	250	100	150
Dancing	275	100	175
Heavy work in factory	450	180	270
Sports	600	250	350

A basic simplified relationship for a preliminary evaluation of heating and cooling loads can be formulated as follows:

$$Q = (c_d + 0.35 \cdot n_v) \cdot VB \cdot \Delta t \pm Q_i$$

where

Q	=	**total load (W)**
c_d	=	**total building dispersion coefficient ($W/m^3 \cdot K$)**
n_v	=	**rate of ventilation in number of building volumes per hour (1/h)**
$0.35\ n_v$	=	**heat capacity flow-rate referred to 1 m^3/h of air flow ($W/m^3 \cdot K$); only the contribution of the sensible heat is here considered (see enthalpy of the mixture in Chapter 13.2)**
VB	=	**volume of the building (m^3)**
Δt	=	**average temperature difference between outside and inside**
Q_i	=	**internal sources contribution to be subtracted from heating consumption in cold seasons and to be added to cooling consumption in hot seasons (W). In cooling mode, the internal contribution from lighting and occupants reaches 10-20% of the total load. Higher values can be found in particular situations.**

Values of the previous coefficients depend on the insulation of the building (see also Chapter 8), on the heat capacity (specific heat multiplied by the mass) of the part of the building structures which is influenced by the indoor temperature, on the climate and on the ratio between the total dispersion surface (ground floor, ceiling, external surfaces with their windows) and the building volume.

The colder the average external temperature, the lower must be the total building coefficient c_d (typical values range from 0.35 to 1 $W/m^3 \cdot K$); the higher the dispersion surface/volume ratio the higher the c_d coefficient.

The n_v coefficient generally does not exceed unity if referred to the building gross volume, so that average values of (c_d + 0.35 n_v) ranging from 0.7 to 1.3 $W/m^3 \cdot K$ can be assumed for the total load calculation. Notice that for particular situations, the n_v coefficient can be even higher than 10.

Latent heat can be introduced in the simplified relationship (Q) by using the difference between the total enthalpy of the outside and inside air-water mixture (see Chapter 13.2).

13.4 TYPICAL HVAC SYSTEMS

In HVAC systems the heating mode is generally operated by means of boilers and heat pumps; occasionally, depending on local conditions, electrically heated systems can be used.

HVAC systems for cooling are basically refrigeration systems and evaporation systems.

Refrigeration systems use compressors or absorption systems (see Chapter 12). The refrigerant evaporates through an evaporator which can be an air/refrigerant or water/refrigerant exchanger. If the exchanger is placed in the room to be cooled, air is cooled directly when it moves across the coils. Condensation can be by water (in large units, with cooling towers) or by air (in smaller units).

In evaporation HVAC systems, only pumps to move water inside the circuit and to force contact with air are required.

Heat pumps, as individual units or centralized plants, provide both heating and cooling (see Chapter 12).

Auxiliary equipment comprises pumps and fans, and control systems.

HVAC systems can be classified as (i) all-air systems; (ii) all-water systems; (iii) air-water systems; (iv) individual units. Class (i) serve the conditioned space primarily by means of a network of ducts. Class (ii) are based on a piping network which delivers energy as water to distributed units, typically fan-coil units. Class (iii) are a combination of the previous systems. Class (iv) are based on independent equipment.

All-air systems

All-air systems may be classified as follows: variable temperature and constant volume; constant temperature and variable volume; variable volume and variable temperature.

Variable temperature and constant volume systems include single zone systems, multizone with centralized blending, re-heating in each zone, dual-duct with mixing boxes in each zone.

Constant temperature and variable volume systems are mainly dual-duct systems, one duct with ventilation air in neutral conditions equal to those of the space to be conditioned and the other for cooling and heating.

Variable volume and variable temperature systems are mainly dual-duct with mixing boxes. Many combinations of the previous systems are possible.

Notice that in the re-heating mode, chilled air is supplied to each zone (for cooling mode) and then the air is re-heated according to the temperature required for each zone. This system has high energy consumption because of the duplication of energy use. Variable air volume, which requires a constant supply air temperature, is based on valve control or on variable-speed fans control (by means of a local thermostat) of the air flow supplied to each zone, to ensure the supply only of the air needed by the real load of each zone. In this way energy consumption is reduced.

All-water systems

These systems use fan coils or unit ventilators, either of the under-the-window type or fixed at the ceiling, fed by piping network. For heating only, radiators are generally installed. In this case, heat release takes place through both convection (80-60%) and radiation (20-40%).

Air-water systems

In mixed systems, air and water, the energy supplied to each room is shared between water systems, such as radiators, fan-coils, radiant systems (for floor or ceiling installation fed by steam or superheated water) or induction units, and the air supply system. A variety of combinations exists.

The water system can supply most of the heating demand of the rooms while the main function of the air system is to meet the air renewal and quality requirements and to control humidity.

In induction systems, centrally conditioned primary air is supplied to the unit plenum at high pressure, which induces secondary air from the room to flow over the secondary coils where it is heated or cooled as the zone may demand. So-called induction units are used in these systems.

Individual units

These are package units for cooling. The condenser is generally air-cooled and expels heat from the room. A direct expansion cooling coil is usually installed. These units allow only a temperature control.

Radiant systems for heating

Radiant energy can be used for heating by means of radiant tubes where hot fluids or hot gases flow. The temperature of hot gases is roughly 300-400°C (572-752°F). Exhausts from processes can conveniently be used.

This system can be economical if heating must be concentrated in defined areas inside the buildings, without heating the whole surface and volume. It avoids the air stratification which occurs in high-ceilinged constructions.

Typical operating consumption of hot gases radiant systems for industrial applications are: 0.02 Sm^3/h of natural gas per square meter of surface heated (175 W/m^2; 17.5 W/m^3 if the height of the building is 10 m).

13.5 COMPUTERCONTROL

Computer control of HVAC systems allows significant savings of energy and thus of cost.

Control strategies for HVAC systems include many features such as: night temperature setback for heating mode, night cooldown with outside air for cooling mode, outside and inside temperature control, minimization of outside air during heating, reduction of excessive ventilation, use of outside air for cooling when possible (free cooling in cooling mode), control of humidity, implementation of cycles with night shutdown and morning warmup.

Computer control systems are based on the acquisition of a great number of parameters (many hundreds or thousands of values of temperature, humidity, pressure) in all parts of the buildings and outside, and on the modelling of the buildings to make the proper control of each installation possible. Of course, off-line and on-line modelling programs are available to improve the design and the operation of HVAC systems.

In addition to the optimum control of temperature and humidity in each zone of the buildings, which is based on the availability of sensors and transducers to regulate the temperature of the air or water supplied to the single zones, significant energy savings can be achieved by simply using external air or by circulating exhaust air whenever possible.

The rate of heat recovery can be controlled on the basis of the dry-bulb temperature of the outside air or of its enthalpy value. The latter approach performs better because it measures the heat content of the outside air and it avoids errors due to the difference between dry- and wet-bulb temperatures; on the other hand, it requires more maintenance.

13.6 ENERGY-SAVING SYSTEMS

Energy saving in HVAC systems is primarily based on monitoring the existing equipment and on operating single components as efficiently as possible. Correct ventilation (an outside air supply plus any recirculated air that has been treated to maintain the desired quality of air within a designated space) may allow significant energy saving (electric energy for fans, heating and cooling energy and additional treatment such as humidification and dehumidification can be reduced).

Reducing loads, by reducing losses from buildings and the negative contributions to cooling operation from machines and lighting, must always be a goal but the retrofitting needed is quite expensive.

Heat recovery from equipment and other sources such as exhaust air from buildings must always be practised when possible.

Energy storage capability, such as chilled water or ice storage for cooling (see Chapter 4.7) and specially designed buildings with high thermal capacity structures (see Chapter 13.3), will improve the efficiency of the system by lowering the demand profile. Storage also makes it possible to shift energy consumption from peak hours to low-rate hours, thus increasing the energy cost saving.

When the ventilation rate is high, systems which allow heat recovery from exhaust air can be installed (see the example in Chapter 13.7).

Economizer systems are used to mix external air with return air (for instance, in cooling mode return air at 24°C (75°F) can be mixed with cold outside air to maintain input air at 10-15°C (50-60°F) in a cooled space. If the outside temperature is around the preset value, 100% outside air can be used. Thus the system works in a so-called free cooling mode. In hot seasons, the integration of outside air is kept as low as possible.

13.7 PRACTICAL EXAMPLES

Example 1
Heat recovery from exhaust air leaving buildings and inlet air entering buildings

Industrial laboratories, such as those in pharmaceutical and chemical factories, generally require high ventilation rates: 10-15 volumes of external air which enters the building and must be heated or cooled to maintain the required temperature and relative humidity.

This example illustrates the energy saving by a group of exchangers (air to air with a water-glycol solution as intermediate fluid) which recovers the energy content of the exhaust outlet air in the cold season and of inlet air in the hot season.

The flow of the air without and with recovery equipment is represented in Fig. 13.5 and in Fig. 13.6. All heating is effected by boiler-fed steam/air exchangers. If recovery exchangers are installed, the system is modified as in the figures. There are two operation modes:

— Cold season mode (see Fig. 13.5)

Exchangers are installed to recover sensible and sometimes also latent heat from the exhaust outlet air. The heat is released to a water-glycol solution, which circulates by means of pumps, and exchanges heat with the inlet air which, otherwise, would have to be heated by boiler-fed steam/air exchangers.

— Hot season mode (see Fig. 13.6)
The inlet air passes through a cooling dehumidification section, followed by a re-heating section where boiler-fed steam/air exchangers are installed. The recovery system extracts heat from the total inlet air and transfers it to a part (40%) of the air leaving the dehumidification section, thus reducing the energy consumption for re-heating the latter. The total inlet air is cooled; thus, the consumption of the compressor refrigerating plant is reduced.

— Intermediate seasons mode
No recovery occurs and all heat is supplied by boiler-fed steam/air exchangers.

The energy saving in cold season and hot season modes is calculated by considering only (i) the energy which the HVAC system (boiler plant and compressor refrigerating plant) would have required without recovered heat and (ii) the additional consumption due to the new system. The calculations are synthesized in Table 13.2.

Economic evaluation is shown in Table 20.3.

Example 2
Computer control of HVAC systems

The computer control of HVAC systems can be applied to industrial, commercial and tertiary sector buildings in order to guarantee comfortable conditions with the minimum energy consumption.

All kinds of software and hardware are on the market; any particular example would require many details and data on the site concerned and on the hardware and software chosen.

As a general rule, a saving of 15% on the previous energy consumption for HVAC can be assumed as an average value.

A building with 45000 m^2 of floor and 100000 m^3 of space to be heated and cooled is considered. The consumption is 600 TOE for heating in the cold season

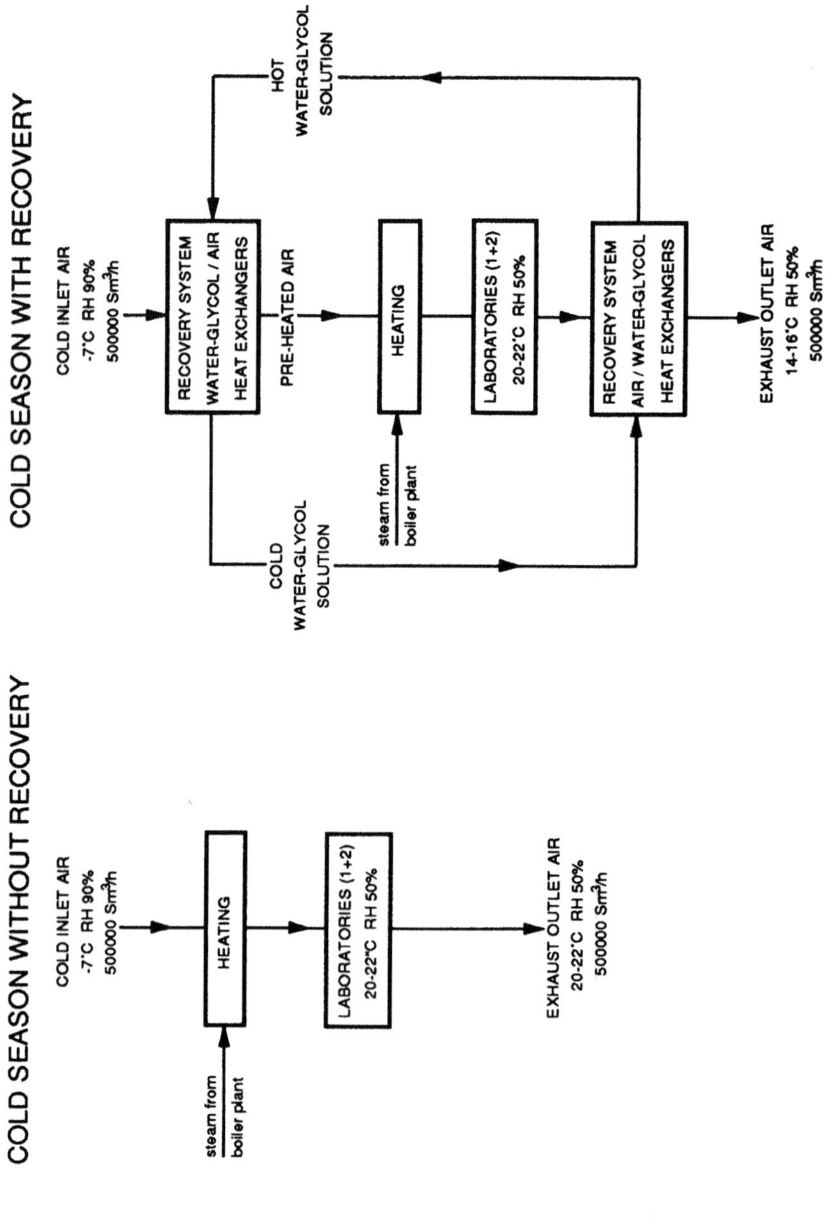

Fig. 13.5 Flow of the air without and with recovery equipment in the cold season

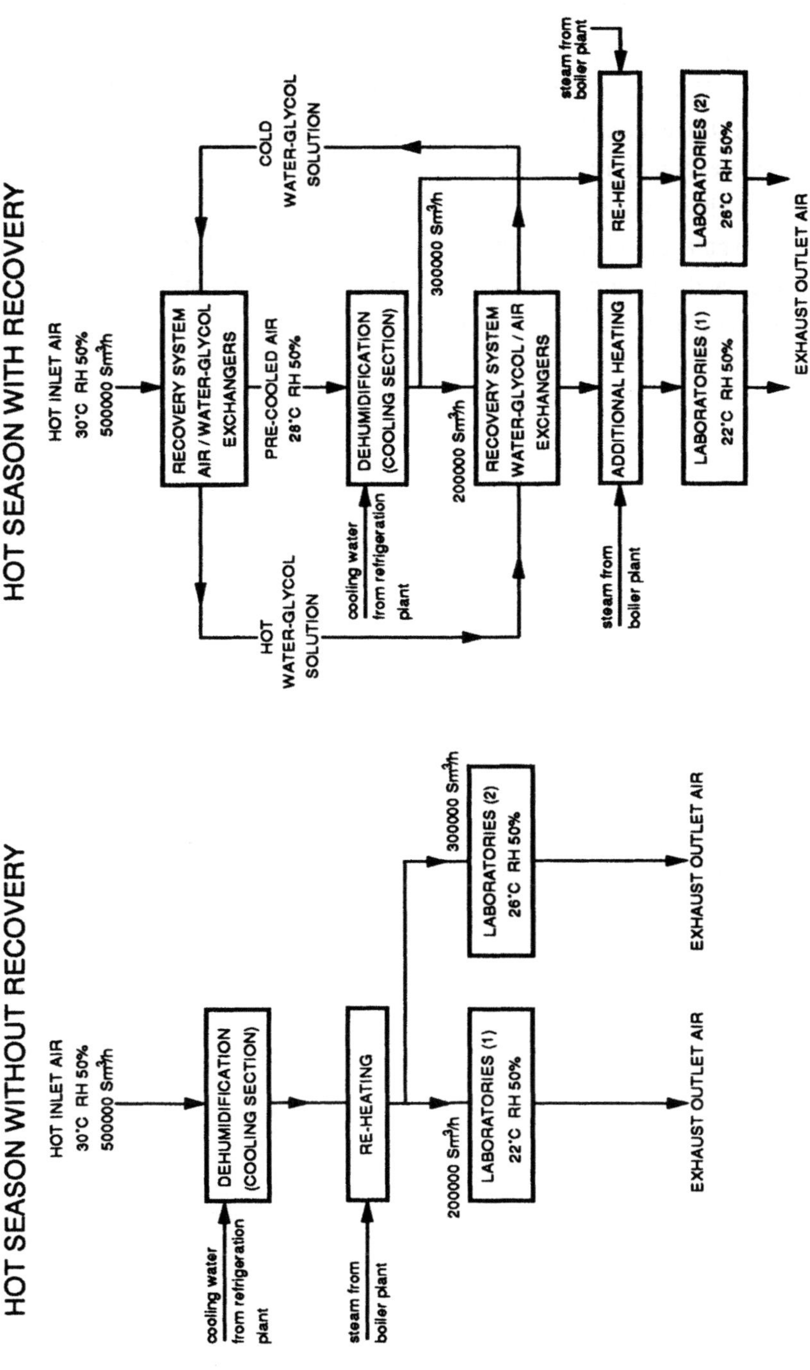

Fig. 13.6 Flow of the air without and with recovery equipment in the hot season

(2500 h/year) and 1600 MWh for cooling in the hot season (2000 h/year). The energy saving associated with the installation of a computer control system, based on 750-1250 control points inside the building, is then estimated to be equal to 90 TOE/year and 240 MWh/year.

Economic evaluation is shown in Table 20.3 where only the contribution of the energy saving is considered.

Other advantages, such as comfort and reduction of working hours for plant maintenance and operation will certainly increase the economic advantages of the control system (see also Chapter 17.10).

Table 13.2 Energy saving by heat recovery from exhaust and inlet air

	COLD SEASON MODE (see Fig. 13.5)		Recovery system	Equivalent fuel-fed boiler
a	exhaust air flow-rate RH 50% 22°C	Sm^3/h	500000	
b = a/0.848	dry air mass flow rate	kg/h	589623	
c = b · 0.00825	moisture flow rate	kg/h	4864	
d	outlet-inlet temperature increase	°C	6	
e = d · (1+0.00825 · 1.8)	enthalpy increase per kg dry air	kJ/kg	6.1	
f = e · b	heat	kJ/h	3590271	3590271
f' = f/3600		kW	997	997
f'' = f/1.055		Btu/h	3403101	3403101
g	boiler efficiency			0.85
h = f/g	boiler consumption (natural gas)	kJ/h		4223849
h' = h/3600		kW		1173
h'' = h/1.055		Btu/h		4003648
i = h/34325		Sm^3/h		123.1
l	additional electrictricity consumption	kW	20	0
m	working hours	h/year	3750	3750
	Energy consumption			
n = l · m	electric energy	kWh/year	75000	0
o = i · m	thermal energy	$Sm^3/year$	0	461454
	HOT SEASON MODE (see Fig. 13.6)		Recovery system	Equivalent fuel-fed boiler
A	after dehumidification RH 100% 10°C	Sm^3/h	200000	
B = A/0.815	dry air mass flow-rate	kg/h	245399	
C = B · 0.0076	moisture flow rate	kg/h	1865	
D	temperature increase	°C	5	
E = D(1+0.0076 · 1.8)	enthalpy increase per kg dry air	kJ/kg	5,07	
F = E · B	heat	kJ/h	1243779	1243779
F' = F/3600		kW	345	345
F'' = F/1.055		Btu/h	1178938	1178938
G	boiler efficiency			0.85
H = F/G	boiler consumption (natural gas)	kJ/h		1463270
H' = H/3600		kW		406
H'' = H/1.055		Btu/h		1386985
I = H/34325		Sm^3/h		42.6
L	additional electrictricity consumption	kW	20	0

	HOT SEASON MODE (continued)		Recovery system	Equivalent refrigeration compressor
M	inlet air flow rate RH 50% 30°C	Sm^3/h	500000	
N = M/0.87	dry air mass flow rate	kg/h	574713	
O = N · 0.0133	moisture flow rate	kg/h	7644	
P	outlet-inlet temperature drop	°C	2	
Q = P · (1+0.013 · 1.8)	enthalpy drop(c) per kg dry air	kJ/kg	2.047	2.047
R = Q · N	heat extracted	kJ/h	1176943	1176943
R' = R/3600		kW	327	327
R = R/1.055		Btu/h	1114997	1114997
S	refrigeration plant COP			4.5
T = R'/S	electric energy consumption	kW		72.6
V	additional electicity consumption	kW	10	
Z	working hours	h/year	1250	1250
	Energy consumption			
X = Z · (L+T+V)	electric energy	kWh/year	37500	90766
Y = Z · I	thermal energy	Sm^3/year	0	53287
	COLD + HOT SEASON MODES		Recovery system	Equivalent boiler+compressor
	Energy consumption		(I)	(II)
EE = X + n	electric energy	kWh/year	112500	90766
ET = Y + o	thermal energy	Sm^3/year	0	514742
	Energy saving (II)-(I)			
SE = EE(II)-EE(I)	electric energy	kWh/year	−21734	
ST = ET(II)-ET(I)	thermal energy	Sm^3/year	514742	
ST' = ST · 34.325/41860		TOE/year	422	

COLD SEASON MODE (Design conditions)	HOT SEASON MODE (design conditions)
Outdoor −7°C, 19.4°F, relative humidity 90%	Outdoor 32°C, 89.6°F, relative humidity 50%
Inlet 22°C, 71.6°F, relative humidity 50%	Inlet 22-26°C, 71.6-78.8°F, relative humidity 50%
COLD SEASON MODE (Average conditions)	HOT SEASON MODE (average conditions)
Outdoor 2°C, 35.6°F, relative humidity 90%	Outdoor 30°C, 86°F, relative humidity 50%
Inlet 22°C, 71.6°F, relative humidity 50%	Inlet 22-26°C, 71.6-78.8°F, relative humidity 50%

Enthalpy of dry air (kJ/kg) = t + humidity ratio · (2500+1.8 · t) where t (°C)
enthalpy increase (e) per kg dry air = 6 · (1 + 0.00825 · 1.8)
enthalpy increase (E) per kg dry air = 5 · (1 + 0.0076 · 1.8)
enthalpy drop (Q) per kg dry air = 2 · (1 + 0.013 · 1.8)
net heating value of natural gas equal to 34325 kJ/Sm^3

CHAPTER 14

FACILITIES—LIGHTING

14.1 INTRODUCTION

Each building, each space in a factory has its own characteristics and should be surveyed carefully in order to ensure the optimum quality and quantity of lighting. Depending on the kind of activity, energy consumption for lighting may be a significant part of the total energy consumption, as in supermarkets and commercial buildings (up to 20-30%). In factories, it is a very small percentage of the total energy consumption.

Energy can be saved both by a proper choice of lighting systems for new installations or when retrofitting and by a control of the energy flow (intensity control, switching on/off).

14.2 BASIC PRINCIPLES OF LIGHTING SYSTEMS AND DEFINITIONS

The basic concepts and definitions regarding lighting are set out briefly below.

Luminous intensity of a lamp expressed in candela (cd; see Table 2.1).

Lumen is the unit of luminous flux output from a lamp. The luminous flux is expressed in lumen (lm) = cd · sr (luminous intensity in candela multiplied by the solid angle in steradian).

Luminance is the amount of light flux per unit of area reflected from or transmitted through a surface. It is expressed in lux (lx) = lm/m^2 (or footcandle if a surface of 1 square foot is assumed).

Lamp efficiency is the ratio between the luminous flux from a lamp (lumen) and the electrical power consumed (watt): it is expressed as lm/W.

Reflectance is the ratio between the light reflected by a surface and the light incident on it.

Chromaticity or color temperature is an index of the light color that includes consideration of its dominant wavelength and purity. It is expressed in Kelvin temperature. The higher the temperature, the cooler the light (typical reference values are: 4100 K for the moon; 5000 K for the sun; 10000-25000 K for blue sky). Chromaticity is defined as the temperature of a black body which radiates light with the same color as the light source.

Color rendering or color rendition is an index of the effect of a light source on the color appearance of objects compared with color appearance under a reference light source (value of color rendering for the reference light source equals 100).

Space distribution diagrams take into account that light emitted from a lamp or a luminaire may vary in luminous intensity depending on the direction. Diagrams are generally made on a plane section and report candle values referred to a flux of 1000 lm to allow comparison among different lighting systems.

Luminaire (or fixture) is defined as a complete lighting unit consisting of a lamp or lamps and components designed to distribute the light (diffuser, reflector, lens and others) and to connect the lamp to the power supply.

Coefficient of Utilization (CU) or luminaire efficiency is the ratio of the luminous flux (lumen) reaching the working plane to the total flux (lumen) generated by the lamp. This coefficient takes into account light absorbed or reflected by walls, ceilings and luminaries. Its values are generally found in charts in the manufacturer's catalogue and permit a rapid estimation of the illumination that can be achieved with a given system.

Lamp and luminaire depreciation factor. This factor relates the initial illumination provided by a clean, new lamp and luminaire to the reduced illumination that they will provide at a particular point in time because of accumulated dirt, and depreciation (see Fig. 14.1).

Task surface is the working area where the recommended levels of illumination specified by international associations should be applied. For the surrounding area the level of luminance should be no more than 1/3 of the average level for the task surface. Table 14.1 shows typical values of luminance for different industrial task surfaces; it ranges from 2000 lx for high precision operations to 50 lx for transit areas.

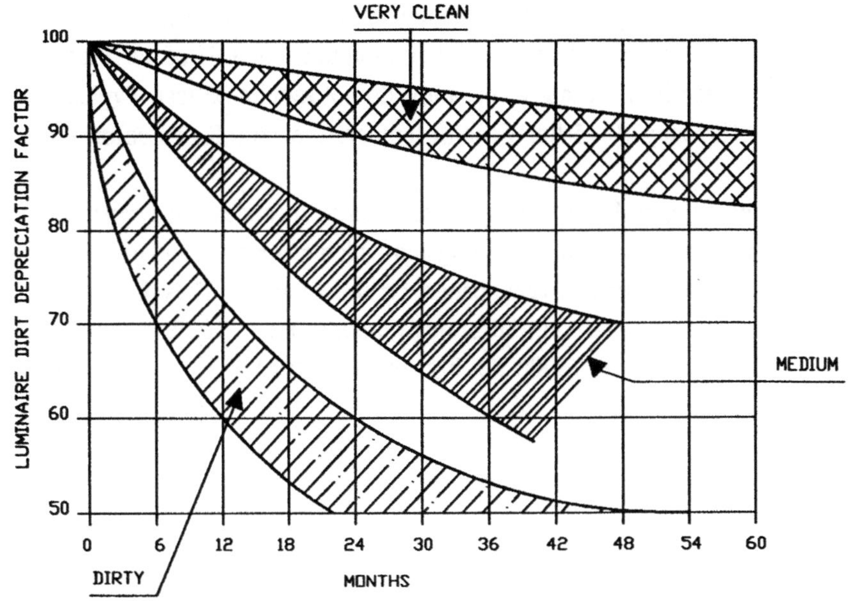

Fig. 14.1 Example of curves of depreciation factor (ratio between the illumination provided at various points in time and that provided by the new system. It is expressed as a percentage)

14.3 TYPICAL LIGHTING SYSTEMS AND RELATED EQUIPMENT

A lighting system may have many components: ballast, lamp, diffuser, reflector, lens, control equipment.

Systems are generally classified according to lamp categories, as follows: incandescent, fluorescent, high-intensity discharge such as mercury vapor, metal halide vapor, sodium vapor (high pressure and low pressure).

Incandescent lamps

Incandescent lamps consist of a resistive filament inside an evacuated glass bulb containing a trace of an inert gas. Light is produced by passing an electric current through the filament. In a typical incandescent lamp about 5% of the energy is converted in the visible spectrum, while the balance is radiated as infrared energy and heat. These lamps have a relatively short life (less than 2000 h) because the filament gradually evaporates, reducing its diameter and increasing its resistance. This also reduces the light output to about 80% of the rated lumens at the end of the life. Incandescent lamps do not require ballast. Typical values are: lamp efficiency 10-15 lm/W, color temperature 2800 K and color rendering 100.

Tungsten halogen lamps are also classified as incandescent lamps; the main operating characteristic is that the tungsten which evaporates and deposits on the bulb walls is removed and redeposited on the filament. They have many advantages such as maintaining the lumen output close to 100% throughout life, higher color temperature (3000 K), higher lamp efficacy 20-25 lm/W and life-span equal to 3000-4000 h.

Fluorescent lamps

Fluorescent lamps require a ballast which provides an initial high voltage pulse to initiate the discharge through a conducting vapor or gas (mercury, argon, krypton, other). The ballast also limits current flow through the lamp. The ballast can be basically an inductor or a high-frequency electronic device. The latter has the advantages of (i) reducing the power consumption of the lighting system by roughly 30% and thus improving the luminaire efficiency, if used with fluorescent lamps designed to work at high frequency, tens of kHz instead of 50-60 Hz, and (ii) dimming from 25% to 100%.

The efficiency of the lamp increases with the arc length; this explains why fluorescent lamps are generally tubular. Typical values are 90-100 lm/W. The life-span is determined by the rate of loss of the electron-emitting material and it is also influenced by the number of times the lamps are switched on (typical life-span is 7000-12000 h instead of 1000-2000 h for incandescent lamps). The lamp depreciation factor reaches 90% after 1000 hours and then slowly decreases to 85% at the end of the life-span. The spectrum of the emitted light is modified by the phosphor coating on the inside of the tube.

Notice that roughly 20% of the input energy is converted to light and 40% to infrared, while 40% is dissipated as heat.

A choice of chromaticity and color rendering (generally 80-90) can be made depending on the phosphor mix of the lamp.

High-intensity discharge lamps

High-intensity discharge lamps include mercury vapor, metal halide and high-pressure sodium lamps. They work in the same way as fluorescent lamps and require starting devices (ballasts).

Mercury vapor lamps provide good color rendition (50-60), color temperature of 3300-4300 K and good efficiency (35-60 lm/W). When the arc is extinguished, it cannot be switched on again until the vapor pressure is lowered to a point suitable for the applied voltage (3-7 min). They have a long life-span (up to 10000 h) and a depreciation factor of more than 75%.

Metal halide lamps have better efficiency (65-85 lm/W), better color rendition (65) and a color temperature of 4500 K, thanks to the inclusion of metal additives. The life-span is generally 6000-7000 h with a depreciation factor of more than 75%.

The high-pressure sodium is the most efficient high-intensity discharge lamp. Its golden-white color quality is a limiting factor for interior use, but improvement in color rendition (from 20 to 80) has made it acceptable also for interiors. The color temperature is around 2000 K and the efficacy is very high, up to 130 lm/W. The life-span is generally 10000-15000 h.

Low-pressure sodium lamps operate as discharge lamps. They need a special ballast. They have the highest efficiency up to 180 lm/W and higher but their monochromatic yellow light limits their use. The life-span is generally 10000-15000 h.

For all lamps, except for the incandescent lamps, the power factor is generally less than 0.5; to improve this value, capacitors (see Chapter 7.3) must be provided.

Tables 14.2, 14.3, 14.4, 14.5 show typical values of the main parameters.

Reflectance varies between 90% and 30%; chromaticity between 2500 K and 3000 K; color rendering between 50 and 90.
As a general rule, with a required task surface luminance ranging between 1000 lx and 200 lx and a lamp efficiency ranging from 50 to 100 lm/W, values of 20-10 W/m^2 are typical. Of course, higher values can be found if incandescent lamps with less than 20 lm/W are used or if high levels of illumination are recommended for particular task surfaces.

Table 14.1 Recommended task surface luminance

Task surface	Recommended luminance	
	lx	fc
Very difficult task	1000-4000	93-372
Commercial surfaces	750-1000	70-93
Material test, drawing, sewing, etc.	750-1500	70-140
Office room	250-750	23-70
Transit area	50-150	5-15

Table 14.2 Average values of reflectance coefficients for different materials and surfaces

Surface	Reflectance range %
Silver	92
Aluminium	70-60
Inox iron	65-55
White paint	90-70
White paper	80-70
White wall	90-70
Wood	30-20
Floor	60-20

Table 14.3 Chromaticity of light sources

Source	Color temperature K
Daylight with blue sky	15000-25000
Overcast sky	5000-15000
Sunlight at noon	5000
Early morning sunlight	4500
Sunlight at sunrise	1700
Incandescent lamps	2500-3500
Fluorescent lamps	3000-6500
High-pressure sodium	2000-2500
High-pressure metal halide	3500-4500

Table 14.4 Suggested color rendering for typical applications

Color rendering	Suggested applications
≥ 90	High-precision tasks, museum
80-90 WARM TONES	House, hospital, offices, school
80-90 COOL TONES	Textile, printing, mechanical engineering
60-80	Manufacturing industries
40-60	Heavy industries
20-40	In cases without particular requirements

Table 14.5 Typical lamp output characteristics

Type		Incandescent	Incandescent halogen	Fluorescent	Fluorescent high-frequency	Mercury vapor	Metal halide	High-pressure sodium	Low-pressure sodium
OUTPUT CHARACTERISTICS									
Power (only lamp)	W	100	500	36	32	250	400	400	180
Lamp lumen	lm	1500	12500	2600	3200	14000	32000	48000	33000
Lamp efficiency	lm/W	15	25	72	100	56	80	120	183
Power (only ballast)	W	0	0	10	4	16	25	35	30
Lamp+ballast efficiency	lm/W	15	25	57	89	52	75	110	157
Typical CU coefficient	%	70	70	70	70	40	40	40	40
Overall efficiency	lm/W	11	18	40	62	21	30	44	63
Life span	h	1000	3000	7500	12500	10000	10000	10000	10000

Luminaires

With all kinds of lamp, luminaires play an important role in lighting systems.

The efficiency of luminaires is highly dependent on the reflectance of their interior surfaces (reflectors with silver films may reflect more than 90% of the incident light), and exterior surfaces such as lenses (with polarizing materials and deflectors). Typical values range between 60% and 75%

Luminaire dirt depreciation is a major factor in maintaining good lighting levels at low energy cost. Typical depreciation factor curves are illustrated in Fig. 14.1 and they are given by manufacturers. A good maintenance program may yield significant energy saving. In addition, other factors of depreciation such as room surface and furniture dirt depreciation, and aging and discoloration of luminaire shield materials must be considered carefully in a lighting conservation program.

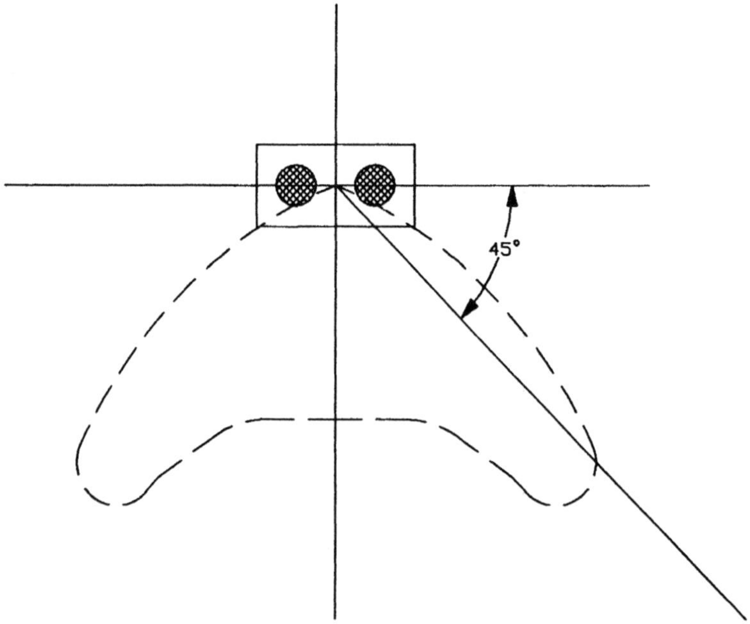

Fig. 14.2 Example of distribution diagram

Luminaires must also avoid uncomfortably bright light (commonly called glare effect) on the task surface. Generally, light emitted within a cone with a 90° angle and with the axis along the vertical line from the luminaire to the floor does not contribute to direct glare. Inside the same cone loss of contrast on the task surface may occur; on the contrary, this effect is minimal outside the cone.

Shielding materials which direct light at particular angles can conveniently be used; distribution such as batwing distribution as in Fig. 14.2 where a high percentage of light is emitted near 45° allows a good compromise among different requirements if the work stations are placed between rows of luminaires.

Lenses with polarizing materials and louvers (series of baffles generally arranged in a geometric pattern) are also used.

Lighting control

The best way to save energy in lighting is to ensure that lights are turned on only when they are needed and that lights are dimmed during periods when lower levels of lighting are required.

Single program clocks, switches, photoelectric cells and dimmers as well as a centralized computer controlled switching system can be introduced depending on the particular situation and working requirements.

Generally, the following rules must be remembered. Light must be turned on at the start of the business day, in specific areas for a watchman's tour of duty, when the burglar alarm is activated and at reduced intensity for cleaning crews. Light must be turned off at the close of the business day, on completion of cleaning, and in predetermined external sunlight conditions.

For more efficient control, each individual working area needs its own switch and large open areas need separate switches for different task zones. All these suggestions must be taken into account in designing new installations; for retrofitting, each case must be considered carefully because the cost of wiring might be too high.

Solid-state dimmers to vary the level between fully on and off give great flexibility which can be valuable if the use of surfaces changes frequently. They reduce the voltage supplied to the lamp from a constant voltage source and they save energy by reducing the light level to the right value for each situation.

Incandescent lamps can be totally dimmed from 100% to zero. Fluorescent lamps and high-intensity discharge lamps cannot be dimmed totally because of their operation mode (at low voltage they cannot remain alight). Specially designed high-frequency fluorescent lamps can be regulated from 25% to 100%.

14.4 HOW TO CHOOSE A LIGHTING SYSTEM

The task-surface level must be created in accordance with local and international recommendations. Efficient lamps with high lumen, high efficacy (lm/W) and good lamp lumen depreciation should be used. Color temperature and color rendition must be checked in view of the kind of activity performed.

The choice of luminaires should be based on a high coefficient of utilization (CU) for the specific application and on the avoidance of depreciation due to dirt in the environment.

A preliminary evaluation of the lighting system can be made as follows:

a) the luminance (lx = lm/m^2) required at the task surface multiplied by the area (m^2) to be lighted is the total luminous flux required (lm);

b) the lumen output for a single lamp multiplied by the Coefficient of Utilization of the luminaire (CU) gives the lumen output that will reach the task surface. The CU coefficient can be found in manufacturers' charts where lamp and luminaire characteristics are taken into account together with the parameters of the surroundings (room dimensions, furniture, etc.). Typical values range between 35% and 60%;

c) the lumen output as at point b) must be reduced by introducing the lamp and luminaire depreciation factors as given by the manufacturers (after a reference period of life of the lighting system);

d) the ratio between the value of the total luminous flux (lm) required as at point a) and the value of lumen per lamp as at step c) gives a preliminary evaluation of the number of lamps and luminaires required and then of the cost of the lighting system;
e) the electric power required is then equal to the product of the single lamp (or lamp with ballast if necessary) power multiplied by the number of lamps as at step d). The energy consumption is thus calculated if the working hours per year are known.

A more detailed approach exploiting knowledge of the lamp and luminaire space distribution diagram will place luminaires to ensure the task-surface level in all the working area, while providing lower lighting levels in surrounding zones. In this way, capital and operating costs are reduced while the illumination requirements are satisfied.

As with any other investment in energy saving, economic analysis must take into account many aspects of the operating (energy and other) and capital costs.

A work-sheet showing the main technical parameters together with capital and operating costs is given in Tab. 14.6 as a suggested guideline for evaluation.

14.5 ENERGY-SAVING OPPORTUNITIES

The foregoing considerations show that the energy-saving opportunities in lighting systems are:

— a proper selection of lamps and luminaires to provide the task surface with the maximum efficiency (lm/W). This opportunity can be exploited both in new installation and in retrofitting;
— a planned maintenance program of replacing lamps, cleaning luminaires, replacing defective components and cleaning surrounding surfaces in order to achieve the maximum exploitation at any time;
— a periodical check-up of the task surface;
— a control of lighting at a degree of complexity determined by the incidence of lighting costs on the energy bill. This control system may interact with other building or factory subsystems, in particular with the HVAC, because of the heat that lamps dissipate in surrounding areas.

Table 14.6 Work-sheet for energy-saving evaluation in lighting

			Reference	New system
	TECHNICAL DATA			
a	Rated initial lumen per lamp	lm		
b	Input electric power per lamp	W		
$c = a/b$	Lamp efficiency	lm/W		
d	Input electric power per lamp+ballast	W		
$e = a/d$	Lamp+ballast efficiency	lm/W		
f	Coefficient of utilization CU	%		
g	Lamp depreciation factor	%		
h	Dirt depreciation factor	%		
$i = e \cdot f \cdot g \cdot h/10^6$	Effective lumen per watt	lm/W		
$l = a \cdot f \cdot g \cdot h/10^6$	Effective lumen per lamp	lm		
m	Average lux required	lx		
n	Surface to be lighted	m^2		
$o = m \cdot n$	Total lumen required	lm		
$p = o/l$	Number of lamps	lamps		
$q = o/i$	Total electric input power	W		
	CAPITAL COST			
r	Cost of one luminaire	U		
s	Auxiliary equipment cost	U		
t	Installation cost	U		
$u = r + s + t$	Total capital cost	U		
	OPERATING COST			
v	Cost of replaced lamps	U/year		
x	Cost for replacing	U/year		
$y = v + x$	Total annual operating cost	U/year		
	ENERGY SAVING			
w	Working hours	h/year		
$EC = w \cdot q/1000$	Annual energy consumption	kWh/year		
$ES = EC(REF) - EC(NEW)$	Annual energy saving	kWh/year		
CA	Electric energy cost	U/kWh		
$CS = ES \cdot CA$	Annual energy cost saving	U/year		

14.6 PRACTICAL EXAMPLES

Example 1

Replacement of standard fluorescent lamps by high frequency systems (HF fluorescent lamps and electronic ballast)

In a shopping center with a shopping surface of 5000 m^2 replace standard fluorescent lamps with HF lamps.

The lighting system existing in the shopping center has 1200 standard F96T12 lamps (input power 75 W per lamp) with two-lamp luminaires and standard ballasts of 10 W per lamp. The input power for each luminaire is 170 W. An average luminance of 600 lx has been measured.

The new lighting system includes two-lamp luminaires with HF 50 W input power per lamp and ballast input power of 6 W per lamp. The initial lamp efficiency is 100 lm/W; the lamp+ballast efficiency is 89.3 lm/W.

The details of the energy-saved evaluation are shown in Table 14.7 which is derived from the work-sheet of Table 14.6.

The annual energy saving is equal to 249321 kWh.

See Table 20.3 for the economic evaluation.

Example 2

Comparison between mercury lamps and high-pressure sodium lamps

A 4000 m^2 working area in a factory, where a new lighting system has to be installed, requires a task surface luminance of 450 lx.

Make a comparison between mercury lamps (lamp efficiency 56 lm/W) and high-pressure sodium lamps (lamp efficiency 93 lm/W), with the same rated initial lumen per each lamp (14000 lm). The sodium lamps can be used if golden-white color is accepted.

Technical data and elaboration are given in Table 14.8, which is derived from the work-sheet of Table 14.6.

The annual energy saving is equal to 178219 kWh.

Economic evaluation is in Table 20.3.

Table 14.7 Replacement of standard fluorescent lamps with HF lamps (Example 1)

			Reference	New system
	TECHNICAL DATA			
a	Rated initial lumen per lamp	lm		5000
b	Input electric power per lamp	W	(*)75	50
c = a/b	Lamp efficiency	lm/W		100
d	Input electric power per lamp+ballast	W	(*)85	56
e = a/d	Lamp+ballast efficiency	lm/W		89.29
f	Coefficient of utilization CU	%		60
g	Lamp depreciation factor	%		90
h	Dirt depreciation factor	%		90
$i = e \cdot f \cdot g \cdot h/10^6$	Effective lumen per watt	lm/W		43.39
$l = a \cdot f \cdot g \cdot h/10^6$	Effective lumen per lamp	lm		2430
m	Average lux required	lx	(*)600	600
n	Surface to be lighted	m^2		5000
$o = m \cdot n$	Total lumen required	lm		3000000
p = o/l	Number of lamps	lamps	(*)1400	1235
q = o/i	Total electric input power	W	(*)119000	69136
	CAPITAL COST			
r	Cost of one luminaire	U		
s	Auxiliary equipment cost	U		
t	Installation cost	U		
u = r + s + t	Total capital cost	U		
	OPERATING COST			
v	Cost of replaced lamps	U/year		
x	Cost for replacing	U/year		
y = v + x	Total annual operating cost	U/year		
	ENERGY SAVING			
w	Working hours	h/year	5000	5000
$EC = w \cdot q/1000$	Annual energy consumption	kWh/year	595000	345679
ES = EC(REF) – EC(NEW)	Annual energy saving	kWh/year		249321
CA	Electric energy cost	U/kWh		
$CS = ES \cdot CA$	Annual energy cost saving	U/year		

Note
(*) INPUT DATA for the existing system (REFERENCE)

Table 14.8 Comparison between mercury and high-pressure sodium lamps (Example 2)

			Reference (Mercury)	New system (HP sodium)
	TECHNICAL DATA			
a	Rated initial lumen per lamp	lm	14000	14000
b	Input electric power per lamp	W	250	150
$c = a/b$	Lamp efficiency	lm/W	56	93
d	Input electric power per lamp+ballast	W	266	170
$e = a/d$	Lamp+ballast efficiency	lm/W	52.63	82.35
f	Coefficient of utilization CU	%	60	62
g	Lamp depreciation factor	%	80	90
h	Dirt depreciation factor	%	90	90
$i = e \cdot f \cdot g \cdot h/10^6$	Effective lumen per watt	lm/W	22.7	41.3
$l = a \cdot f \cdot g \cdot h/10^6$	Effective lumen per lamp	lm	6048	7030.8
m	Average lux required	lx	450	450
n	Surface to be lighted	m^2	4000	4000
$o = m \cdot n$	Total lumen required	lm	1800000	1800000
$p = o/l$	Number of lamps	lamps	298	256
$q = o/i$	Total electric input power	W	79167	43523
	CAPITAL COST			
r	Cost of one luminaire	U		
s	Auxiliary equipment cost	U		
t	Installation cost	U		
$u = r + s + t$	Total capital cost	U		
	OPERATING COST			
v	Cost of replaced lamps	U/year		
x	Cost for replacing	U/year		
$y = v + x$	Total annual operating cost	U/year		
	ENERGY SAVING			
w	Working hours	h/year	5000	5000
$EC = w \cdot q/1000$	Annual energy consumption	kWh/year	395833	217614
$ES = EC(REF) - EC(NEW)$	Annual energy saving	kWh/year		178219
CA	Electric energy cost	U/kWh		
$CS = ES \cdot CA$	Annual energy cost saving	U/year		

CHAPTER 15

HEAT EXCHANGE AND RECOVERY IN PROCESS AND FACILITIES

15.1 OPPORTUNITIES FOR HEAT RECOVERY FROM PROCESS AND FACILITY PLANTS

A great part of the total energy consumed all over the world is discharged as waste heat into the environment; part of this waste heat can be recovered by proper equipment if user demand matches it. The quality, defined mainly by the temperature value and the quantity of the waste heat, and their profiles versus time must match the user demand if energy recovery is to be significant from both the technical and economic points of view. Storage of waste heat is an expensive and not very effective approach with limited capability, and little used (water and steam storage, refractory material storage systems, etc.; see Chapter 4.7).

Waste heat is available mainly as fluid, gas or liquid streams at different temperatures. The streams can be more or less clean or corrosive, and this determines the possibility and cost of using recovery systems. Firstly, the temperature level qualifies the waste stream; the higher the temperature the greater the quantity of heat recoverable down to the ambient temperature which is the minimum level of recoverable heat. Below or near this temperature, heat pumps can be used, but their application involves additional energy consumption, so their installation is not always opportune.

Table 15.1 lists the commonest sources of waste heat and their temperatures classified as high range (>925 K; >625°C; >1160°F), medium range (900-500 K; 625-225°C; 1160-440°F) and low range (500-300 K; 225-25°C; 440-80°F). The first two ranges belong to direct fired uses of fuel or other reaction and they represent a good source of energy still usable for cogeneration plants and for the direct replacement of process heat. Attention must be paid, in the higher temperature ranges, to the equipment for heat recovery (proper material for high temperature, corrosion, etc.).

Table 15.1 Commonest sources of waste heat and their range of temperature

	K	°C	°F
LOW-TEMPERATURE RANGE	500-300	225-25	440-80
Steam condensate	330-365	56-92	134-198
Cooling water from			
process machines	305-370	32-96	90-206
air compressors	300-325	27-52	80-126
internal combustion engines	340-395	66-122	152-252
refrigeration condensers	305-315	32-42	90-107
ovens and similar	350-505	76-232	170-450
hot-processed materials	340-505	66-232	152-450
	K	°C	°F
MEDIUM-TEMPERATURE RANGE	900-500	625-225	1160-440
Steam boiler stack exhausts	500-750	226-375	440-890
Gas turbine exhausts	650-800	376-526	710-980
Internal engine exhausts	525-850	251-576	485-1070
Furnace stack gases	700-900	426-625	800-1160
Catalytic crackers	700-900	426-625	800-1160
	K	°C	°F
HIGH-TEMPERATURE RANGE	>925	>625	>1160
Glass melting furnace	1250-1800	977-1526	1791-2780
Cement kiln	900-1000	626-726	1160-1340
Refining furnace	925-1350	651-1076	1205-1970
Steel heating furnace	1200-1300	928-1026	1702-1880
Waste incinerator	925-1250	652-977	1206-1791

The low-range heat sources, generally air and water cooling streams from process and facility equipment, are available for producing or preheating hot water.

The quantity of heat available in any stream can generally be quantified in terms of enthalpy:

$$Q = m \cdot h$$

where

Q = total heat flow-rate of the stream (W; Btu/h)
m = mass flow-rate of the stream (kg/s; lb/h)
h = specific enthalpy of the stream (J/kg; Btu/lb)

The mass flow-rate can be calculated from the volume flow-rate by introducing the density of the stream (m = mass density · volume flow-rate).

It is important that every heat recovery system include auxiliary equipment for discharging heat, generally into the environment, when the waste heat demand from the user is interrupted or reduced. Otherwise, unacceptable operating conditions or interruptions would be imposed on the plant producing the waste heat.

15.2 HEAT EXCHANGE AND RECOVERY SYSTEMS

Heat exchangers are devices which transfer energy between fluids at different temperatures by heat transfer modes such as those discussed in Chapter 8.

Heat exchangers effect both heat recovery from waste heat and heat transfer from heating media to end users (see Chapter 8.4).

The exchange between two streams, a hot stream which has to be cooled and a cold stream which has to be heated, is induced either by mixing the two fluid streams to form a third stream at intermediate temperature, or by a system which separates the two streams, thus allowing heat exchange through separating surfaces and sometimes by an intermediate fluid.

The first system, such as a vessel where hot and cold streams are mixed directly, has limited application due to the problems of contamination and of different pressures which often make it impossible to operate.

The second system is based on the use of heat exchangers often called closed heat exchangers or recuperators, of which many different forms exist. Such characteristics of streams as chemical composition, physical phase, temperature and pressure affect the choice of heat exchangers and auxiliary equipment.

The following sections cover the basic principles of operation and the selection criteria for the commonest heat exchangers.

15.3 BASIC PRINCIPLES OF HEAT EXCHANGER OPERATION

The parameters which govern the functioning of a heat exchanger are:

- maximum pressure. Notice that many exchanger types work only at low pressure; thus the choice among different types must be made with careful regard to this;
- temperature range;
- fluid compatibility both between fluids and materials of construction and between fluid streams in the case of failure;
- range and size available for each unit. Obviously, many units can be assembled together but this may involve economic and technical problems.

Table 15.2 lists the more commonly used types with observations on the main parameters listed above. All heat exchangers are classified mainly by flow arrangement: parallel flow, counterflow, crossflow, mixed flow (see Fig. 15.1 and Fig. 15.2).

In order to choose among the different types available, the same basic principles must be followed, from which a few suggested methods are derived.

The rate of heat transfer Q (W) from fluid 1 (hot side fluid) to fluid 2 (cold side fluid) can be calculated by the following expression:

$$Q = m_1 \cdot (h_{1i} - h_{1o}) = m_2 \cdot (h_{2o} - h_{2i}) \qquad \text{(W)} \qquad \text{(Btu/h)}$$

where

stream 1 = hot stream; stream 2 = cold stream

m_1, m_2 = mass flow-rate of the two fluids (kg/s; lb/h)

h_{1i}, h_{2i} = inlet specific enthalpy of the two fluids (J/kg; Btu/lb)

h_{1o}, h_{2o} = outlet specific enthalpy of the two fluids (J/kg; Btu/lb)

If the two fluids have a constant specific heat (this can generally be assumed if no change of phase occurs), then:

$$Q = m_1 \cdot c_1 \cdot (T_{1i}-T_{1o}) = m_2 \cdot c_2 \cdot (T_{2o}-T_{2i}) =$$
$$= C_1 \cdot (T_{1i}-T_{1o}) = C_2 \cdot (T_{2o}-T_{2i})$$

where

T_{1i}, T_{2i} = inlet temperatures of the two fluids (K, °C, °F)

T_{1o}, T_{2o} = outlet temperatures of the two fluids (K, °C, °F)

c_1, c_2 = specific heat capacity of the two fluids (J/kg · K, Btu/lb · °F)

$m_1 \cdot c_1 = C_1$ = heat capacity flow-rate of fluid 1 (W/K, Btu/h · °F)

$m_2 \cdot c_2 = C_2$ = heat capacity flow-rate of fluid 2 (W/K, Btu/h · °F)

Table 15.2 Typical parameters for the commonest heat exchangers

Exchanger type	Max. pressure	Range of temperature		Normal size	Notes
	MPa	°C	°F	m^2	
SHELL AND TUBE	25-30	-200 to 600	-328 to 1112	10 to 1000	General purposes
DOUBLE PIPE	>30 (shell) >150 (tube)	-100 to 600	-148 to 1112	0.25 to 200	No limitation
GASKETED-PLATE	1-2	-25 to 175	-13 to 347	1 to 1250	Limitations due to gaskets
WELDED-PLATE	3	>400	>752	>1000	Differential pressure less than 3 MPa
CONVECTION BANK(gas/liquid)	0.1	<700	<1292	<500	Usually for water heating
GAS TO GAS	0.1(shell) >0.1(tube)	250 to 400	482 to 752	5 to 100 1000 to 3000	For waste gas stream recovery
AIR COOLED	>0.1(tube)	no limitation		5 to 200	Heat stream from process

The effectiveness of the heat exchanger is then defined as the ratio of the heat actually transferred to the maximum heat transfer possible, which is ideally obtained when the input temperature of the hot fluid equals the output temperature of the cold fluid ($T_{1i} = T_{2o}$; thus, $T_{2o} - T_{1o} = T_{1i} - T_{2i}$). The maximum heat transfer is equal to the minimum heat capacity flow-rate of the two fluids ($m_1 \cdot c_1$ or $m_2 \cdot c_2$) multiplied by the difference between input temperatures of the two fluids (T_{1i}-T_{2i}); the actual heat transfer is given by the previous heat transfer relationships.

In a parallel or counterflow arrangement (Fig. 15.1 and Fig. 15.2):

Effectiveness $E = (T_{2o}-T_{2i})/(T_{1i}-T_{2i})$
if the cold fluid (2) has the minimum heat capacity;

Effectiveness $E = (T_{1i}-T_{1o})/(T_{1i}-T_{2i})$
if the hot fluid (1) has the minimum heat capacity.

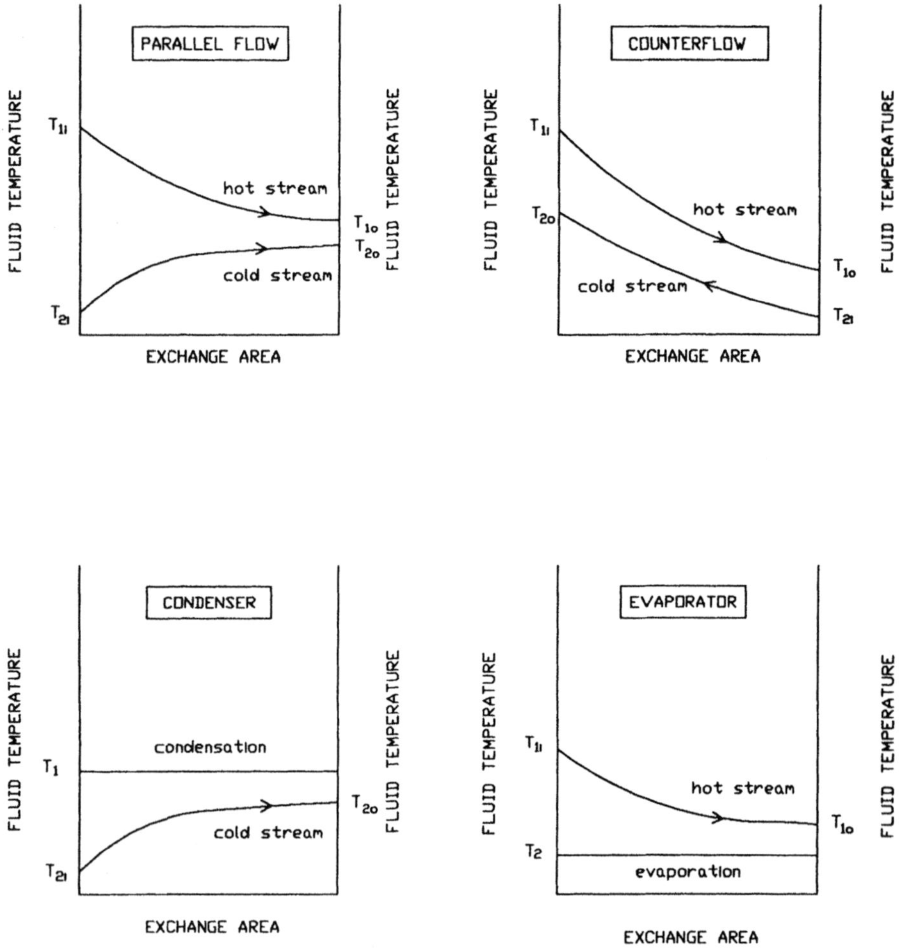

Fig. 15.1 Temperature profiles for different types of exchanger

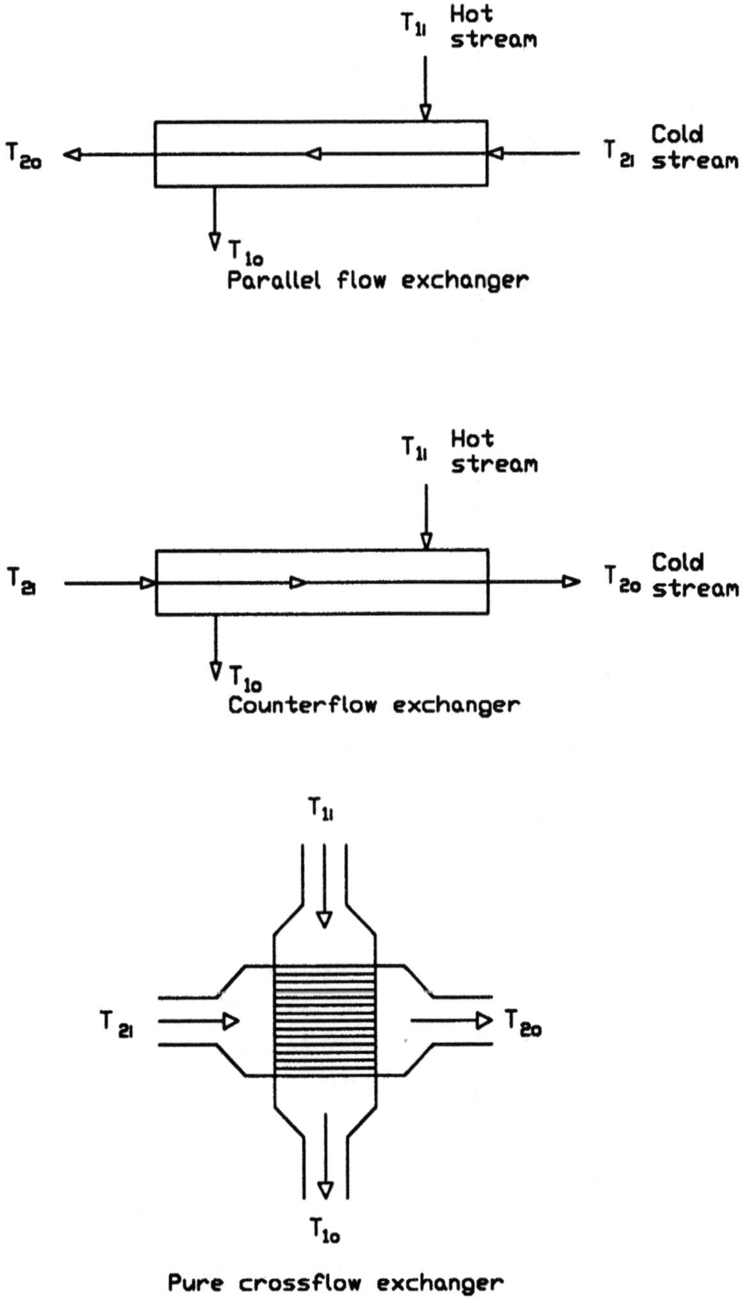

Fig. 15.2 Basic arrangements of different types of exchanger

A general expression of the heat transferred from one fluid to the other through the heat exchanger separating surfaces can be formulated as follows (see Chapter 8.3):

$Q = U \cdot A \cdot \Delta T$ (W) (Btu/h)

where

U = overall heat-transfer coefficient = $1/R_{th}$ ($W/m^2 \cdot K$; $Btu/h \cdot ft^2 \cdot °F$)

A = surface area of the exchanger, generally taken as the surface in contact with one fluid or stream

ΔT = effective temperature difference between the streams, assumed constant in any point of the exchanger surface, generally expressed as the log-mean difference between the temperatures of the fluids at the two sides of the exchanger. Note that these differences are not the same for parallel and for counterflow arrangements.

With a pure parallel flow exchanger, which is the simplest heat exchanger, assuming constant specific heats and constant overall coefficient U along the exchanger, Δ T must be introduced as follows (see Fig. 15.1 for fluid temperatures at both sides of the exchanger):

$$\Delta T = \Delta T_{LM} = \frac{(T_{1i} - T_{2i}) - (T_{1o} - T_{2o})}{\ln((T_{1i} - T_{2i})/(T_{1o} - T_{2o}))}$$

With a pure countercurrent flow exchanger, which is the most efficient heat exchanger, assuming constant specific heats and constant overall coefficient U along the exchanger, ΔT must be introduced as follows (see Fig. 15.1 for fluid temperatures at both sides of the exchanger):

$$\Delta T = \Delta T_{LM} = \frac{(T_{1i} - T_{2o}) - (T_{1o} - T_{2i})}{\ln((T_{1i} - T_{2o})/(T_{1o} - T_{2i}))}$$

In practice, many exchangers can be derived from a counterflow device, so that the effective temperature difference is defined as follows:

$\Delta T = FT \cdot \Delta T_{LM}$ (pure counterflow)

where

FT is a correction factor of the log-mean difference of temperature, generally obtainable from curves (see Fig. 15.3) where FT value is expressed as functions of R or P parameters:

$R = (T_{1i}-T_{1o})/(T_{2o}-T_{2i})$

$P = (T_{2o}-T_{2i})/(T_{1i}-T_{2i}) = (T_{1i} - T_{1o}) / (T_{1i} - T_{2i})$

The parameter R is the ratio between the difference of input/output temperatures of hot fluid (1) and cold fluid (2), that is, the ratio between the heat capacity flow-rates of cold fluid (2) and hot fluid (1) on the assumption that the cold and hot fluid have the same heat capacity. The parameter P is the effectiveness already defined on the same assumption as for parameter R.

The overall heat-transfer coefficient U is not constant along the heat exchangers and it depends on many factors such as the geometry of the exchanger, the stream speed, the temperatures of the streams, phase changes, etc.. An exact calculation generally requires an iterative method and it is made by the manufacturer. For a preliminary evaluation, a constant value can conveniently be taken from reference tables where a set of values is generally given. Table 15.3 reports some typical values of the U coefficient for a wide range of applications. Values of U for the commonest heat exchangers are reported in Table 15.4 and in Table 15.5.

Attention must be paid to the surface condition of the exchanger. This varies during the operation because of (i) film deposits, such as oil films; (ii) surface scaling due to the precipitation of solid compounds from solution; (iii) corrosion of the surface; (iv) deposit of solid in liquid or gaseous streams. This phenomenon is called fouling and a fouling factor is introduced to take it into account. The fouling factor decreases with increased fouling and effectiveness drops correspondingly.

In the case of heat exchange between many hot and cold streams inside the same process, a single composite curve of all hot streams and a single composite curve of all cold streams can be described in the temperature-heat content (Q) diagram. Heat exchanger networks can be designed or revamped by means of specialized techniques such as the Pinch method to obtain the best energy exchange.

Table 15.3 Typical range of values of overall heat-transfer coefficients U

Application	Range of values of U coefficient	
	$W/m^2 \cdot K$	$Btu/h \cdot ft^2 \cdot °F$
Steam condenser	1000 - 5000	176 - 882
Water to water	500 - 2000	88 - 353
Water to heavy organic fluid	100 - 400	18 - 70
Water to medium organic fluid	200 - 500	35 - 88
Water to light organic fluid	300 - 600	53 - 106
Steam to heavy organic fluid	60 - 300	11 - 53
Steam to medium organic fluid	300 - 400	53 - 70
Steam to light organic fluid	400 - 1000	70 - 176
Water to air with finned tube	30 - 60	5 - 11
Steam to air with finned tube	30 - 300	5 - 53
	(a)	(b) = a · 0.1761

Notes
heavy organic fluid: oil, etc.
medium organic fluid: gasoil, etc.
light organic fluid: freon, etc.

Table 15.4 Shell and tube parameters (Q/ΔT ranges between 1000 W/K and 100000 W/K)

SHELL AND TUBE	Cold side operation	U Coefficient	Hot side operation			
			Gas (0.1 MPa)	Gas (1 MPa)	Process water	Condensing steam
	Gas (0.1 Mpa)	$W/m^2 \cdot K$ (a)	55	93	102	107
		$Btu/h \cdot ft^2 \cdot {}^\circ F$ (b)	9.7	16.4	18	18.8
	Gas (2 MPa)	$W/m^2 \cdot K$	93	300	429	530
		$Btu/h \cdot ft^2 \cdot {}^\circ F$	16.4	52.8	75.5	93.3
	Cooling water	$W/m^2 \cdot K$	105	484	938	1607
		$Btu/h \cdot ft^2 \cdot {}^\circ F$	18.5	85.2	165.2	283
	Boiling water	$W/m^2 \cdot K$	105	467	875	1432
		$Btu/h \cdot ft^2 \cdot {}^\circ F$	18.5	82.2	154.1	252.2

Note
(b) = (a) · 0.1761

Table 15.5 Double pipe finned tube parameters (Q/ΔT ranges between 5000 W/K and 100000 W/K)

DOUBLE PIPE FINNED TUBE	Cold side operation	U Coefficient	Hot side operation			
			Gas (0.1 MPa)	Gas (1 MPa)	Process water	Condensing steam
	Gas (0.1 Mpa)	$W/m^2 \cdot K$ (a)	21	56	77	70
		$Btu/h \cdot ft^2 \cdot {}^\circ F$ (b)	3.7	9.9	13.6	12.3
	Gas (2 MPa)	$W/m^2 \cdot K$	56	97	181	182
		$Btu/h \cdot ft^2 \cdot {}^\circ F$	9.9	17.1	31.9	32.1
	Cooling water	$W/m^2 \cdot K$	87	238	325	534
		$Btu/h \cdot ft^2 \cdot {}^\circ F$	15.3	41.9	57.2	94
	Boiling water	$W/m^2 \cdot K$	84	216	284	511
		$Btu/h \cdot ft^2 \cdot {}^\circ F$	14.8	38	50	90

Notes
(b) = (a) · 0.1761
Q/ΔT range from 5000 W/K to 100000 W/K and higher
The parameter cost per unit of Q/ΔT decreases as Q/ΔT value increases.

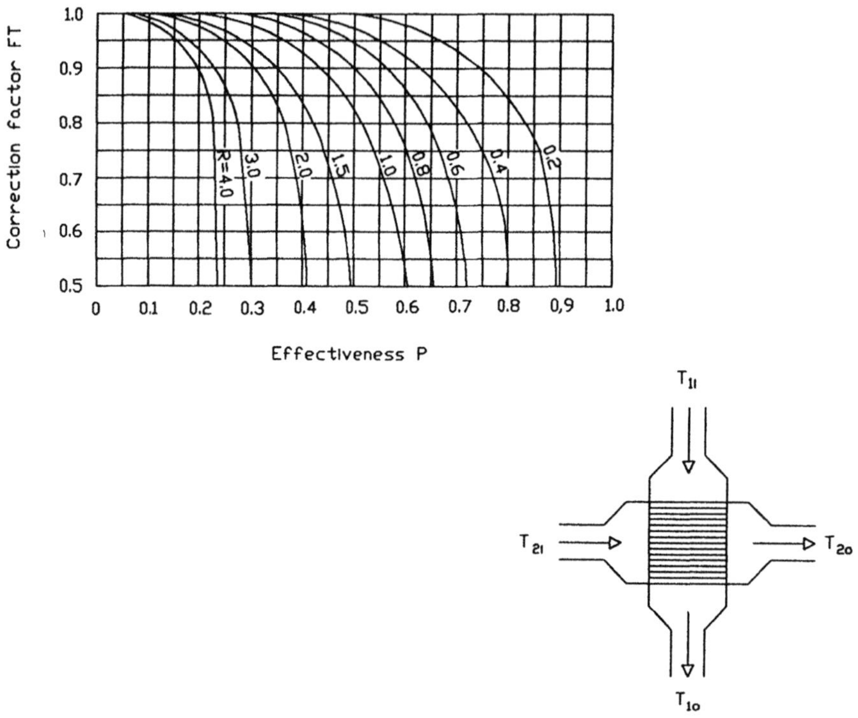

Fig. 15.3 Correction factor FT for a crossflow exchanger (P and R parameters are calculated on the assumption that the cold and hot fluid have the same heat capacity)

15.4 PRELIMINARY EVALUATION CRITERIA FOR HEAT EXCHANGERS

A preliminary evaluation of the heat exchanger surface can be made by simplified approaches such as the log-mean temperature method and the effectiveness-NTU method.

The log-mean temperature method

Preliminary sizing and pricing of heat exchangers is possible if the U coefficient is estimated, preferably by introducing also a fouling factor, and if inlet/outlet temperatures are known. The heat exchanger area may be calculated as:

$$A = (Q/\Delta T) \cdot (1/U)$$

The expressions given above can be applied to different flow-arrangement types of exchanger: parallel flow, counterflow, crossflow and mixed flow.

The ratio ($Q/\Delta T$) is an index of the heat exchanger duty and from this value a preliminary evaluation of the exchanger cost can be made by multiplying the area A by a cost per unit area generally taken from tables.

The procedure for evaluating various types of heat exchanger involves the following steps:

1) the heat flow-rate Q of the stream is calculated using the relationships given above ($m \cdot \Delta h$ or $m \cdot c \cdot \Delta T$);
2) the log-mean difference of temperatures ΔT_{LM} is calculated on the basis of the inlet/outlet temperatures and then corrected by factor FT;
3) the ratio $Q/\Delta T_{LM}$ should be calculated for each possible configuration;
4) if U is estimated (see Tables 15.3, 15.4), the exchanger surface is obtained and then the cost of the exchanger. Alternatively, curves or tables which give the cost per unit of W/K (= W/°C) or of W/°F corresponding to different values of U can conveniently be used to determine the cost of the exchanger.

Effectiveness-NTU Method

This approach is based on the introduction of a parameter defined as Number of Transfer Units $NTU = A \cdot U/C_{min}$ where A is the effective heat transfer area, U is the effective overall heat-transfer coefficient and C_{min} is the minimum heat-capacity flow-rate. The parameter NTU, which is an adimensional quantity, is indicative of the size of the heat exchanger.

The effectiveness of typical heat exchangers versus the parameter NTU is expressed by means of curves. Different curves can be described for different values of C_{min}/C_{max} (Fig. 15.4 and Fig. 15.5). These curves are plotted using the log-mean difference expression.

Notice that a linear relationship exists between effectiveness and the parameter NTU until the value of the latter reaches 1; after this point the curve begins to knee, which means that effectiveness can be increased with a corresponding higher increase of the NTU parameter. This involves paying more and more for exchangers with high effectiveness.

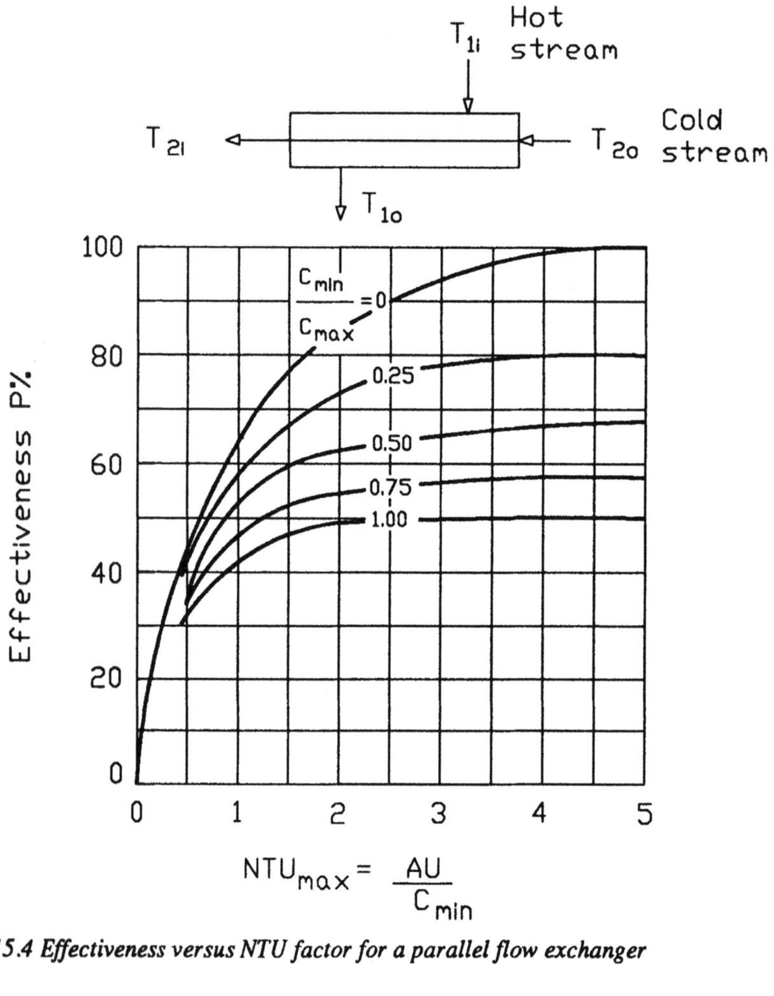

Fig. 15.4 Effectiveness versus NTU factor for a parallel flow exchanger

This method is very useful when the inlet and outlet temperatures in the heat exchanger are not known.

A short outline of a preliminary design for a heat exchanger made by the NTU method, if only the input temperatures and the heat capacity flow-rates of the fluids are known, includes the following steps:

1) a value of the overall heat-transfer coefficient U is estimated;
2) the heat capacity flow-rates of the two fluids are known ($m \cdot c$);

3) for a given effectiveness, which corresponds to a given output temperature of one fluid, the value of the NTU parameter is derived from curves like those in Fig. 15.4 and Fig. 15.5;
4) the exchange surface value can be calculated from the NTU expression (NTU $= A \cdot U/C_{min}$) and the unknown outlet temperature is derived from the effectiveness expression.

Alternatively

3') for a given heat exchanger surface the NTU parameter is calculated;

4') the effectiveness is derived from the same curves as before; then the actual heat flow-rate is calculated (by means of the effectiveness definition) and the outlet temperatures derive from an energy balance.

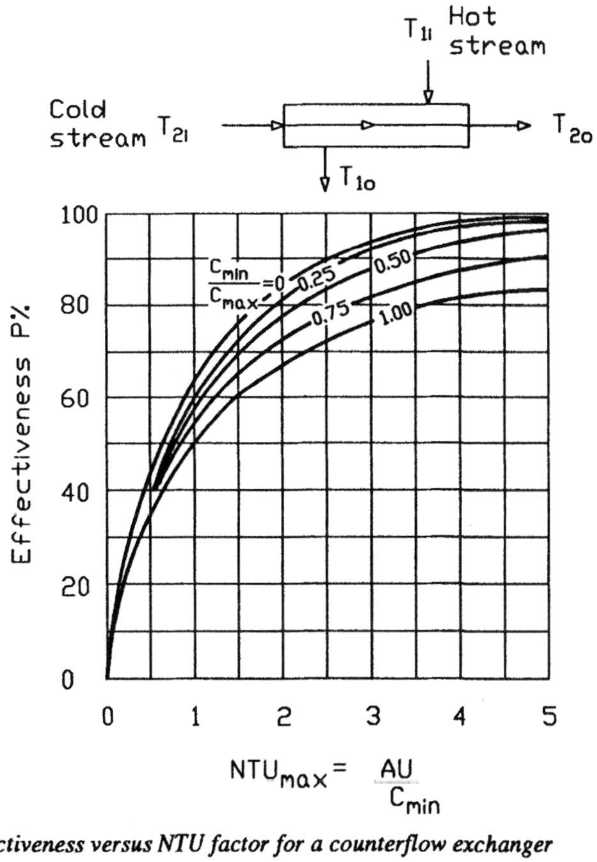

Fig. 15.5 Effectiveness versus NTU factor for a counterflow exchanger

15.5 A CLASSIFICATION OF HEAT EXCHANGERS AND THEIR SELECTION CRITERIA

The capital cost of a heat exchanger depends on many factors, such as
— type of exchanger
— tube diameter
— number of passes
— pressure
— temperature
— material

Typical values of the main parameters are listed in Table 15.2.

These factors can also be used to group heat exchangers as is shown below.

Shell and tube heat exchanger

The shell and tube heat exchanger is the most widely used in industry because of its flexibility both in operating parameters (temperatures, pressures) and in construction materials. It can be classified as fixed tube plate, U-tube type or floating head. The exchangers are generally defined in terms of a standard nomenclature introduced by the Tubular Exchanger Manufacturers Association (TEMA) of the USA, based on the front end, shell and rear end type.

U coefficient values vary from 50 to 2000 $W/m^2 \cdot K$ (see Table 15.4).

Double pipe heat exchanger

The double pipe heat exchanger has the advantage of offering almost perfect countercurrent flow. It can be a single tube within an outer tube, or a group of tubes.

U coefficient values vary from 20 to 600 $W/m^2 \cdot K$ (see Table 15.5).

Gasketed-plate heat exchanger

This exchanger is based on plates of stainless steel or titanium or other materials, which are packed together with gaskets. It can be enlarged at little cost by adding further plates, according to user requirements. Temperature and pressure limitations and the need for integrity of sealing are the major drawbacks of this type of exchanger. U coefficient values vary from 300 $W/m^2 \cdot K$ when high-viscosity liquid is on one side of the exchanger to 3500-4000 $W/m^2 \cdot K$ with water or low-viscosity organic fluid.

Welded-plate exchanger

This exchanger, which has been designed to overcome the problems of the gasketed exchanger, has welded plates. If it is mounted within a shell, high pressures and high temperatures, equivalent to those in shell and tube exchangers, can be achieved. A large exchange area, more than 1000 m^2, can be exploited.

U coefficient values are similar to those of gasketed-plate heat exchangers.

Hot-gas-to-liquid convective bank system

This type of exchanger is used when waste heat from process is transferred to a liquid phase, generally water. It consists of groups of tubes with fins of cast iron or carbon steel or stainless steel. The working temperature ranges between 1000 and 500 K (727-227°C; 1340-440°F), depending on the location of the tubes, on their material and on the gas characteristics. Attention must be paid to the condensation of acidic moisture. U coefficient values vary from 15 to 30 $W/m^2 \cdot K$, with water or other low-viscosity liquid on one side of the tube.

Gas-to-gas recuperative heat exchanger

This type of exchanger is widely used in industry in many applications such as the use of combustion gases to preheat combustion air, heat exchange in chemical reactors, heat recovery from exhaust air in buildings, etc.. Many different configurations with tubes and plates and different materials (steel, cast iron, glass, plastic tube) are used.

U coefficient values vary from 10 to 35 $W/m^2 \cdot K$.

Heat pipe exchanger

The heat pipe exchanger uses an intermediate fluid to transfer heat from a hot gas to a cold fluid. The basic principle is as follows: heat from hot gases is transferred to the evaporation section. The intermediate fluid changes its state into vapor and moves along the heat pipe to the condensation section where it condenses by releasing heat to the cold stream to be heated. Water can be used as heat pipe fluid. Higher temperatures can be reached with organic fluid.

Large surfaces can be used for both sides.

A typical value of the U coefficient is 20 $W/m^2 \cdot K$.

Rotary regenerator

This consists of a disk, fabricated of materials having a significant heat capacity, which rotates between cold gas and hot gas ducts, side by side. The disk acts as an intermediate medium: it stores heat from the hot gas and it transfers heat when it moves in the area of the cold gas duct. The effectiveness of this system is very high, more than 90%. Because the high temperatures in contact with hot gases are uniformly distributed, the danger of corrosion is quite insignificant.

Attention must be paid to cross-contamination between the gas streams; this can be reduced by special devices such as labyrinth seals.

15.6 PRACTICAL EXAMPLES

Example 1
Cooling water-to-process water exchanger (log-mean method)

A process water flowing at 8 kg/s must be cooled from a temperature of 110°C to a temperature of 40°C using cooling water at the same flow-rate. The inlet temperature of the cooling water is 15°C and the desired outlet temperature is 85°C. The pressure of the two fluids is around 0.3 MPa. A preliminary evaluation of possible heat exchangers is required (see Chapter 15.4 for the log-mean temperature method).

Given temperatures

$T_{1i} = 110°C$ $T_{1o} = 40°C$

$T_{2i} = 15°C$ $T_{2o} = 85°C$

Step 1

Calculate the heat flow-rate Q of the two streams

— stream 1 (hot water)

$Q = 8 \cdot 4186 \cdot (110 - 40) = 2344000 \text{ W} = 2344 \text{ kW}$

— stream 2 (cold water)

$Q = 8 \cdot 4186 \cdot (85 - 15) = 2344000 \text{ W} = 2344 \text{ kW}$

Step 2

Calculate the difference between the temperatures of the fluids at the two sides of a counterflow exchanger

$T_{1i} - T_{2o} = 110 - 85 = 25°C$

$T_{1o} - T_{2i} = 40 -- 5 = 25°C$

$\Delta T = 25 \text{ K} = 25°C$ (in this case, it is constant along the counterflow exchanger)

$R = (110-40)/(85-15) = 1$

$P = (85-15)/(110-15) = 0.73$

Correction factor FT is preliminarily assumed equal to 1

Step 3

Calculate $Q/\Delta T$ (W/K)

$Q/\Delta T = 2344000/25 = 93700 \text{ W/K}$

Step 4

Calculate the exchange surface for different kinds of heat exchanger

— from Table 15.4 (shell and tube exchanger)

$U = 938 \text{ W/m}^2 \cdot \text{K}$ (typical value)

$\text{area} = Q/(U \cdot \Delta T) = 2344 \cdot 1000/(938 \cdot 25) = 100 \text{ m}^2$

— for plate exchanger U is assumed equal to $4000 \text{ W/m}^2 \cdot \text{K}$

$\text{area} = Q/(U \cdot \Delta T) = 2344 \cdot 1000/(4000 \cdot 25) = 23.4 \text{ m}^2$

In both cases, if the parameter cost per unit of W/K is known, the cost of the heat exchangers can be found:

$C \cdot (Q/\Delta T)$ = cost of exchanger

Although costs are not here included, the plate exchanger, which has the smaller surface, seems to be the more economical.

The energy saving is calculated with reference to a boiler plant which will heat the cold water flow from 15°C to 85°C as required from end users with efficiency equal to 85% (Net Heating Value as reference; 41860 kJ/kg of oil).

In the case of 2000 h/year of recovery, it follows that:

ENERGY SAVING = (2344 · 3600) · 2000/(41860 · 0.85)
= 474320 kgoil/year = 474.32 TOE/year

The economic evaluation is shown in Table 20.3.

Example 2

Cooling water-to-process water exchanger (effectiveness-NTU method)

The case is similar to that in Example 1; outlet temperatures of both fluids are supposed unknown (see effectiveness-NTU method in Chapter 15.4).

Given temperatures
$T_{1i} = 110°C$
$T_{2i} = 15°C$

Step 1
Estimate a value of U coefficient
$U = 938\ W/m^2 \cdot K$ (shell and tube)

Step 2
Calculate the heat capacity flow-rate of the two fluids
$m_1 \cdot c_1 = 8 \cdot 4186 = 33488\ W/K$
$m_2 \cdot c_2 = 8 \cdot 4186 = 33488\ W/K$

Step 3

— the output temperature of fluid 1 is assumed T_{1o} = 40°C; the effectiveness is calculated as follows:

$P = (110\text{-}40)/(110-15) = 0.736$

— $C_{min}/C_{max} = 1$; from Fig. 15.5 it follows that $NTU_{max} = 2.9$

Step 4

Calculate the exchange surface and the outlet temperature of fluid 2

$NTU = A \cdot U/C_{min}$

$A = NTU \cdot C_{min}/U = 2.9 \cdot 33488/938 = 103\ m^2$

$P = (T_{20}-15)/(110-15) = (100 - 40)/(110 - 15) = 0.736$

$T_{20} = P \cdot (110-15)+15 = 0.736 \cdot 95+15 = 85°C$

Example 2 has the same energy saving as Example 1.

Example 3

Air-to-exhaust stream exchanger (log-mean method)

A process requires 3720 Sm^3/h of air at 145°C, which can be obtained by heating air from the ambient temperature of 20°C to the desired value by means of an exhaust gas stream. The exhaust is at 1000°C and has a flow rate equal to 1860 Sm^3/h. A preliminary evaluation of the heat exchanger surface, which is a gas-to-gas exchanger with a crossflow arrangement, is required (see Chapter 15.4 for the log-mean temperature method).

Operating data at both sides of the heat exchanger are shown in Fig. 15.7.

Given temperatures

— T_{1i} = 1000°C T_{1o} = 750°C

— T_{2i} = 20°C T_{2o} = 145°C

Step 1

Calculate the heat flow-rate of the two streams (specific heat of air and exhaust is assumed equal to 1 kJ/kg · K)

— stream 1 (exhaust)

$Q = (1860/3600) \cdot 1.29 \cdot 1 \cdot (1000-750) = 166\ kW$

— stream 2 (air)

$Q = (3720/3600) \cdot 1.29 \cdot 1 \cdot (145-20) = 166\ kW$

Step 2

Calculate the difference between the temperatures of the fluids at the two sides of the exchanger, by means of the log-mean difference and the correction factor FT

$T_{1i}-T_{2o} = 1000-145 = 855°C$
$T_{1o}-T_{2i} = 750-20 = 730°C$
$\Delta T_{LM} = 791\ K = 791°C$
$R = (1000-750)/(145-20) = 2$
$P = (1000-750)/(1000-20) = 0.25$
$FT = 0.95$ (see Fig. 15.3)
$\Delta T = \Delta T_{LM} \cdot FT = 791 \cdot 0.95 = 751\ K = 751°C$

Step 3

Calculate $Q/\Delta T$ (W/K)

$Q/\Delta T = 166 \cdot 1000/751 = 221$ (W/K)

Step 4

Make a preliminary evaluation of a gas-to-gas exchanger surface and cost. U is assumed equal to 20 $W/m^2 \cdot K$; it follows that:

area $= Q/(U \cdot \Delta T) = 166 \cdot 1000/(20 \cdot 751) = 11\ m^2$

If the parameter cost per unit of W/K is known, the cost of the heat exchangers can be found:

$C \cdot (Q/\Delta T)$ = cost of exchanger

The energy saving is calculated with reference to a boiler plant which will heat the cold air stream (stream 2) as required by the process with efficiency equal to 85% (Net Heating Value as reference; 41860 kJ/kg of oil).

For 5000 h/year of recovery, it follows that:

ENERGY SAVING = $(166 \cdot 3600) \cdot 5000/(41860 \cdot 0.85) = 83977\ kg_{oil}$/year
= 84 TOE/year

The economic evaluation is shown in Table 20.3.

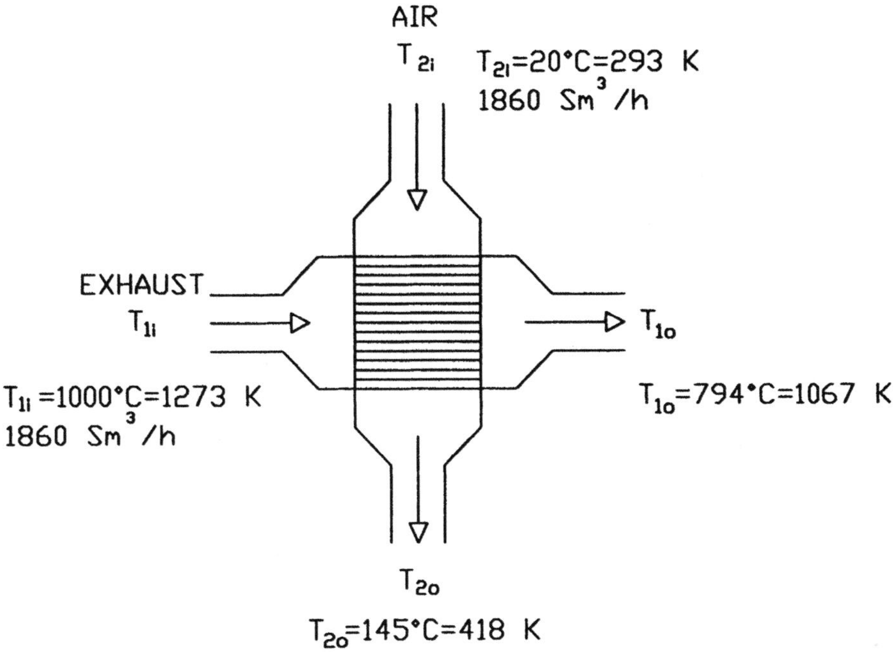

Fig. 15.6 Air/exhaust gas crossflow exchanger. Operating data as for Example 2

CHAPTER 16

WASTE FROM INDUSTRIAL PROCESSES

16.1 INTRODUCTION

Industry produces solid, liquid and gaseous waste.

The re-use of industrial waste inside the factory itself as raw material and in heat-recovery plants are possibilities that must always be investigated. The main aspects to point out are: the chemical composition of the waste, the possibility of re-use for internal process, the technology available for treatment, waste produced by treatment plants and the consequent air and water pollution and also the economic viability, taking into account environmental, energy and operating costs.

Treatment of liquid and gaseous waste is widely practised, to varying extents determined by local regulations, sensitivity to the problem and the kind of pollution concerned. Liquid and gaseous waste can be classified on the basis of pollutant content; specialized literature and practical experience are available to solve most cases either on-site or outside the factory. The energy content in liquid and gaseous emissions can be partially recovered by means of heat exchangers (see Chapter 15), to be used in process or in cogeneration plants (see Chapter 9), depending on the temperatures and on the pollutant content.

Solid waste from industry is classified in each country according to local regulations; nevertheless, the various systems have naturally much in common. In an industrial country, such as the example illustrated in Fig. 3.1, with a gross energy consumption of 160 MTOE/year (of which 50 MTOE/year in industry), the amount of industrial waste is roughly 35 million t per year; it follows that, generally speaking, one can find 0.7-1 million t of industrial waste per TOE of gross energy consumption in industry. In addition, note that in an industrial country each person disposes of about 1.5-2 kg/day (0.5-0.7 t/year) of urban waste; if all kinds of waste are considered, then in an industrial country such as the example mentioned before with a population of 50 million, the total waste can reach 100 million t, that is, 60% in weight of the total TOE consumption.

Four categories of industrial solid waste are widely recognized:

- **inert waste from extractive, brick and similar industries, to which can also be assimilated debris from building demolition. This can easily be re-used for road construction and public works;**
- **waste assimilable to urban waste. This includes plastic, paper, rubber, and, in limited quantities, glass. These can be treated as urban waste;**
- **industrial waste includes all industrial waste with the exceptions of the categories mentioned above (inert and assimilable to urban waste) and hazardous waste. It presents some risks for the environment, but it can be rendered less harmful by appropriate treatment or re-used in processes. It includes organic and inorganic materials, liquids, sludges and solids with limited concentrations of pollutant;**
- **hazardous and infectious waste, which is industrial waste with high concentrations of toxics or pollutants.**

Urban waste is commonly landfilled, but this is no longer an acceptable disposal method because of environmental problems and the scarcity of space. Utilities are very interested in burning waste to produce electricity and steam, but air pollution must be avoided. Urban waste treatment is not discussed here in detail, our main concern being industrial waste and its on-site utilisation.

Non-hazardous industrial waste can be treated as urban waste, but the quantity produced by a single medium-sized factory is generally too small for on-site re-use in heat-recovery plants. As a general rule, on-site treatment is economically viable if the quantity of waste is not less than hundreds of t/year (500-1000). This depends on the type of waste and on particular situations, but it can be assumed as a general indication.

16.2 WASTE MANAGEMENT

Waste-management strategies comprise a few main sections, some of which exist also in energy management (see Chapter 18): analysis of historical data, energy audits and accounting, analysis of local regulations and environments, engineering analyses and investment proposals based on feasibility studies, training and information of personnel.

Several factors must be considered:

— incidence of waste costs on turnover and added value;
— waste as a percentage of production leaving the factory;
— identification and quantification of waste following the flow of raw materials and energy inside the factory;
— selection of significant indexes valid for the whole factory and for single production lines;
— possibility of recycling inside the factory or outside;
— level of pollution and compatibility with local regulations.

Waste management involves so many technical and economic as well as environmental aspects that it requires a separate approach, although correlated with several topics from other fields. Here, only correlation with energy management is treated and the basic data useful for preliminary evaluations are provided.

Development of waste-management strategies at different levels depends on the importance of the waste cost in comparison with other production factors and on local regulations which may require specific procedures to be followed, regardless of the quantity of waste involved.

As a general rule, waste management is performed in three main ways: (i) waste reduction by using clean technologies; (ii) waste elimination; (iii) recycling of waste as raw material or energy.

The clean-technologies approach is the most appealing but the most expensive and time-consuming because it may require drastic transformations in all phases of the process, from raw-material input to packaging and delivery. Technologists must be involved in the process modification. Also important and not to be overlooked are the effects of the packaging phase on urban and industrial waste, due to the plastic and wood waste produced. In addition, the use of combustibles such as natural gas, liquid and solid fuels with low sulfur concentrations (less than 1% in weight) can be included in this area of waste management; it can be called clean technology in facilities, because gaseous pollutants from combustion are reduced without abatement plants at the end of the combustion process.

Elimination of solid and liquid waste by employing external agencies is the simplest but generally the most expensive policy for the single factory. Quite often it is the only one practicable because the quantity of waste produced is too small to justify an independent plant. Nevertheless, local regulations may require stockage and transportation in accordance with quite strict rules which may increase the final cost of disposal. It must be pointed out that the difficulties in finding suitable sites for this operation are growing dramatically, particularly in crowded areas such as the most industrialized part of Europe.

In the case of liquid waste, water if discharged to external agencies is charged for, according to the pollutant content. If not, liquid waste should be treated before discharge by means of mechanical, chemical and aerobic systems, which consume energy.

In a similar way, the treatment of gaseous streams at high temperature must be performed inside the factory, depending on local regulations (by burning the pollutant streams and by keeping them for a given time at temperatures higher than 900-1000°C (1650-1832°F) and then cooling them to a lower temperature before discharge from the stack). Notice that the cooling phase can economically be effected by recovering heat from the exhaust by means of gas-to-gas exchangers, thus reducing the energy consumed in treatment.

Recycling as raw material or energy, whenever possible, is always the most attractive approach.

The use of waste as raw material inside the process itself remains the optimum form of waste management because the waste does not cross the factory boundary. In the same way, the use of waste from other processes can be equally successful, in a wider waste-management approach. It is a method practised in foundries, and with wood, plastic and rubber; it requires separation of material flows inside the factory in order to facilitate the recycling, whenever possible, at different steps of the process.

Incineration of waste (RDF, sludges and others) inside the factory is an always attractive possibility but quite often it is not practicable because of environmental pollution problems and the corrosion of equipment by substances such as polyvinyl chloride, etc. (see combustion principles in Chapter 6.11).

In addition, pyrolysis, distillation and anaerobic treatment of solid waste, sludges and liquids produce a liquid or a gas suitable for use as fuel. Many of these treatments, such as anaerobic digestion processes, although often mentioned as economical treatments, are confined mostly to small plants. The energy recovery from these plants is quite low and often the energy recovered only equals the input energy for auxiliaries. Nevertheless, these treatments offer a possibility of saving energy as compared with other systems already mentioned which consume energy.

16.3 ENERGY RECOVERY FROM WASTE

Energy recovery from waste can be classified on the basis of the kinds of treatment and of waste available.

Heat recovery from hot liquid and gaseous streams by means of heat exchangers has been discussed in Chapter 15. It is the simplest and most economical method of recovery, but it requires that the recoverable stream match the demand profile of the end user in quantity and in quality. In addition, the pollutant content of the stream, which might provoke corrosion, must be considered carefully.

Many types of solid and liquid waste can be stored when they are produced and then transformed into energy as liquid, solid and gaseous fuels according to the end-user demand.

Table 16.1 lists the main treatment technologies for energy recovery and a few basic operating data to be considered in a preliminary evaluation. These systems are subject to the following considerations:

— urban and industrial solid waste can be burned in incinerators to produce hot water, steam and electricity. So can industrial and urban sludges. Solid waste can be prepared as RDF (Refuse-Derived Fuel) or as solid fuel pellets. Net heating values range from 16000 to 21000 kJ/kg;

— liquid and solid waste can be treated by means of anaerobic digestion to produce biogas. The main parameter to be considered for liquid waste is the COD content (Chemical Oxygen Demand) to which the production of biogas is generally referred;

— other systems, such as pyrolysis, distillation and gasification are possible in particular applications, which must be investigated individually and evaluated carefully from the economic and technical points of view.

16.4 WASTE AND ENERGY MANAGEMENT FUNCTIONS IN INDUSTRY

Waste management is a strictly energy-related function which concerns all factories and the activities of every person at work and in the home.

If energy must be considered with an eye to the limited amount available and the search for new sources must always be continued, waste must be viewed in the perspective of its huge presence everywhere in the world. On the other hand, waste means energy: (i) energy can be produced from disposal at the end of a production process; (ii) energy is consumed both during the transformation of raw materials into the end product and during the waste treatment.

Although energy can properly be saved by means of individual investment, waste reduction is more difficult to achieve individually and certainly requires the contributions of both the public and the industrial sectors and sensitivity to the waste problem all over the world.

Cost of waste treatment is certainly one of the factors to be considered in economic analysis of investments (see Chapter 20).

Table 16.1 Main technologies for energy recovery in waste treatment

Type of technology	Energy recovery		Note
INCINERATION	Net Heating Value kJ/kg		
RDF fluff	16000-21000		
RDF pellet	16000-21000		
Urban solid waste	>4000		
Industrial solid waste	20000-40000		Depends on the type of refuse
Sludges	10000		Moisture content 5%
ANAEROBIC DIGESTION (BIOGAS)	Net Heating Value kJ/kg	Production Sm^3/kg	
Urban solid waste landfilled	18000-22000	0.35-.45	Per kg of dry mass
Sludges (urban, agricultural)	15000-20000	0.6-1.5	Per kg of solid substance gasified
Liquid waste	20000-25000	0.2-0.4	Per kg of COD (liquid Chemical Oxygen Demand); COD>5000
Biomass	15000-20000	variable	Depends on the type of biomass

CHAPTER 17

USE OF COMPUTERS FOR ENERGY AND WASTE MANAGEMENT

17.1 INTRODUCTION

The use of computerized systems to control process and facilities produces many benefits which mainly affect four important functions of the factory: Operation, Maintenance, Planning and Management. Energy is involved in each function; thus, energy is saved through a more complex approach which involves every aspect of the factory's life.

A detailed analysis of different control systems and their architecture is here omitted; specialized books and papers on information science and technical documentation from system manufacturers should be consulted to ensure that the chosen system is up-to-date and is that best suited to the user's needs.

A basic scheme is shown in Fig. 17.1.

17.2 HOW COMPUTER CONTROL MAY AFFECT THE FUNCTIONING OF THE FACTORY

The main operations performed by a computer control system can be summarized as follows:

— sampling of the operation parameters selected as the most significant;
— checking that the operative functional limits of the individual machines or systems are respected, with immediate indication of anomalies;
— choosing the most opportune control actions. These actions can be optimized from several points of view (energy saving, maintenance, production of process lines, etc.);

— implementing of the control instructions (start-stop, position control, etc.) by means of actuators. Verification that the instructions have been carried out is generally performed;
— memorizing on paper and/or other support (magnetic, etc.) individual operating profiles, power-demand peaks, daily-weekly-monthly consumptions, etc.. The quantity recorded can be real (flows, temperatures, pressures, currents, ..) or result from elaborations such as efficiency, percentages, etc..

The following sections review briefly the influence of computer control on factory functions.

Operation

The introduction of a computerized control system generally leads to increased productivity. This is due both to the fact that the installations are now working in conditions quite close to the designed ideal, and also to the rationalization which is a collateral effect of the exhaustive analysis preceding any introduction of a computerized system. In consequence, a "re-layout" or "retrofitting" of plants is generally effected with such improvement of productivity as often to postpone the need to enlarge the installations.

A second benefit is the reduction of the specific cost of the service supplied, if facilities are controlled. This is due to the implementation of energy strategies and to the reduction of wasteful functional anomalies.

A last and not negligible benefit is the reduction of the staff needed for operation, their remaining function being only to supervise the automated system and to intervene as it indicates.

Maintenance

Automated operation has proved able to reduce the number of machine faults; in any case, efficient diagnosis reduces almost to zero the time needed to identify the nature of a fault and the time which elapses between the first indication of a fault and its correction.

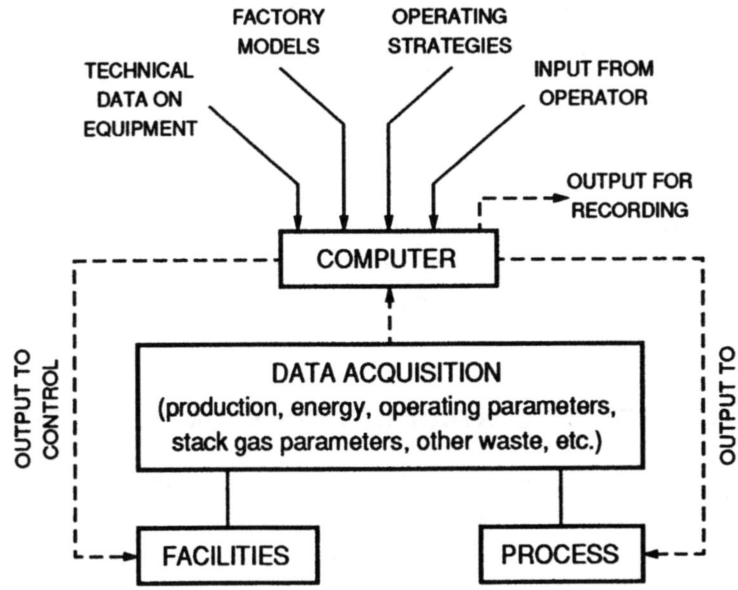

Fig. 17.1 A basic scheme for energy and waste management by computerized systems

Periodical scheduled maintenance is also facilitated by taking into account a predetermined number of working hours for each machine.

Maintenance is organized by computer control in many industries and specialized technical literature can be found on this subject. Here, the possibility is outlined of interaction between computer control of maintenance and of energy.

Planning

In the absence of data regarding real past functioning, the plant planner tends to make the installation too big, as a precaution. The acquisition and computer-processing of real data on plant operation help the engineering staff to avoid excessive size and a consequent investment burden. In addition, mathematical models of the plant can be made in order to simulate, in simplified form, the functioning of the real installations. Thus, it is possible to study the influence of variations in the parameters on the process being checked, and so to discover the optimal values to apply in operation strategies.

Management

Availability of memorized data permits a more exhaustive analysis of past operating modes of the plant and more detailed energy audits. These together are the basis for the implementation of energy-management strategies (see Chapter 18).

17.3 HOW COMPUTER CONTROL CAN WORK FOR ENERGY SAVING

A computer control system for energy saving, whatever its architecture and the HW and SW available, collects data from the plants, processes and facilities, throughout the factory. It analyzes and elaborates these data so as to activate operation strategies defined by the user and designed to maintain optimum operating conditions with the minimum expenditure of energy. Sensors and transducers play an important role and very often represent the major part of the installation cost.

As a general rule, a standard SW program performs the following main functions:

— it acquires data from sensors by scanning on individual frequencies of acquisition for various quantities (temperature, electric energy, pressure, flow of steam, flow of natural gas, flow of water, flow of exhausts, etc.). Subsequent processing converts the readings into engineering values (K or °C or °F, kW, t of steam, etc.). The maximum and minimum limits of the reading and the functioning of the instrumentation can usually be checked. The on/off state of equipment such as valves, switches, etc. and also of alarms is detected. If required by the energy program, other quantities such as efficiency, enthalpy, coefficient of performance, etc., can periodically be calculated. The data collected and subsequently elaborated are generally memorized and preserved for 24 hours or longer, according to the archive space available;

— it chooses the operating modes on the basis of standard strategies already present or of strategies introduced by the users in particular situations and it gives the output signals to activate the machines.
  Basically, strategies can be classified as operation strategies and optimization strategies. The first usually serve to bring an anomalous situation back to within correct operational limits or to conduct an installation according to predetermined optimum criteria. The second, optimization strategies, are generally more complex and cause variables to interact with one another in such a way that the whole plant (total plant, individual production line, individual facility, etc.) is always working at maximum efficiency;
— it gives the operator, periodically and on request, information on past data, current data, daily-monthly-yearly data, maintenance reports, and other reports for particular needs. Communication between operator and system must be as simple as possible in order to facilitate the full exploitation of the control system which is generally under-used because of difficulties of interaction with operator.

The following sections review the main facility-control strategies. No indications are given about process because the great variety of cases would require specialized knowledge of multifarious sectors.

17.4 CONTROL OF BOILER PLANTS AND ELECTRICAL SUBSTATIONS

The control of boiler plants to achieve the maximum efficiency and the lowest emission levels is one of the most attractive ways of using a computer control system for energy management, because of the great quantity of energy passing through the boiler plants.

As explained in Chapter 6, a control system must ensure at the same time minimum fuel consumption, minimum pollution and safe operation. A gain of a few points in efficiency, at least 1-2% or more depending on the previous efficiency, is always obtainable, but one of the main advantages is certainly the possibility of maintaining this saving throughout the operation thus avoiding departures from the optimum operating mode, such as generally occur with manual operation or with frequent changes of the load profile demand.

On the contrary, the control of electrical substations, though important from the safety and maintenance point of view, is scarcely justified by the amount of energy saved by the optimum exploitation of the transformers and by the reduction of the no-load losses if unloaded transformers are switched off.

17.5 AIR-COMPRESSOR PLANT CONTROL

In many factories, where compressed air is widely used for power purposes, the air-compressor plant is one of the most important facilities from the energy point of view.

Control strategies are implemented by the use of analogic signals, which indicate pressure and temperature, as well as of digital signals which indicate the on/off state of compressors and related equipment. Depending on the type of compressor (reciprocating compressor, screw compressor, etc.) on/off control or continuous control can be used. Cooling system operation can also be optimized.

17.6 PUMP AND FAN CONTROL

Pumps and fans, if controlled by means of non-dissipative systems such as variable-speed drives, allow significant energy saving (see Chapter 10.3). A centralized control of pumps and fans can increase the energy saved through a correct choice of the operating conditions for the whole factory, while the best working point of each system depends on the individual control of pump or fan.

17.7 REFRIGERATION-PLANT CONTROL

Refrigeration-plant control can yield significant energy saving in large plants, both for process and conditioning uses.

A centralized control requires the collection of thousands of parameters all over the site to be refrigerated (internal and external temperatures, pressure in evaporator equipment, oil temperature, cooling water or air temperature, etc.) and the correlation of these values with the demand profile in order to achieve the range of working points which perform best from the energy point of view. If a variable-speed drive compressor is installed, the centralized system delivers input signals to the control of the power electronic converter of the electrical drive.

17.8 HVAC SYSTEMS AND LIGHTING CONTROL

In residential and in commercial and industrial buildings, depending on their specific use, HVAC and lighting may represent a significant part of the energy consumption.

The centralized control must be included in the original design and installation because introducing it at a later stage would require retrofitting of the lighting and HVAC systems which quite often would necessitate an expenditure little less than the original capital cost of the plant.

The basic principles of the control strategies have been presented briefly in Chapters 13 and 14. Details can be suggested by specialists; here the stress is on the importance of planning centralized control before plant installation, since after this phase it would become extremely costly and technically less satisfactory.

This system is different from other facilities controls in industry, where equipment is generally concentrated in specific rooms to render data collection and control cheaper and easier.

17.9 CORRELATION BETWEEN PROCESS, ENERGY AND WASTE MANAGEMENT

Process, energy and waste are closely interconnected throughout the existence of the factory.

Production control, of both quality and quantity, means (i) a better choice of raw materials and of their treatment procedure; (ii) reduction of rejects and waste material from production; (iii) reduction of environmental pollution due to the reduction of emission from combustion during the process.

Energy control, of both electric and thermal energy, means a set of actions performed to reduce the energy consumption per unit of production either by modifying the operation of the plant or by installing new plants designed to be easily controlled. A reduction of primary energy consumption is then achieved; inside the factory it means reduction of energy costs and reduction of the pollution due to fuel combustion.

Control of solid, liquid and gaseous waste flow can reduce energy consumption if waste is recovered as hot stream or as combustible. Reduction of waste means also reduction of energy consumption for its treatment, and reduction of environmental pollution. In controlling the operating parameters of processes, care must be taken to choose the most appropriate raw materials and to avoid undesirable reactions.

17.10 PRACTICAL EXAMPLES

An example of the application of the computer to energy management in a factory would require a great many details and data on the process and facility equipment concerned and on the hardware and software chosen.

Many kinds of software and hardware are on the market, and many levels of control are possible inside a factory.

Example

A computerized system to control factory facilities

In order to give a general example of application in industry, in what follows the factory is represented by its electric and thermal model, as detailed in Table 17.1, which gives estimates of energy saving in facilities on the basis of practical experience.

Other advantages, such as reduction of working hours for plant maintenance and operation will certainly increase the economic advantages of the control system (see also Example 2 in Chapter 13.7), but they are not here considered.

Economic evaluation is shown in Table 20.3.

Table 17.1 Energy saving due to computer control of facilities

Electric model	Average absorbed power	Working hours	Annual consumption	ESTIMATED ENERGY SAVING	
	kW	h/year	kWh/year	%	kWh/year
Air-compressor plant	250	5000	1250000	7	87500
Pump plant	75	5000	375000	15	56250
HVAC refrigerating plant	500	1750	875000	7.5	65625
HVAC auxiliaries	100	3500	350000	12.5	43750
Lighting	100	5000	500000	5	25000
Process	800	5000	4000000	0	0
Others	250	5000	1250000	0	0
TOTAL	2075		8600000		278125
	a	b	c = a · b	d	e = c · d/100

Thermal model	Average absorbed power	Working hours	Annual consumption	ESTIMATED ENERGY SAVING	
	kW	h/year	TOE/year	%	TOE/year
Process boiler plant	7500	5000	3225	3	96.751
HVAC boiler plant	1000	1750	150.5	3	4.515
TOTAL	8500		3375.5		101.266
	a	b	c(*)	d	e = c · d/100

Notes
(*)$c = a \cdot b \cdot 3600/(41860 \cdot 10^3)$
41860 kJ/kg as net heating value

CHAPTER 18

ENERGY MANAGEMENT STRATEGIES FOR CONTROL AND PLANNING

18.1 BASIC PRINCIPLES OF ENERGY MANAGEMENT STRATEGIES

Energy management strategies traditionally include four main sections: analysis of historical data; energy audits and accounting; engineering analysis and investment proposals based on feasibility studies; training and information of personnel. Several factors must be considered:

— incidence of energy costs on turnover and added value;

— percentages of thermal and electric energy both as used in process and as primary energy;

— identification and quantification of thermal and electric energy flows from transformation plants to users;

— selection of significant energy indexes valid for the whole factory and for single production lines;

— possibilities of heat recovery;

— comparative incidence on total consumption of energy used for process and services.

> **Energy strategies can be implemented at different levels depending on the importance of energy in comparison with other production factors. Of course, also the absolute cost of energy consumed in the factory is a parameter to be considered before starting any energy saving program.**

It is suggested that energy strategies should be correlated with waste strategies because of their close interdependence.

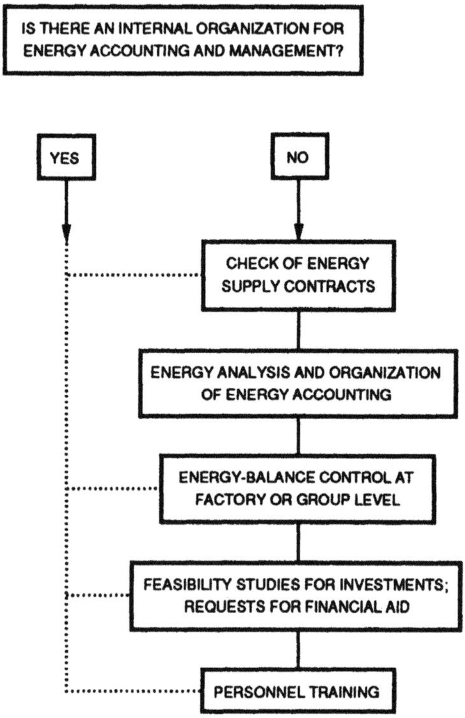

Fig. 18.1 A suggested basic approach to energy management: the program can be initiated at different levels

18.2 THE DIFFERENT STEPS OF AN ENERGY MANAGEMENT APPROACH

Fig. 18.1 shows a suggested basic approach to energy management. To initiate the energy management program, it is necessary to know the organization and functioning of the factory.

If the energy management problem has not been treated before, all the phases shown in Fig. 18.1 must be carried out. In the opposite case, each phase of the program can be developed separately.

18.3 THE CHECKING OF ENERGY SUPPLY CONTRACTS WITH UTILITIES

This is a preliminary check based on the analysis of the contracts with utilities in order to reduce costs without any significant change in the process and service operation modes. Depending on the local tariffication, the suitability of the contract for the factory demand profile (kilowatt load profiles with 15-min or 30-min integrated mean values, kilowatt-hour energy consumption during peak and off-peak hours and load factor) must be investigated carefully in order to obtain the cheapest available rate.

A more detailed analysis can be made when the energy situation of the factory is completely known.

18.4 ENERGY ANALYSIS AND AUDIT FOR PROCESS AND FACILITIES

Energy analysis and audit include the following five steps which are summarized in the block of Tables 18.A (from Table A.0 to Table A.22):

1) data collection regarding at least the last three years, if no significant changes occurred in this period. For cost of unit of energy see Chapter 20.3. The data can be organized on a monthly basis (from Table A.10 to Table A.20):
   - electrical energy consumption in kWh and power consumed in kW during different rate hours, if rate hours tariffication (hours a, b, c) is applied, and related costs (Table A.10). When different contracts exist, several tables can be used to report data referring to a single contract. The power factor charge can also be reported if applied
   - natural gas consumption in Sm^3 and related costs (Table A.11)
   - oil consumption in kg and related costs (Table A.12)
   - gasoil consumption in liter (L) and related costs (Table A.13)
   - LPG consumption in liter (L) and related costs (Table A.14)
   - coal consumption in kg and related costs (Table A.15)
   - other combustibles and related costs (Table A.16)
   - water consumption from both utilities and factory wells in m^3 and related costs (Table A.17)

- production grouped in main categories (in this example: productions A, B and total A+B) expressed in t or number of pieces or some other significant unit. A unifying unit must also be employed in order to have a measure of the total production; otherwise, the working hours can be assumed as an index of the total production (Table A.18)
- waste production grouped in main categories expressed in t or in m^3 (Table A.19). Energy content can be detailed if significant
- turnover of each year as in Table A.20;
- conversion factors used in the energy analysis are shown in Table A.21;

2) construction of energy and mass flow charts from input to output of the factory. Process and facilities energy flow charts must be correlated with each other for a preliminary test of congruence. They must be congruent with factory models as defined at step 4. An example of a mass and energy flow chart is shown in Table A.22;

3) elaboration of data collected in steps 1) and 2) and correlation among them in order to extrapolate parameters significant for assessment of the factory from the energy point of view (from Table A.1 to Table A.9). Annual data of the last three years are used here, derived from the monthly-based data as in step 1:
   - synthesis of energy consumption and costs as in Table A.1. Percentage distribution of different forms of energy
   - yearly variations of energy consumption and related costs (Table A.2)
   - yearly variations of different types of production (Table A.3)
   - incidence of the total energy cost on the turnover for the last three years (Table A.4)
   - attribution of the total energy consumption, electric and thermal, to the different categories of production (Table A.6.1, A.6.2). This attribution can be made in different degrees of detail, depending on the data available and on their mode of elaboration (see Tables A.5.1, A.5.2, A.5.3, A.5.4 in step 4)
   - yearly specific energy consumption and cost, both electric and thermal, referred to parameters which have been assumed as the most significant for production as discussed in step 1 (Table A.7)

- comparison of the specific energy consumption and cost for the factory under examination with the same parameters selected as standard values for the same sector or for the factory group. To avoid misleading comparisons, care must be taken to use the same methods of collecting and elaborating data (Table A.8)
- Table A.9 is used for comparison of consumption and cost among individual end users (process and facility), depending on particular situations inside the single factory or the group of factories;

4) modelling of the factory as electric, thermal, refrigeration, water models as shown in Tables A.5.1, A.5.2, A.5.3, A.5.4.
   The approach for the models is:
   - list of users with their installed power or rated consumption per hour (electric, thermal, cooling, water consumed per hour)
   - percentage load factor depending on the mode of operation in the period to which the model refers. As this parameter is multiplied by the installed power to obtain the consumed power, the efficiency of the system (motors, distribution lines, etc.) is included in the percentage load factor
   - consumed power equal to the installed power multiplied by the percentage load factor
   - working hours per day, working days per year and working hours per year as product of these two parameters
   - yearly energy consumption as the product of the consumed power multiplied by the working hours per year
   - codification of end users by introducing several identification codes, generally four are enough, to indicate the user categories (process or facilities; areas of production; etc.) or their attribution to boiler plants and electrical substations around the factory;

5) investment classification as shown in Table A.0. Investment cost, savings (saving in energy and operating cost minus additional costs) and a preliminary economic analysis by means of the pay-back parameter are detailed. The state of each investment evaluation must be shown.

The different models must be cross-checked to verify the consistency of different modes of water and energy consumption (electric, thermal, etc.) by the same end user. In this way, anomalies due to variation from standards can be detected.

The results of the model elaboration can also be used to attribute the energy consumption to the different categories of production.

Energy flow measurements, which are made systematically in many parts of chemical, petrochemical and other energy-intensive factories, can also be implemented in medium energy intensive manufacturing industries by a proper selection of the main flows to be measured in order to reduce investment for instrumentation and related maintenance costs. The results of these measurements can, moreover, be used to improve the validity of the energy models and the precision with which energy consumption is attributed to the different categories of production (as shown in Tables A.6.1, A.6.2).

The modelling, which evidences the relative importance of different users and facilities from the energy point of view, is a preliminary phase for any energy accounting, whether off-line or on-line, and for any centralized energy management system.

18.5 PRELIMINARY TECHNICAL AND ECONOMIC EVALUATION OF ENERGY-SAVING INVESTMENTS

The chief investments aimed at saving costs and energy must be analyzed, technically and economically, on the basis of the results of the previous steps. This is one of the main goals of an energy analysis and is often the phase in which many people in the factory are most interested. Various methods of evaluation, such as those shown in Chapter 20, can be applied. It is worth noticing that quite often energy-saving investments involve many other factors, such as reduction of maintenance costs and air pollution, the incidence of which is not easy to evaluate, but which often improve the return on the investment.

Investments can be classified as in Table A.0, which is generally presented as the first part of an energy analysis report to stress the importance of these items.

Note that investments on process plants must be checked carefully with the production technologists because they can involve important changes in process both in quality and in quantity. Except in chemical, petrochemical and other energy-intensive industries, it is difficult to justify investment on process plant from the energy point of view alone, but plant automation, improvement in production and replacement of obsolete plants are often the key factors in such decisions.
On the other hand, energy-saving investments on facilities may be agreed more easily, since they do not affect process nor involve factory overall strategies directly. Therefore the energy factor alone is often enough to account for investments on facilities which may prove profitable and at the same time meet production needs.

18.6 OFF-LINE AND ON-LINE PROCEDURES FOR ENERGY ACCOUNTING

The energy-consumption trends are checked periodically by examining the energy accounting so as to verify data consistency and detect variance from standard values and company goals (both for the whole process and for single lines).

This procedure can be entrusted either to the company's internal structures or to independent organizations by means of on-line procedures (as in Chapter 17) or off-line procedures.

As a rule, these procedures include:

— periodical elaboration (daily, weekly, monthly) of energy consumption and production data to update significant consumption parameters;
— meetings with staff for critical evaluation of consumption trends and to check the efficiency and development of company policies;

— controller's reports providing an up-to-date picture of the energy situation and comments on observed anomalies. These reports can be diversified to suit the company functions they are intended for.

Energy accounting ensures a continuously high level of energy awareness and allows operating anomalies to be detected very quickly.

The depth and detail of accounting procedures depend on the size and complexity of the company's energy use and on the main motive determining the collection of data (basically: to attribute only the energy cost to different categories of production or to achieve the greatest efficiency in energy use).

The block of Tables 18.B (from Table B.1 to Table B.10) shows an example of energy accounting for a medium energy intensive factory or group of factories (each center of cost can be part of the factory or a whole factory in a group of factories belonging to the same company):

— Tables B.6-B.10 are the input data tables (thermal energy, electricity, water, production) which are collected periodically (monthly in most cases) or measured and then elaborated. These data refer to each center of cost (single production line or single factory from NO.1 to NO.5 or more);
— Tables B.1-B.5 refer singly to each center of cost (single production line or single factory) and elaborate data given in Tables B.6-B.10. The following data are introduced: production (units of production, working hours, others), electric energy consumption and related cost, thermal energy consumption and related cost, water consumption and related cost. Elaboration of these data yields: specific energy consumption per production unit, specific energy cost per production unit, specific water consumption per production unit (see Chapter 20.3 for cost of unit of energy);
— Table B.0 summarizes all the centers of cost by giving a picture of the energy situation of the whole factory (as sum of the production lines) or of a group of factories (as sum of the factories). Specific consumption related to a unifying production parameter can be assumed as a reference energy consumption value.

These procedures can be implemented off-line or on-line depending on the availability of centralized control for process and facilities plants. The achievement of better operation and maintenance is the main reason for having centralized control equipment; however, if existing, it can conveniently be used for accounting procedures.

18.7 FEASIBILITY STUDIES AND FINANCING

Feasibility studies are a necessary step, accompanying any proposal for plant investments, which of course involve a technical and economic evaluation of the solutions put forward in relation to the present and future needs of the company.

Should it be decided to execute a project, it is necessary to choose the best financing channels and go through all the bureaucratic, energy and environmental steps.

In many industrial sectors, besides the classic heat recovery from water or hot gas streams, investment in centralized plant control (see Chapter 17), particularly in facilities, is a quite significant approach to energy management which requires a preliminary study of specific energy parameters, measurement methods and control algorithms. For this purpose, it is absolutely indispensable to know the energy flows of the whole factory as well as of single process lines: hence the need for combined strategies like those described in the previous paragraphs.

Non-energy additional costs and savings must be included in feasibility studies, particularly for the replacement of process and facility plants and for computer control systems. Very often, only the contribution of these savings can justify the investment.

Economic evaluation must always be made. The principal methods are reviewed in Chapter 20.

18.8 PERSONNEL TRAINING

This is a further aspect of energy management which must not be overlooked, because energy-saving policies require responsive and motivated staff (see also Chapter 19).

This is fundamental at all levels of company functioning among plant operators, production planning and accounting operators, who often do not receive sufficient information or fail to see their own role and tasks in the context of an economical energy policy.

18.9 A SUCCESSFUL ENERGY MANAGEMENT PROGRAM

The success of an energy management program depends on many activities which must bé promoted by the energy management staff and accepted and developed by others working in the factory.

A thorough knowledge of the whole factory, in terms of process and facilities plants and of their operation and maintenance, is the first requirement. Thus, any opportunities to save energy, whether in conducting or retrofitting the existing plants or in installing new plants, must be listed, and any correlation with production and environmental conservation must be pointed out by the energy management staff. The results of these investigations must be checked with other departments (production, maintenance, engineering, finance) before the presentation of a final report.

Advantages and disadvantages must be clearly indicated and an economic evaluation must always be included.

A periodical account of energy flows related to production must be organized before promoting any investment on plants. In this way, the results will be easy to check and the success of the energy strategies will be better appreciated.

Information on energy strategies and results must be made known throughout the factory to motivate people to maintain the level of efficiency already reached and to improve it in the future.

Table 18.A Energy analysis and audit report — Output

A.0 SUMMARY OF THE INVESTMENTS — USER: FACTORY 1

TYPE OF INVESTMENT	INVESTMENT MU	SAVING MU/year	PAYBACK year	INVESTMENT STATE	INVESTMENT TYPE
OPTIMIZATION OF ENERGY SUPPLY CONTRACT	25	115.3	0.22	C	5
HEAT RECOVERY FROM PROCESS WATER	45	25	1.80	C-F	2
INSTALLATION OF RADIANT TUBES FOR HEATING	850	215	3.95	C-NF	6
GAS TURBINE COGENERATION (1000 kW)	1500	225	6.67	C-TD	1
ALTERNATIVE MOTOR COGENERATION (1000 kW)	1000	300	3.33	C-TD	1

NQ = Not quantifiable
C = Completed
S = Under scrutiny
TD = Technical description to be completed
NF = Financial aid not available
F = Financial aid available

Type of Investment:

1 Cogeneration
2 Heat recovery
3 Power station retrofitting/substitution of fuels
4 Process equipment retrofitting
5 Optimization of energy supply contracts
6 Others
7 Process waste recovery

Table 18.A Energy analysis and audit report — Output

A.1 YEARLY CONSUMPTION AND COSTS — USER: FACTORY 1

			CONSUMPTION TOE	%	COSTS MU	%	MU/TOE
YEAR 199x							
TOTAL ENERGY	——————>		6397.18	100.00%	2043.336	100.00%	0.319
ELECTRICITY	13415	MWh	3353.76	52.43%	1318.078	64.51%	0.393
THERMAL ENERGY	127397	GJ	3043.42	47.57%	725.258	35.49%	0.238
- Natural Gas	3688990	Sm^3	3043.42	47.57%	725.258	35.49%	0.238
- Gasoil	0	L	0.00	0.00%	0	0.00%	
- Oil	0	kg	0.00	0.00%	0	0.00%	
- LPG	0	L	0.00	0.00%	0	0.00%	
YEAR 199y			CONSUMPTION TOE	%	COSTS MU	%	MU/TOE
TOTAL ENERGY	——————>		6314.25	100.00%	2050.131	100.00%	0.325
ELECTRICITY	13078	MWh	3269.60	51.78%	1486.482	72.51%	0.455
THERMAL ENERGY	127449	GJ	3044.65	48.22%	563.649	27.49%	0.185
- Natural Gas	3690479	Sm^3	3044.65	48.22%	563.649	27.49%	0.185
- Gasoil	0	L	0.00	0.00%	0	0.00%	
- Oil	0	kg	0.00	0.00%	0	0.00%	
- LPG	0	L	0.00	0.00%	0	0.00%	
YEAR 199z			CONSUMPTION TOE	%	COSTS MU	%	MU/TOE
TOTAL ENERGY	——————>		6469.06	100.00%	2285.725	100.00%	0.353
ELECTRICITY	12766	MWh	3191.61	49.34%	1559.682	68.24%	0.489
THERMAL ENERGY	137194	GJ	3277.45	50.66%	726.043	31.76%	0.222
- Natural Gas	3909360	Sm^3	3225.22	49.86%	703.890	30.80%	0.218
- Gasoil	57724	L	48.83	0.75%	20.952	0.92%	0.429
- Oil	0	kg	0.00	0.00%	0	0.00%	
- LPG	5459	L	3.39	0.05%	1.201	0.05%	0.354

Table 18.A Energy analysis and audit report — Output

A.2 TRENDS OF YEARLY ENERGY CONSUMPTION AND COSTS — USER: FACTORY 1

ENERGY CONSUMPTION

YEAR	TOTAL ENERGY		ELECTRICITY			THERMAL ENERGY	
	TOE	VAR %	TOE	MWh	VAR %	TOE	VAR %
199x	6397.18		3353.76	13415		3043.42	
199y	6314.25	-1.30%	3269.60	13078	-2.51%	3044.65	0.04%
199z	6469.06	2.45%	3191.61	12766	-2.39%	3277.45	7.65%

YEAR	NATURAL GAS			GASOIL			OIL			LPG		
	TOE	Sm^3	VAR %	TOE	L	VAR %	TOE	kg	VAR %	TOE	L	VAR %
199x	3043.42	3688990		0.00	0		0.00	0		0.00	0	
199y	3044.65	3690479	0.04%	0.00	0		0.00	0		0.00	0	
199z	3225.22	3909360	5.93%	48.83	57724		0.00	0		3.39	5459	

ENERGY COSTS

YEAR	TOTAL ENERGY		ELECTRICITY		THERMAL ENERGY	
	MU	VAR %	MU	VAR %	MU	VAR %
199x	2043.336		1318.078		725.258	
199y	2050.131	0.33%	1486.482	12.78%	563.649	-22.28%
199z	2285.725	11.49%	1559.682	4.92%	726.043	28.81%

YEAR	NATURAL GAS		GASOIL		OIL		LPG	
	MU	VAR %	MU	VAR %	MU	VAR %	MU	VAR %
199x	725.258		0		0		0	
199y	563.649	-22.28%	0		0		0	
199z	703.890	24.88%	20.952		0		1.201	

Table 18.A Energy analysis and audit report						Output
A.3 YEARLY PRODUCTION DATA					USER: FACTORY 1	
	PRODUCTION A		PRODUCTION B		TOTAL PRODUCTION	
	t	VAR %	t	VAR %	t	VAR %
199x	17830		6629		24459	
199y	17917	0.49%	6705	1.15%	24622	0.67%
199z	16160	-9.81%	8169	21.82%	24328	-1.19%

Table 18.A Energy analysis and audit report						Output
A.4 INCIDENCE OF TOTAL ENERGY COSTS ON TURNOVER					USER: FACTORY 1	
YEAR	ENERGY COSTS		TURNOVER		ENERGY COSTS/TURNOVER	
	MU	VAR %	MU	VAR %	%	VAR %
199x	2043.34		43525		4.69%	
199y	2050.13	0.33%	45675	4.94%	4.49%	-4.39%
199z	2285.73	11.49%	50245	10.01%	4.55%	1.35%

Table 18.A Energy analysis and audit report — Input

A.5.1 ELECTRIC MODEL OF THE FACTORY (Reference year 199y) USER: FACTORY 1

End users	Installed power kW	Load factor %	Absorbed power kW	Working hours h/d	Working days d/year	Working hours h/year	Absorbed energy MWh/year	Percentage energy %	Codification (4 codes)
PRODUCTION LINES									
Line 1	1176	25	294	16	220	3520	1035	7.91	
Line 2	845	25	211	16	220	3520	744	5.69	
Line 3	760	25	190	8	330	2640	502	3.84	
Line 4	1250	30	375	24	330	7920	2970	22.71	
Line 5	900	25	225	24	220	5280	1188	9.08	
Line 6	1525	20	305	16	220	3520	1074	8.21	
Line 7	750	35	263	24	220	5280	1386	10.60	
FACILITIES								0.00	
Compressed air	220	40	88	24	330	7920	697	5.33	
Offices	120	40	48	16	220	3520	169	1.29	
Boiler plant	230	65	150	24	330	7920	1184	9.05	
Indoor lighting	75	80	60	16	220	3520	211	1.61	
Outdoor lighting	25	70	18	8	365	2920	51	0.39	
Waste treatment	150	70	105	24	330	7920	832	6.36	
Water wells	150	60	90	24	365	8760	788	6.03	
Others	210	15	32	24	330	7920	249	1.91	
TOTAL	8386		2452				13079	100.00	
NOTE FOR TABLE A.6.1:									
PRODUCTION A	75% FACILITIES						3136		
	Line 4/5/6/7						6618		
	Total Production A						9754	74.00	
PRODUCTION B	25% FACILITIES						1045		
	Line 1/2/3						2280		
	Total Production B						3325	26.00	

Table 18.A Energy analysis and audit report — Input

A.5.2 THERMAL MODEL OF THE FACTORY (Reference year 199y) USER: FACTORY 1

End users	Installed power kW	Load factor %	Absorbed power kW	Working hours h/d	Working days d/year	Working hours h/year	Absorbed energy TOE/year	Percentage energy %	Codification (4 codes)
PRODUCTION LINES									
Line 4	4650	40	1860	24	220	5280	845	27.75	
Line 5	3780	35	1323	24	330	7920	901	29.61	
Line 6/7	1100	40	440	12	220	2640	100	3.28	
HEATING									
Factory	9750	40	3900	24	140	3360	1127	37.03	
Offices	1050	35	368	16	140	2240	71	2.33	
TOTAL	20330		7891				3043	100	
NOTE FOR TABLE A.6.1:									
PRODUCTION A		100% PRODUCTION LINES (4/5/6/7)					1846		
		75% HEATING					899		
		Total Production A					2745	90.2	
PRODUCTION B		25% HEATING					298	9.8	

Table 18.A Energy analysis and audit report — Input

A.5.3 REFRIGERATION MODEL OF THE FACTORY (Reference year 199y) USER: FACTORY 1

End users	Installed power kW	Load factor %	Absorbed power kW	Working hours h/d	Working days d/year	Working hours h/year	Absorbed energy TOE/year	Percentage energy %	Codification (4 codes)
Line	0	.....	0	.....	.....	.....	0	.....	.....
Line	0	.....	0	.....	.....	.....	0	.....	.....
Line	0	.....	0	.....	.....	.....	0	.....	.....
Facility	0	.....	0	.....	.....	.....	0	.....	.....
Facility	0	.....	0	.....	.....	.....	0	.....	.....
Offices	0	.....	0	.....	.....	.....	0	.....	.....
TOTAL	0		0				0	.....	

NOTE
No consumption for refrigeration in Factory 1

Table 18.A Energy analysis and audit report — Input

A.5.4 WATER CONSUMPTION MODEL OF THE FACTORY (Reference year 199y) USER: FACTORY 1

End users	Rated consumption m^3/h	Load factor %	Actual consumption m^3/h	Working hours h/d	Working days d/year	Working hours h/year	Annual consumption $m^3/year$	Percentage consumption %	Codification (4 codes)
PRODUCTION LINES									
Line 1	70	35	25	16	220	3520	86240	10.14	
Line 2	55	40	22	16	220	3520	77440	9.11	
Line 3	55	40	22	8	330	2640	58080	6.83	
Line 4	75	35	26	24	330	7920	207900	24.44	
Line 5	55	35	19	24	220	5280	101640	11.95	
Line 6	115	35	40	16	220	3520	141680	16.66	
Line 7	60	40	24	24	220	5280	126720	14.90	
FACILITIES							0	0.00	
Compressed air	6	75	5	24	330	7920	35640	4.19	
Boiler plant	2	65	1	24	330	7920	10296	1.21	
Others	1	62	1	24	330	7920	4871	0.57	
TOTAL	494		185				850507	100.00	

NOTE			
PRODUCTION A	75% FACILITIES	38105	
	Line 4/5/6/7	577940	
	Total Production A	616045	72.50
PRODUCTION B	25% FACILITIES	12702	
	Line 1/2/3	221760	
	Total Production B	234462	27.50

Table 18.A Energy analysis and audit report			Input
A.6.1 PERCENTAGE ATTRIBUTION OF TOTAL ENERGY CONSUMPTION TO THE PRODUCTION			USER: FACTORY 1
YEAR 199x	PRODUCTION UNIT u	ELECTRICITY %	THERMAL ENERGY %
PRODUCTION A	t	71.00%	87.00%
PRODUCTION B	t	29.00%	13.00%
TOTAL PRODUCTION	t	100.00%	100.00%
TOTAL PRODUCTION		100.00%	100.00%
YEAR 199y	PRODUCTION UNIT u	ELECTRICITY %	THERMAL ENERGY %
PRODUCTION A	t	74.00%	90.00%
PRODUCTION B	t	26.00%	10.00%
TOTAL PRODUCTION	t	100.00%	100.00%
TOTAL PRODUCTION		100.00%	100.00%
YEAR 199z	PRODUCTION UNIT u	ELECTRICITY %	THERMAL ENERGY %
PRODUCTION A	t	74.00%	90.00%
PRODUCTION B	t	26.00%	10.00%
TOTAL PRODUCTION	t	100.00%	100.00%
TOTAL PRODUCTION		100.00%	100.00%

Table 18.A Energy analysis and audit report Output

A.6.2 YEARLY PRODUCTION DATA AND BREAKING DOWN OF CONSUMPTION BY PRODUCTION USER: FACTORY 1

YEAR 199x	PRODUCTION UNIT	PRODUCTION	ELECTRICITY	THERMAL EN.	TOTAL ENERGY	ELECTR. COSTS	THERM. COSTS	TOTAL COSTS
	u	QUANTITY	TOE	TOE	TOE	MU	MU	MU
			3353.76	3043.42	6397.18	1318.078	725.258	2043.336
PRODUCTION A	t	17830	2381.17	2647.77	5028.94	935.835	630.975	1566.810
PRODUCTION B	t	6629	972.59	395.64	1368.23	382.243	94.284	476.526
TOTAL PRODUCTION	t	24459	3353.76	3043.42	6397.18	1318.078	725.258	2043.336
YEAR 199y	PRODUCTION UNIT	PRODUCTION	ELECTRICITY	THERMAL EN.	TOTAL ENERGY	ELECTR. COSTS	THERM. COSTS	TOTAL COSTS
	u	QUANTITY	TOE	TOE	TOE	MU	MU	MU
			3269.60	3044.65	6314.25	1486.482	563.649	2050.131
PRODUCTION A	t	17917	2419.50	2740.18	5159.68	1100.00	507.28	1607.28
PRODUCTION B	t	6705	850.10	304.46	1154.56	386.49	56.36	442.85
TOTAL PRODUCTION	t	24622	3269.60	3044.65	6314.25	1486.48	563.65	2050.13
YEAR 199z	PRODUCTION UNIT	PRODUCTION	ELECTRICITY	THERMAL EN.	TOTAL ENERGY	ELECTR. COSTS	THERM. COSTS	TOTAL COSTS
	u	QUANTITY	TOE	TOE	TOE	MU	MU	MU
			3191.61	3277.45	6469.06	1559.68	726.04	2285.73
PRODUCTION A	t	16160	2361.79	2949.70	5311.49	1154.16	653.44	1807.60
PRODUCTION B	t	8169	829.82	327.74	1157.56	405.52	72.60	478.12
TOTAL PRODUCTION	t	24328	3191.61	3277.45	6469.06	1559.68	726.04	2285.73

Table 18 A energy analysis and audit report — Output

A.7 CONSUMPTION AND SPECIFIC COSTS BY PRODUCTION — USER: FACTORY 1

YEAR 199x	PRODUCTION UNIT	PRODUCTION	Electricity Consumption		Thermal Cons.	Total Cons.	Electr.Costs	Therm.Costs	Total Costs
	u	t	1000kWh/u	TOE/u	TOE/u	TOE/u	U/u	U/u	U/u
PRODUCTION A	t	17830	0.534	0.134	0.149	0.282	52487	35388	87875
PRODUCTION B	t	6629	0.587	0.147	0.060	0.206	57662	14223	71885
TOTAL PRODUCTION	t	24459	0.548	0.137	0.124	0.262	53889	29652	83541
YEAR 199y	**PRODUCTION UNIT**	**PRODUCTION**	**Electricity Consumption**		**Thermal Cons.**	**Total Cons.**	**Electr.Costs**	**Therm.Costs**	**Total Costs**
	u	t	1000kWh/u	TOE/u	TOE/u	TOE/u	U/u	U/u	U/u
PRODUCTION A	t	17917	0.540	0.135	0.153	0.288	61395	28313	89708
PRODUCTION B	t	6705	0.507	0.127	0.045	0.172	57638	8406	66044
TOTAL PRODUCTION	t	24622	0.531	0.133	0.124	0.256	60372	22892	83264
YEAR 199z	**PRODUCTION UNIT**	**PRODUCTION**	**Electricity Consumption**		**Thermal Cons.**	**Total Cons.**	**Electr.Costs**	**Therm.Costs**	**Total Costs**
	u	t	1000kWh/u	TOE/u	TOE/u	TOE/u	U/u	U/u	U/u
PRODUCTION A	t	16160	0.585	0.146	0.183	0.329	71423	40437	111860
PRODUCTION B	t	8169	0.406	0.102	0.040	0.142	49644	8888	58532
TOTAL PRODUCTION	t	24328	0.525	0.131	0.135	0.266	64111	29844	93955

Table 18.A Energy analysis and audit report — Output

A.8 COMPARISON WITH SECTOR MEAN VALUES OF SPECIFIC CONSUMPTION — USER: FACTORY 1

REFERENCE PRODUCTION	REFERENCE UNIT u	SPECIFIC CONSUMPTION TOE/u	FACTORY	REFERENCE UNIT u	199x TOE/u	199y TOE/u	199z TOE/u
PRODUCTION A	t	0.275	PRODUCTION A	t	0.282	0.288	0.328
PRODUCTION B	t	0.15	PRODUCTION B	t	0.206	0.172	0.142
TOTAL PRODUCTION	t	0.26	TOTAL PRODUCTION	t	0.261	0.256	0.266

REFERENCE DATA FROM DATA BANK OR HISTORICAL DATA

Table 18.A Energy analysis and audit report — Output

A.9 COMPARISON WITH REFERENCE MEAN VALUES OF SPECIFIC CONSUMPTION OF SINGLE PRODUCTION LINE OR FACILITY — USER: FACTORY 1

END USER	REFERENCE UNIT	SPECIFIC CONSUMPTION		FACTORY END USER	REFERENCE UNIT	199x		199y		199z	
	u	kWh/u	TOE/u		u	kWh/u	TOE/u	kWh/u	TOE/u	kWh/u	TOE/u
LINE ...	u1	.....	.....	LINE ...	u1	....	....	....	....	....	....
LINE...	u2	.....	.....	LINE...	u2	....	....	....	....	....	....
FACILITY...	u3	.....	.....	FACILITY...	u3	....	....	....	....	....	....

REFERENCE DATA FROM DATA BANK OR HISTORICAL DATA

Table 18.A Energy analysis and audit report — Input

A.10 MONTHLY CONSUMPTION AND COSTS OF ELECTRICITY 199x-199z USER: FACTORY 1

	HOURS a	HOURS b	HOURS c	HOURS a	HOURS b	HOURS c	ENERGY	COSTS	COST/unit
YEAR 199x	kW	kW	kW	kWh	kWh	kWh	kWh	U	U/kWh
JAN	2838	2847	2298	181760	495360	423040	1100160	126904199	115.35
FEB	2841	2769	2435	188160	504320	484480	1176960	128791173	109.43
MAR	2808	2944	2445	198400	543360	471680	1213440	135981181	112.06
APR	2848	2846	2576	164480	449920	504320	1118720	122322999	109.34
MAY		2777	2734		718080	497920	1216000	86851310	71.42
JUN		2861	2509		735360	485120	1220480	86944785	71.24
JUL		3107	2694		707840	504960	1212800	87679842	72.30
AUG		2880	2628		382080	228480	610560	64818619	106.16
SEPT		2962	2728		763520	512640	1276160	98786762	77.41
OCT	2743	2677	2687	192000	536320	511360	1239680	124907686	100.76
NOV	2824	2853	2683	194560	536320	476800	1207680	137285356	113.68
DEC	2825	2965	2547	136960	375680	309760	822400	116804052	142.03
YEAR 199y									
JAN	2699	2645	2410	188800	512640	421120	1122560	129939743	115.75
FEB	2715	2610	2442	168320	471680	379520	1019520	123839500	121.47
MAR	2745	2834	2645	187520	538240	457600	1183360	139813081	118.15
APR	2671	2706	2419	162560	457600	354560	974720	124823112	128.06
MAY		2560	2256		717440	387840	1105280	102475338	92.71
JUN		2626	2623		673920	453120	1127040	102596954	91.03
JUL		2817	2676		696320	437760	1134080	105103946	92.68
AUG		2850	2612		400000	232960	632960	83937223	132.61
SEPT		2849	2757		706560	479360	1185920	108018132	91.08
OCT	2906	2896	2755	214400	595200	458240	1267840	161515537	127.39
NOV	2834	2777	2195	224400	615200	488240	1327840	171515537	129.17
DEC	2800	2750	2225	174320	487520	335440	997280	132904364	133.27
YEAR 199z									
JAN	2823	2736	2074	205440	571520	404480	1181440	144781055	122.55
FEB	2716	2750	1859	186880	519040	364800	1070720	138640494	129.48
MAR	2609	2624	1910	185600	513920	371200	1070720	139691302	130.46
APR	2790	2875	2147	168960	470400	341760	981120	135400000	138.01
MAY		2990.7	2147		775680	390400	1166080	112928062	96.84
JUN		2958	2202		705920	392320	1098240	109069592	99.31
JUL		2800	2676		836813	470707	1307520	120663000	92.28
AUG		2800	2612		309658	174182	483840	74351000	153.67
SEPT		2800	2757		799539	449741	1249280	120997000	96.85
OCT	2850	2727	2075	205400	572535	398000	1175935	167735675	142.64
NOV	2715	2755	1863	187000	521035	364975	1073010	157860525	147.12
DEC	2620	2650	1980	162000	445300	301230	908530	137564530	151.41

YEAR	HOURS a	HOURS b	HOURS c	ENERGY	COSTS	COST/unit
	kW	kW	kW	kWh	MU	U/kWh
199x	2818.1	2874	2580.3	13415040	1318.078	98.25
199y	2767.1	2743.3	2501.3	13078400	1486.482	113.66
199z	2734.5	2822.3	2056.5	12766435	1559.682	122.17

Table 18.A Energy analysis and audit report				Input
A.11 MONTHLY NATURAL GAS CONSUMPTION AND COSTS			199x-199z	USER: FACTORY 1
	QUANTITY	COSTS	COST/unit	MEAN CONSUMPTION
YEAR 199x	Sm^3	U	U/Sm^3	Sm^3/d (22 d)
JAN	535385	75694250	141.38	24335.68
FEB	472000	92350000	195.66	21454.55
MAR	400375	76825670	191.88	18198.86
APR	321000	55983540	174.40	14590.91
MAY	242258	45605345	188.25	11011.73
JUN	225562	49387825	218.95	10252.82
JUL	172525	39857650	231.03	7842.05
AUG	55675	22750860	408.64	2530.68
SEPT	172545	38675450	224.15	7842.95
OCT	182546	47500875	260.21	8297.55
NOV	411254	80750950	196.35	18693.36
DEC	497865	99875640	200.61	22630.23
YEAR 199y				
JAN	520381	80692616	155.06	23653.68
FEB	461117	73778720	160.00	20959.86
MAR	422250	62598563	148.25	19193.18
APR	330455	48576885	147.00	15020.68
MAY	232257	33677265	145.00	10557.14
JUN	228591	33831468	148.00	10390.50
JUL	175431	26340965	150.15	7974.14
AUG	58146	13033426	224.15	2643.00
SEPT	170859	25654479	150.15	7766.32
OCT	185525	29173806	157.25	8432.95
NOV	409141	61841662	151.15	18597.32
DEC	496326	74448900	150.00	22560.27
YEAR 199z				
JAN	524634	78695000	150.00	23847.02
FEB	524634	86565000	165.00	23847.02
MAR	524634	81318000	155.00	23847.02
APR	334950	72545670	216.59	15225.00
MAY	289516	65748976	227.10	13159.82
JUN	196693	46750875	237.68	8940.59
JUL	189972	42535480	223.90	8635.09
AUG	40155	9875650	245.94	1825.23
SEPT	160096	32450765	202.70	7277.09
OCT	185675	42756076	230.27	8439.77
NOV	450750	68897650	152.85	20488.64
DEC	487650	75750985	155.34	22165.91
YEAR	QUANTITY	COSTS	COST/unit	MEAN CONS.
	Sm^3	MU	U/Sm^3	Sm^3/d (22 d)
199x	3688990	725.258	196.60	13973.45
199y	3690479	563.649	152.73	13979.09
199z	3909360	703.890	180.05	19744.24

Table 18.A Energy analysis and audit report — Input

A.12 MONTHLY OIL CONSUMPTION AND COSTS 199x-199y USER: FACTORY 1

	QUANTITY	COSTS	COST/unit	MEAN CONS.
YEAR 199x	kg	U	U/kg	kg/d (22 d)
JAN	0	0	.....	0.00
FEB	0	0	.....	0.00
MAR	0	0	.....	0.00
APR	0	0	.....	0.00
MAY	0	0	.....	0.00
JUN	0	0	.....	0.00
JUL	0	0	.....	0.00
AUG	0	0	.....	0.00
SEPT	0	0	.....	0.00
OCT	0	0	.....	0.00
NOV	0	0	.....	0.00
DEC	0	0	.....	0.00
YEAR 199y				
JAN	0	0	.....	0.00
FEB	0	0	.....	0.00
MAR	0	0	.....	0.00
APR	0	0	.....	0.00
MAY	0	0	.....	0.00
JUN	0	0	.....	0.00
JUL	0	0	.....	0.00
AUG	0	0	.....	0.00
SEPT	0	0	.....	0.00
OCT	0	0	.....	0.00
NOV	0	0	.....	0.00
DEC	0	0	.....	0.00
YEAR 199z				
JAN	0	0	.....	0.00
FEB	0	0	.....	0.00
MAR	0	0	.....	0.00
APR	0	0	.....	0.00
MAY	0	0	.....	0.00
JUN	0	0	.....	0.00
JUL	0	0	.....	0.00
AUG	0	0	.....	0.00
SEPT	0	0	.....	0.00
OCT	0	0	.....	0.00
NOV	0	0	.....	0.00
DEC	0	0	.....	0.00

YEAR	QUANTITY	COSTS	COST/unit	MEAN CONS.
	kg	MU	U/kg	kg/d (22 d)
199x	0	0	.....	0.00
199y	0	0	.....	0.00
199z	0	0	.....	0.00

Table 18.A Energy analysis and audit report — Input

A.13 MONTHLY GASOIL CONSUMPTION AND COSTS — 199x-199z — USER: FACTORY 1

	QUANTITY	COSTS	COST/unit	MEAN CONS.
YEAR 199x	L	U	U/L	L/d (22 d)
JAN	0	0	.....	0.00
FEB	0	0	.....	0.00
MAR	0	0	.....	0.00
APR	0	0	.....	0.00
MAY	0	0	.....	0.00
JUN	0	0	.....	0.00
JUL	0	0	.....	0.00
AUG	0	0	.....	0.00
SEPT	0	0	.....	0.00
OCT	0	0	.....	0.00
NOV	0	0	.....	0.00
DEC	0	0	.....	0.00
YEAR 199y				
JAN	0	0	.....	0.00
FEB	0	0	.....	0.00
MAR	0	0	.....	0.00
APR	0	0	.....	0.00
MAY	0	0	.....	0.00
JUN	0	0	.....	0.00
JUL	0	0	.....	0.00
AUG	0	0	.....	0.00
SEPT	0	0	.....	0.00
OCT	0	0	.....	0.00
NOV	0	0	.....	0.00
DEC	0	0	.....	0.00
YEAR 199z				
JAN	5032	1761320	350.00	228.74
FEB	5032	1761320	350.00	228.74
MAR	5032	1761320	350.00	228.74
APR	2717	991705	365.00	123.50
MAY	4832	1763680	365.00	219.64
JUN	4832	1763680	365.00	219.64
JUL	4832	1763680	365.00	219.64
AUG	4832	1763680	365.00	219.64
SEPT	4832	1715360	355.00	219.64
OCT	5250	1968750	375.00	238.64
NOV	5250	1968750	375.00	238.64
DEC	5250	1968750	375.00	238.64

YEAR	QUANTITY	COSTS	COST/unit	MEAN CONS.
	L	MU	U/L	L/d (22 d)
199x	0	0	.....	0.00
199y	0	0	.....	0.00
199z	57724	20.952	362.97	291.54

Table 18.A Energy analysis and audit report				Input
A.14 MONTHLY LPG CONSUMPTION AND COSTS			199x-199z	USER: FACTORY 1
	QUANTITY	COSTS	COST/unit	MEAN CONS.
YEAR 199x	L	U	U/L	L/d (22 d)
JAN	0	0	.....	0.00
FEB	0	0	.....	0.00
MAR	0	0	.....	0.00
APR	0	0	.....	0.00
MAY	0	0	.....	0.00
JUN	0	0	.....	0.00
JUL	0	0	.....	0.00
AUG	0	0	.....	0.00
SEPT	0	0	.....	0.00
OCT	0	0	.....	0.00
NOV	0	0	.....	0.00
DEC	0	0	.....	0.00
YEAR 199y				
JAN	0	0	.....	0.00
FEB	0	0	.....	0.00
MAR	0	0	.....	0.00
APR	0	0	.....	0.00
MAY	0	0	.....	0.00
JUN	0	0	.....	0.00
JUL	0	0	.....	0.00
AUG	0	0	.....	0.00
SEPT	0	0	.....	0.00
OCT	0	0	.....	0.00
NOV	0	0	.....	0.00
DEC	0	0	.....	0.00
YEAR 199z				
JAN	450	90000	200.00	20.45
FEB	450	90000	200.00	20.45
MAR	450	90000	200.00	20.45
APR	754	165880	220.00	34.27
MAY	410	90200	220.00	18.64
JUN	410	90200	220.00	18.64
JUL	420	94500	225.00	19.09
AUG	287	64575	225.00	13.05
SEPT	403	90675	225.00	18.32
OCT	475	111625	235.00	21.59
NOV	475	111625	235.00	21.59
DEC	475	111625	235.00	21.59
YEAR	QUANTITY	COSTS	COST/unit	MEAN CONS.
	L	MU	U/L	L/d (22 d)
199x	0	0	.....	0.00
199y	0	0	.....	0.00
199z	5459	1.201	219.99	27.57

Table 18.A Energy analysis and audit report — Input

A.15 MONTHLY COAL CONSUMPTION AND COSTS — 199x-199y — USER: FACTORY 1

	QUANTITY	COSTS	COST/unit	MEAN CONS.
YEAR 199x	kg	U	U/kg	kg/d (22 d)
JAN	0	0	.....	0.00
FEB	0	0	.....	0.00
MAR	0	0	.....	0.00
APR	0	0	.....	0.00
MAY	0	0	.....	0.00
JUN	0	0	.....	0.00
JUL	0	0	.....	0.00
AUG	0	0	.....	0.00
SEPT	0	0	.....	0.00
OCT	0	0	.....	0.00
NOV	0	0	.....	0.00
DEC	0	0	.....	0.00
YEAR 199y				
JAN	0	0	.....	0.00
FEB	0	0	.....	0.00
MAR	0	0	.....	0.00
APR	0	0	.....	0.00
MAY	0	0	.....	0.00
JUN	0	0	.....	0.00
JUL	0	0	.....	0.00
AUG	0	0	.....	0.00
SEPT	0	0	.....	0.00
OCT	0	0	.....	0.00
NOV	0	0	.....	0.00
DEC	0	0	.....	0.00
YEAR 199z				
JAN	0	0	.....	0.00
FEB	0	0	.....	0.00
MAR	0	0	.....	0.00
APR	0	0	.....	0.00
MAY	0	0	.....	0.00
JUN	0	0	.....	0.00
JUL	0	0	.....	0.00
AUG	0	0	.....	0.00
SEPT	0	0	.....	0.00
OCT	0	0	.....	0.00
NOV	0	0	.....	0.00
DEC	0	0	.....	0.00

YEAR	QUANTITY	COSTS	COST/unit	MEAN CONS.
	kg	MU	U/kg	kg/d (22 d)
199x	0	0	.....	0.00
199y	0	0	.....	0.00
199z	0	0	.....	0.00

Table 18.A Energy analysis and audit report — Input

A.16 MONTHLY CONSUMPTION AND COSTS OF OTHER COMBUSTIBLES — 199x-199y — USER: FACTORY 1

	QUANTITY	COSTS	COST/unit	MEAN CONS.
YEAR 199x	kg	U	U/kg	kg/d (22 d)
JAN	0	0	.....	0.00
FEB	0	0	.....	0.00
MAR	0	0	.....	0.00
APR	0	0	.....	0.00
MAY	0	0	.....	0.00
JUN	0	0	.....	0.00
JUL	0	0	.....	0.00
AUG	0	0	.....	0.00
SEPT	0	0	.....	0.00
OCT	0	0	.....	0.00
NOV	0	0	.....	0.00
DEC	0	0	.....	0.00
YEAR 199y				
JAN	0	0	.....	0.00
FEB	0	0	.....	0.00
MAR	0	0	.....	0.00
APR	0	0	.....	0.00
MAY	0	0	.....	0.00
JUN	0	0	.....	0.00
JUL	0	0	.....	0.00
AUG	0	0	.....	0.00
SEPT	0	0	.....	0.00
OCT	0	0	.....	0.00
NOV	0	0	.....	0.00
DEC	0	0	.....	0.00
YEAR 199z				
JAN	0	0	.....	0.00
FEB	0	0	.....	0.00
MAR	0	0	.....	0.00
APR	0	0	.....	0.00
MAY	0	0	.....	0.00
JUN	0	0	.....	0.00
JUL	0	0	.....	0.00
AUG	0	0	.....	0.00
SEPT	0	0	.....	0.00
OCT	0	0	.....	0.00
NOV	0	0	.....	0.00
DEC	0	0	.....	0.00

YEAR	QUANTITY	COSTS	MEAN COST	MEAN CONS.
	kg	MU	U/kg	kg/d (22 d)
199x	0	0	.....	0.00
199y	0	0	.....	0.00
199z	0	0	.....	0.00

Table 18.A Energy analysis and audit report				Input
A.17 MONTHLY WATER CONSUMPTION AND COSTS			199x-199z	USER: FACTORY 1
	QUANTITY	COSTS	COST/unit	MEAN CONS.
YEAR 199x	m^3	U	U/m^3	m^3/d (22 d)
JAN	61500	2890500	47.00	2795.45
FEB	61500	2890500	47.00	2795.45
MAR	61500	2890500	47.00	2795.45
APR	78000	3510000	45.00	3545.45
MAY	78000	3510000	45.00	3545.45
JUN	78000	3510000	45.00	3545.45
JUL	82500	3712500	45.00	3750.00
AUG	75000	3375000	45.00	3409.09
SEPT	82500	3960000	48.00	3750.00
OCT	61000	2928000	48.00	2772.73
NOV	61000	2928000	48.00	2772.73
DEC	61000	2928000	48.00	2772.73
YEAR 199y				
JAN	62500	3000000	48.00	2840.91
FEB	62500	3000000	48.00	2840.91
MAR	62500	3000000	48.00	2840.91
APR	76000	3648000	48.00	3454.55
MAY	76000	3648000	48.00	3454.55
JUN	76000	3648000	48.00	3454.55
JUL	82000	3936000	48.00	3727.27
AUG	82000	3936000	48.00	3727.27
SEPT	82000	3936000	48.00	3727.27
OCT	63000	3024000	48.00	2863.64
NOV	63000	3024000	48.00	2863.64
DEC	63000	3024000	48.00	2863.64
YEAR 199z				
JAN	63333	3166667	50.00	2878.79
FEB	63333	3166667	50.00	2878.79
MAR	63333	3166667	50.00	2878.79
APR	75000	3750000	50.00	3409.09
MAY	75000	3750000	50.00	3409.09
JUN	75000	3750000	50.00	3409.09
JUL	80000	4000000	50.00	3636.36
AUG	75000	3750000	50.00	3409.09
SEPT	75000	3750000	50.00	3409.09
OCT	62500	3250000	52.00	2840.91
NOV	62500	3250000	52.00	2840.91
DEC	62500	3250000	52.00	2840.91
YEAR	QUANTITY	COSTS	COST/unit	MEAN CONS.
	m^3	MU	U/m^3	m^3/d (22 d)
199x	841500	39.033	46.39	3187.50
199y	850500	40.824	48.00	3221.59
199z	832500	42	50.45	4204.55

Table 18.A Energy analysis and audit report				Input
A.18 MONTHLY PRODUCTION DATA			199x-199z	USER: FACTORY 1
	PRODUCTION A	PRODUCTION B	PRODUCTION A+B	....
YEAR 199x	t	t	t	
JAN	1347	615	1962	
FEB	1486	400	1886	
MAR	1525	725	2250	
APR	1650	487	2137	
MAY	1765	500	2265	
JUN	1735	723	2458	
JUL	1568	729	2297	
AUG	778	300	1078	
SEPT	1680	657	2337	
OCT	1921	723	2644	
NOV	1500	335	1835	
DEC	875	435	1310	
YEAR 199y				
JAN	1354	585	1939	
FEB	1518	408	1926	
MAR	1575	731	2306	
APR	1619	486	2105	
MAY	1751	518	2269	
JUN	1754	737	2491	
JUL	1593	761	2354	
AUG	758	303	1061	
SEPT	1726	658	2384	
OCT	1939	747	2686	
NOV	1545	345	1890	
DEC	785	427	1212	
YEAR 199z				
JAN	1075	852	1926	
FEB	1258	769	2027	
MAR	1372	774	2146	
APR	1340	467	1808	
MAY	1740	748	2488	
JUN	1396	735	2131	
JUL	1631	913	2544	
AUG	420	274	694	
SEPT	1581	907	2488	
OCT	1876	750	2626	
NOV	1546	545	2091	
DEC	925	435	1360	
YEAR	PRODUCTION A	PRODUCTION B	PRODUCTION A+B	....
	t	t	t	
199x	17830	6629	24459	
199y	17917	6705	24622	
199z	16160	8169	24328	

Table 18.A Energy analysis and audit report — Input

A.19 MONTHLY WASTE PRODUCTION — 199x-199z — USER: FACTORY 1

	WASTE no.1	WASTE no.2	WASTE no.3	WASTE no.4	WASTE no.5	WASTE no.6
YEAR 199x	t	t	t	m^3	...	...
JAN	1.35	0.43	0.25	12350	0	0
FEB	1.42	0.23	0.21	11000	0	0
MAR	1.87	0.76	0.12	13457	0	0
APR	1.67	0.34	0.23	11345	0	0
MAY	2.14	0.25	0.23	12450	0	0
JUN	1.35	0.56	0.15	15678	0	0
JUL	1.76	0.57	0.15	13465	0	0
AUG	1.12	0.14	0.04	7500	0	0
SEPT	1.76	0.56	0.17	11346	0	0
OCT	1.98	0.56	0.13	14532	0	0
NOV	1.25	0.33	0.11	10250	0	0
DEC	1.12	0.27	0.08	8750	0	0
YEAR 199y						
JAN	1.27	0.45	0.22	11240	0	0
FEB	1.35	0.36	0.16	10000	0	0
MAR	1.67	0.65	0.23	11234	0	0
APR	1.85	0.43	0.24	14567	0	0
MAY	1.84	0.46	0.19	13280	0	0
JUN	1.67	0.87	0.56	12389	0	0
JUL	1.89	0.82	0.44	11345	0	0
AUG	0.87	0.23	0.12	8760	0	0
SEPT	1.88	0.65	0.35	15670	0	0
OCT	1.78	0.47	0.22	17658	0	0
NOV	1.34	0.27	0.12	16790	0	0
DEC	0.87	0.12	0.07	9800	0	0
YEAR 199z						
JAN	1.34	0.43	0.22	13450	0	0
FEB	1.34	0.86	0.23	13250	0	0
MAR	1.45	0.55	0.27	17860	0	0
APR	1.44	0.45	0.18	11435	0	0
MAY	1.89	0.36	0.16	11279	0	0
JUN	1.56	0.76	0.35	14567	0	0
JUL	1.78	0.33	0.13	13480	0	0
AUG	0.75	0.14	0.07	8890	0	0
SEPT	1.58	0.25	0.13	12356	0	0
OCT	1.87	0.43	0.23	13834	0	0
NOV	1.55	0.45	0.25	12223	0	0
DEC	0.78	0.22	0.11	11638	0	0

	WASTE no.1	WASTE no.2	WASTE no.3	WASTE no.4	WASTE no.5	WASTE no.6
YEAR	t	t	t	m^3	...	...
199x	18.79	5.00	1.87	142123	0	0
199y	18.28	5.78	2.92	152733	0	0
199z	17.33	5.23	2.33	154262	0	0

Table 18.A Energy analysis and audit report			Input
A.20 TURNOVER			USER: FACTORY 1
YEAR	TURNOVER		
	MU	VAR %	
199x	43525		
199y	45675	4.94%	
199z	50245	10.01%	

Table 18.A Energy analysis and audit report						Input
A.21 CONVERSION FACTORS USED IN THE ENERGY ANALYSIS						
THERMAL ENERGY (fuel)	Net Heating Value					
Oil	kJ/kg	41860				
Natural gas	kJ/Sm^3	34534.5				
Gasoil	kJ/L	35162.4				
LPG	kJ/L	26016				
TOE conversion factors						
Oil	1 TOE =	1000	kg			
Natural gas	1 TOE =	1212.1	Sm^3			
Gasoil	1 TOE =	1190.5	L	=	952.4 kg	
LPG	1 TOE =	1609.0	L	=	909.1 kg	
Electric energy	1 TOE =	4000	kWh			

Table 18.A Energy analysis and audit report — Input

A.22 MASS AND ENERGY FLOW CHART (WITH REFERENCE TO A GIVEN QUANTITY OF PRODUCTION) AND INDIVIDUAL END-USER ENERGY REQUIREMENTS

		INLET				OUTLET				
Production phase	Reference (see blocks)	Flow description (mass and energy)	Temperature K	Mass kg	Energy MJ	Flow description (mass and energy)	Temperature K	Mass kg	Energy losses MJ	Energy recovery MJ
....	1	MA material input	....	....	....	MA material output	....	....	....	....
		Electric energy			....					
....	2	MA material input	....	....	....	MC material output	....	....	....	....
		MB material input	....	....	....					
		Electric energy			....					
		Steam input as fuel	....	....	....					
		Natural gas (burners)	....	....	....	Exhaust	....	....	....	....
		Cooling water	....	....	....	Hot water	....	....	....	....
....	3	MC material input	....	....	....	MC material output	....	....	....	....
		Electric energy			....					
Other facilities	4	Electric energy			....					
		Natural gas		....	....	Exhaust	....	....	....	....

Notes

hourly production value can conveniently be introduced

all streams of energy (steam, cooling water, etc.) must be charged with the losses of transformation plants inside the factory before entering the Phase block. Mass flow outlet from facilities are not shown in the diagram

fuels burned inside the block must be introduced by means of their net heating values

1 kWh of electric energy = 3.6 MJ

enthalpy of material flows is calculated by means of the specific heat (enthalpy = mass flow-rate * specific heat)

Energy consumed = (individual end-user consumption) - (waste and energy recovery inside the process) + (balance between the energy contained in raw materials entering the factory and that in products leaving the factory) + (energy losses including transformation and distribution losses)

A.22 (continued)

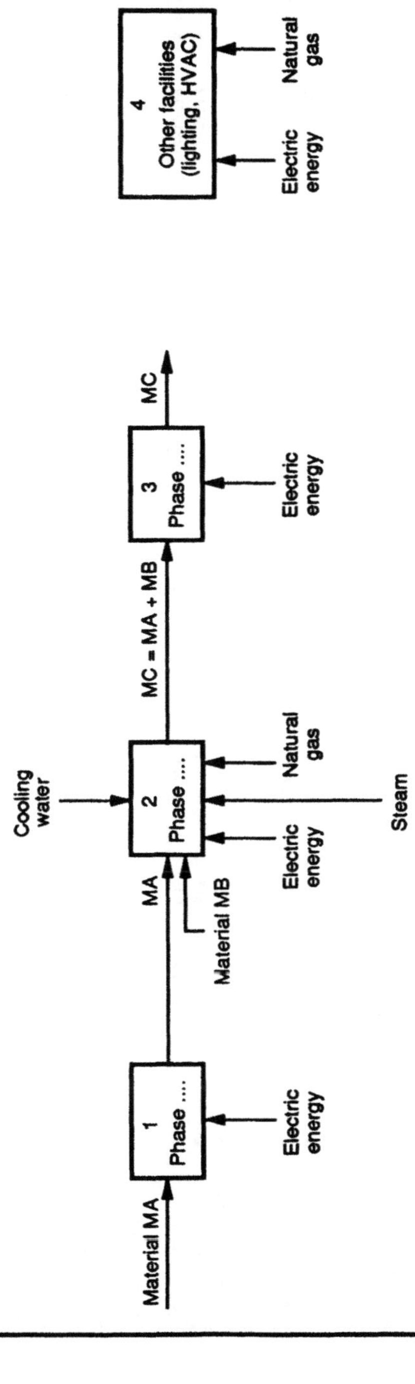

Table 18.B Energy account report — Output

B.0 FACTORY OR GROUP OF FACTORIES

PRODUCTION - TOTAL ENERGY CONSUMPTION 199z (JAN-MAR)

		JAN	FEB	MAR	APR	MAY	JUN	JUL	AUG	SEPT	OCT	NOV	DEC	TOTAL 199z	TOTAL 199y Ref.Year
PRODUCTION															
PRODUCTION	t	4818	5045	5412	....	....	....	....	....	....	....	....	....	15275.32	59981
WORKING HOURS	h	2856	2544	2832	....	....	....	....	....	....	....	....	....	8232	31725
prod/hours	t/l.	1.687	1.983	1.911	....	....	....	....	....	....	....	....	....	1.856	1.891
TOTAL ENERGY CONSUMPTION															
ELECTRICITY	kWh	3485500	3232520	3256920	....	....	....	....	....	....	....	....	....	9974940	36554920
OIL	kg	353239	357622	231365	....	....	....	....	....	....	....	....	....	942226	2421051
NATURAL GAS	Sm^3	524634	524634	524634	....	....	....	....	....	....	....	....	....	1573902	3690479
GASOIL	L	31740	31982	29324	....	....	....	....	....	....	....	....	....	93046	107244
LPG	L	870	1062	1135	....	....	....	....	....	....	....	....	....	3067	8284
WATER	m^3	224843	221281	247392	....	....	....	....	....	....	....	....	....	693516	1766487
TOTAL THERMAL ENERGY															
Equiv.Therm.En.	kg Oil	813455	818162	689702	....	....	....	....	....	....	....	....	....	2321318	5560930
Equiv.Therm.En.	Sm^3Gas	986006	991711	836002	....	....	....	....	....	....	....	....	....	2813719	6740521
ENERGY PER PRODUCTION UNIT															
ELECTRIC ENERGY															
	kWh/h	1220.413	1270.645	1150.042	....	....	....	....	....	....	....	....	....	1211.727	1152.243
	kWh/t	723.385	640.725	601.807	....	....	....	....	....	....	....	....	....	653.010	609.442
THERMAL ENERGY															
Equiv.Therm.Oil	kg/h	284.823	321.605	243.539	....	....	....	....	....	....	....	....	....	281.987	175.285
Equiv.Therm.Gas	Sm^3/h	345.240	389.824	295.198	....	....	....	....	....	....	....	....	....	341.803	212.467
Equiv.Therm.Oil	kg/t	168.825	162.170	127.442	....	....	....	....	....	....	....	....	....	151.965	92.712
Equiv.Therm.Gas	Sm^3/t	204.637	196.569	154.475	....	....	....	....	....	....	....	....	....	184.200	112.378
WATER															
	m^3/h	78.727	86.982	87.356	....	....	....	....	....	....	....	....	....	84.246	55.681
	m^3/t	46.664	43.861	45.713	....	....	....	....	....	....	....	....	....	45.401	29.451

B.0 (continued)

		199z (JAN-MAR)												TOTAL	TOTAL
		JAN	FEB	MAR	APR	MAY	JUN	JUL	AUG	SEPT	OCT	NOV	DEC	199z	199y Ref.Year
ENERGY COSTS															
Electricity	U/kWh	128.0	137.8	136.0	....	....	....	....	....	....	....	....	....	133.8	113.6
Electricity	MU	446.133	445.292	442.914	....	....	....	....	....	....	....	....	....	1334.339	4151.076
Thermal-Oil	U/kg	169.2	163.2	171.1	....	....	....	....	....	....	....	....	....	167.4	163.2
Thermal-Oil	MU	59.755	58.347	39.590	....	....	....	....	....	....	....	....	....	157.692	395.116
Thermal-Nat.Gas	U/Sm^3	150.0	165.0	155.0	....	....	....	....	....	....	....	....	....	156.7	152.0
Thermal-Nat.Gas	MU	78.695	86.565	81.318	....	....	....	....	....	....	....	....	....	246.578	560.953
Thermal-Gasoil	U/L	331.8	323.4	337.3	....	....	....	....	....	....	....	....	....	330.7	310.0
Thermal-Gasoil	MU	10.531	10.344	9.892	....	....	....	....	....	....	....	....	....	30.766	33.246
Thermal-LPG	U/L	224.1	220.2	233.2	....	....	....	....	....	....	....	....	....	226.1	215.0
Thermal-LPG	MU	0.195	0.234	0.265	....	....	....	....	....	....	....	....	....	0.693	1.781
Thermal-Total	MU	149.176	155.489	131.065	....	....	....	....	....	....	....	....	....	435.729	991.095
Energy-Total	MU	595.309	600.781	573.979	....	....	....	....	....	....	....	....	....	1770.068	5142.171
Electricity	U/t	92591	88262	81841	....	....	....	....	....	....	....	....	....	87353	69207
Thermal energy	U/t	30960	30820	24218	....	....	....	....	....	....	....	....	....	28525	16523
Total	U/t	123551	119082	106059	....	....	....	....	....	....	....	....	....	115878	85730
Water	U/m^3	61.1	64.2	54.2	....	....	....	....	....	....	....	....	....	59.6	53.0
Water	MU	13.742	14.196	13.404	....	....	....	....	....	....	....	....	....	41.342	93.624
Water	U/t	2852.0	2813.8	2476.8	....	....	....	....	....	....	....	....	....	2706.4	1560.9

Table 18.B Energy account report — Output

B.1...B.5 SINGLE FACTORY OR SINGLE PRODUCTION LINE (NO.1.....NO. 5)

PRODUCTION - ENERGY CONSUMPTION 199z (JAN-MAR)

		JAN	FEB	MAR	APR	MAY	JUN	JUL	AUG	SEPT	OCT	NOV	DEC	TOTAL 199z	TOTAL 199y Ref.Year
PRODUCTION															
PRODUCTION	t	1074.5	1258	1371.8	....	....	....	....	....	....	....	....	....	3704.3	17917
WORKING HOURS	h	744	672	744	....	....	....	....	....	....	....	....	....	2160	8125
prod/hours	t/h	1.444	1.872	1.844	....	....	....	....	....	....	....	....	....	1.715	2.205
TOTAL ENERGY CONSUMPTION															
ELECTRICITY	kWh	1181400	1070720	1070720	....	....	....	....	....	....	....	....	....	3322840	13078400
OIL	kg	0	0	0	....	....	....	....	....	....	....	....	....	0	0
NATURAL GAS	Sm^3	524634	524634	524634	....	....	....	....	....	....	....	....	....	1573902	3690479
GASOIL	L	5032	5032	5032	....	....	....	....	....	....	....	....	....	15096	0
LPG	L	450	450	450	....	....	....	....	....	....	....	....	....	1350	0
WATER	m^3	63333	63333	63333	....	....	....	....	....	....	....	....	....	189999	850500
TOTAL THERMAL ENERGY															
Equiv.Therm.En.	kg Oil	437360	437360	437360	....	....	....	....	....	....	....	....	....	1312079	3044645
Equiv.Therm.En.	Sm^3Gas	530133	530133	530133	....	....	....	....	....	....	....	....	....	1590399	3690479
ENERGY PER PRODUCTION UNIT															
ELECTRIC ENERGY															
	kWh/h	1587.903	1593.333	1439.140	....	....	....	....	....	....	....	....	....	1538.352	1609.649
	kWh/t	1099.488	851.129	780.522	....	....	....	....	....	....	....	....	....	897.022	729.944
THERMAL ENERGY															
Equiv.Therm.Oil	kg/h	587.849	650.833	587.849	....	....	....	....	....	....	....	....	....	607.444	374.726
Equiv.Therm.Gas	Sm^3/h	712.544	788.889	712.544	....	....	....	....	....	....	....	....	....	736.296	454.213
Equiv.Therm.Oil	kg/t	407.036	347.663	318.822	....	....	....	....	....	....	....	....	....	354.204	169.931
Equiv.Therm.Gas	Sm^3/t	493.377	421.409	386.451	....	....	....	....	....	....	....	....	....	429.339	205.976
WATER															
	m^3/h	85.125	94.246	85.125	....	....	....	....	....	....	....	....	....	87.963	104.677
	m^3/t	58.942	50.344	46.168	....	....	....	....	....	....	....	....	....	51.291	47.469

B.1....B.5 (continued)

		199z (JAN-MAR)													
		JAN	FEB	MAR	APR	MAY	JUN	JUL	AUG	SEPT	OCT	NOV	DEC	TOTAL 199z	TOTAL 199y Ref.Year
ENERGY COSTS															
Electricity	U/kWh	122.6	129.4	130.5	....	....	....	....	....	....	....	....	....	127.4	113.7
Electricity	MU	144.880	138.600	139.700	....	....	....	....	....	....	....	....	....	423.180	1486.482
Thermal-Oil	U/kg	0	0	0	....	....	....	....	....	....	....	....	....	0	0
Thermal-Oil	MU	0	0	0	....	....	....	....	....	....	....	....	....	0	0
Thermal-Nat.Gas	U/Sm^3	150.0	165.0	155.0	....	....	....	....	....	....	....	....	....	156.7	152.7
Thermal-Nat.Gas	MU	78.695	86.565	81.318	....	....	....	....	....	....	....	....	....	246.578	563.6
Thermal-Gasoil	U/L	350.0	350.0	350.0	....	....	....	....	....	....	....	....	....	350.0	0
Thermal-Gasoil	MU	1.761	1.761	1.761	....	....	....	....	....	....	....	....	....	5.284	0
Thermal-LPG	U/L	200.0	200.0	200.0	....	....	....	....	....	....	....	....	....	200.0	0
Thermal-LPG	MU	0.090	0.090	0.090	....	....	....	....	....	....	....	....	....	0.270	0
Thermal-Total	MU	80.546	88.416	83.169	....	....	....	....	....	....	....	....	....	252.132	563.649
Energy-Total	MU	225.426	227.016	222.869	....	....	....	....	....	....	....	....	....	675.312	2050.131
Electricity	U/t	134835	110175	101837	....	....	....	....	....	....	....	....	....	114240	82965
Thermal energy	U/t	74962	70283	60628	....	....	....	....	....	....	....	....	....	68065	31459
Total	U/t	209796	180458	162465	....	....	....	....	....	....	....	....	....	182305	114424
Water	U/m^3	50.0	50.0	50.0	....	....	....	....	....	....	....	....	....	50.0	48.0
Water	MU	3.167	3.167	3.167	....	....	....	....	....	....	....	....	....	9.500	40.824
Water	U/t	2947.1	2517.2	2308.4	....	....	....	....	....	....	....	....	....	2564.6	2278.5

Table 18.B Energy account report — Input

B.6 THERMAL CONSUMPTION OF SINGLE FACTORIES OR SINGLE PRODUCTION LINES

OIL kg 199z

FACTORY/ PRODUCTION LINE	JAN	FEB	MAR	APR	MAY	JUN	JUL	AUG	SEPT	OCT	NOV	DEC	TOTAL	%
NO.1	0	0	0	....	....	....	....	....	....	....	....	....	0	0.00%
NO.2	159677	146455	114396	....	....	....	....	....	....	....	....	....	420528	44.63%
NO.3	147050	157600	70520	....	....	....	....	....	....	....	....	....	375170	39.82%
NO.4	46512	53567	46449	....	....	....	....	....	....	....	....	....	146528	15.55%
NO.5	0	0	0	....	....	....	....	....	....	....	....	....	0	0.00%
TOTAL	353239	357622	231365	0	0	0	0	0	0	0	0	0	942226	100%

NATURAL GAS Sm^3 199z

FACTORY/ PRODUCTION LINE	JAN	FEB	MAR	APR	MAY	JUN	JUL	AUG	SEPT	OCT	NOV	DEC	TOTAL	%
NO.1	524634	524634	524634	....	....	....	....	....	....	....	....	....	1573902	100.00%
NO.2	0	0	0	....	....	....	....	....	....	....	....	....	0	0.00%
NO.3	0	0	0	....	....	....	....	....	....	....	....	....	0	0.00%
NO.4	0	0	0	....	....	....	....	....	....	....	....	....	0	0.00%
NO.5	0	0	0	....	....	....	....	....	....	....	....	....	0	0.00%
TOTAL	524634	524634	524634	0	0	0	0	0	0	0	0	0	1573902	100%

B.6 (continued)

			GASOIL		L								199z	
FACTORY/ PRODUCTION LINE	JAN	FEB	MAR	APR	MAY	JUN	JUL	AUG	SEPT	OCT	NOV	DEC	TOTAL	%
NO.1	5032	5032	5032	....	....	....	....	....	....	....	....	....	15096	16.22%
NO.2	5897	6440	5648	....	....	....	....	....	....	....	....	....	17985	19.33%
NO.3	8190	8204	5290	....	....	....	....	....	....	....	....	....	21684	23.30%
NO.4	5621	5306	3634	....	....	....	....	....	....	....	....	....	14561	15.65%
NO.5	7000	7000	9720	....	....	....	....	....	....	....	....	....	23720	25.49%
TOTAL	31740	31982	29324	0	0	0	0	0	0	0	0	0	93046	100%

			LPG		L								199z	
FACTORY/ PRODUCTION LINE	JAN	FEB	MAR	APR	MAY	JUN	JUL	AUG	SEPT	OCT	NOV	DEC	TOTAL	%
NO.1	450	450	450	....	....	....	....	....	....	....	....	....	1350	44.02%
NO.2	0	0	0	....	....	....	....	....	....	....	....	....	0	0.00%
NO.3	420	612	685	....	....	....	....	....	....	....	....	....	1717	55.98%
NO.4	0	0	0	....	....	....	....	....	....	....	....	....	0	0.00%
NO.5	0	0	0	....	....	....	....	....	....	....	....	....	0	0.00%
TOTAL	870	1062	1135	0	0	0	0	0	0	0	0	0	3067	100%

Table 18.B Energy account report — Input

B.7 ELECTRICITY CONSUMPTION OF SINGLE FACTORIES OR SINGLE PRODUCTION LINES

	ELECTRICITY		kWh										199z	
	JAN	FEB	MAR	APR	MAY	JUN	JUL	AUG	SEPT	OCT	NOV	DEC	TOTAL	%
FACTORY/ PRODUCTION LINE														
NO.1	1181400	1070720	1070720	....	....	....	....	....	....	....	....	....	3322840	33.31
NO.2	936000	857200	878400	....	....	....	....	....	....	....	....	....	2671600	26.78
NO.3	823500	778800	820200	....	....	....	....	....	....	....	....	....	2422500	24.29
NO.4	401600	372800	353600	....	....	....	....	....	....	....	....	....	1128000	11.31
NO.5	143000	153000	134000	....	....	....	....	....	....	....	....	....	430000	4.31
TOTAL	3485500	3232520	3256920	0	0	0	0	0	0	0	0	0	9974940	100.00

Table 18.B Energy account report

Input

B.8 WATER CONSUMPTION OF SINGLE FACTORIES OR SINGLE PRODUCTION LINES

WATER m^3

FACTORY/ PRODUCTION LINE	JAN	FEB	MAR	APR	MAY	JUN	JUL	AUG	SEPT	OCT	NOV	DEC	199z TOTAL	%
NO.1	63333	63333	63333	....	....	....	....	....	....	....	....	....	189999	27.40%
NO.2	152960	150820	176020	....	....	....	....	....	....	....	....	....	479800	69.18%
NO.3	4320	4355	4650	....	....	....	....	....	....	....	....	....	13325	1.92%
NO.4	4230	2773	3389	....	....	....	....	....	....	....	....	....	10392	1.50%
NO.5	0	0	0	....	....	....	....	....	....	....	....	....	0	0.00%
TOTAL	224843	221281	247392	0	0	0	0	0	0	0	0	0	693516	100%

Table 18.B Energy account report — Input

B.9 ENERGY COSTS OF SINGLE FACTORIES OR SINGLE PRODUCTION LINES

FACTORY/ PRODUCTION LINE		JAN	FEB	MAR	APR	MAY	JUN	JUL	AUG	SEPT	OCT	NOV	DEC	199z TOTAL
NO.1														
Electricity	MU	144.880	138.600	139.700	....	....	....	....	....	....	....	....	....	423.180
Thermal-Oil	MU	0	0	0	....	....	....	....	....	....	....	....	....	0
Thermal-Nat.Gas	MU	78.695	86.565	81.318	....	....	....	....	....	....	....	....	....	246.578
Thermal-Gasoil	MU	1.761	1.761	1.761	....	....	....	....	....	....	....	....	....	5.284
Thermal-LPG	MU	0.090	0.090	0.090	....	....	....	....	....	....	....	....	....	0.270
Water	MU	3.167	3.167	3.167	....	....	....	....	....	....	....	....	....	9.500
NO.2														
Electricity	MU	124.083	124.148	127.925	....	....	....	....	....	....	....	....	....	376.157
Thermal-Oil	MU	26.347	22.701	18.303	....	....	....	....	....	....	....	....	....	67.351
Thermal-Nat.Gas	MU	0	0	0	....	....	....	....	....	....	....	....	....	0.000
Thermal-Gasoil	MU	1.858	1.996	1.864	....	....	....	....	....	....	....	....	....	5.718
Thermal-LPG	MU	0	0	0	....	....	....	....	....	....	....	....	....	0
Water	MU	9.942	10.557	9.681	....	....	....	....	....	....	....	....	....	30.181
NO.3														
Electricity	MU	88.839	90.966	91.033	....	....	....	....	....	....	....	....	....	270.8388
Thermal-Oil	MU	25.734	26.004	12.694	....	....	....	....	....	....	....	....	....	64.431
Thermal-Nat.Gas	MU	0	0	0	....	....	....	....	....	....	....	....	....	0
Thermal-Gasoil	MU	2.867	2.666	1.878	....	....	....	....	....	....	....	....	....	7.411
Thermal-LPG	MU	0.105	0.144	0.175	....	....	....	....	....	....	....	....	....	0.423
Water	MU	0.324	0.283	0.302	....	....	....	....	....	....	....	....	....	0.909

B.9 (continued)

FACTORY/ PRODUCTION LINE		JAN	FEB	MAR	APR	MAY	JUN	JUL	AUG	SEPT	OCT	NOV	DEC	199z TOTAL
NO.4														
Electricity	MU	67.271	66.879	61.663	....	....	....	....	....	....	....	....	....	195.8138
Thermal-Oil	MU	7.674	9.642	8.593	....	....	....	....	....	....	....	....	....	25.910
Thermal-Nat.Gas	MU	0	0	0	....	....	....	....	....	....	....	....	....	0
Thermal-Gasoil	MU	1.771	1.645	1.181	....	....	....	....	....	....	....	....	....	4.597
Thermal-LPG	MU	0	0	0	....	....	....	....	....	....	....	....	....	0
Water	MU	0.309	0.189	0.254	....	....	....	....	....	....	....	....	....	0.752
NO.5														
Electricity	MU	21.060	24.698	22.592	....	....	....	....	....	....	....	....	....	68.3498
Thermal-Oil	MU	0	0	0	....	....	....	....	....	....	....	....	....	0
Thermal-Nat.Gas	MU	0	0	0	....	....	....	....	....	....	....	....	....	0
Thermal-Gasoil	MU	2.275	2.275	3.2076	....	....	....	....	....	....	....	....	....	7.758
Thermal-LPG	MU	0	0	0	....	....	....	....	....	....	....	....	....	0
Water	MU	0	0	0	....	....	....	....	....	....	....	....	....	0
TOTAL														
Energy	MU	595.309	600.781	573.979	0	0	0	0	0	0	0	0	0	1770.068
Water	MU	13.742	14.196	13.404	0	0	0	0	0	0	0	0	0	41.342

Table 18.B Energy account report — Input

B.10 PRODUCTION DATA OF SINGLE FACTORIES OR SINGLE PRODUCTION LINES

														199z
		JAN	FEB	MAR	APR	MAY	JUN	JUL	AUG	SEPT	OCT	NOV	DEC	TOTAL
FACTORY/PRODUCTION LINE														
NO.1														
	t	1074.5	1258	1371.8	....	....	....	....	....	....	....	....	....	3704
	h	744	672	744	....	....	....	....	....	....	....	....	....	2160
NO.2														
	t	1837	2045	1938	....	....	....	....	....	....	....	....	....	5820
	h	744	672	744	....	....	....	....	....	....	....	....	....	2160
NO.3														
	t	1071	908	1232	....	....	....	....	....	....	....	....	....	3211
	h	744	672	744	....	....	....	....	....	....	....	....	....	2160
NO.4														
	t	661	660	678	....	....	....	....	....	....	....	....	....	1999
	h	416	352	400	....	....	....	....	....	....	....	....	....	1168
NO.5														
	t	174.82	174.1	192.1	....	....	....	....	....	....	....	....	....	541
	h	208	176	200	....	....	....	....	....	....	....	....	....	584
TOTAL PRODUCTION	t	4818.32	5045.1	5411.9	0	0	0	0	0	0	0	0	0	15275
TOTAL HOURS	h	2856	2544	2832	0	0	0	0	0	0	0	0	0	8232

CHAPTER 19

EDUCATION FOR ENERGY MANAGEMENT

19.1 EDUCATION FROM PRIMARY SCHOOL TO PROFESSIONAL TRAINING

Awareness of the role of energy in all aspects of life — home, industry, transport and services — is a primary step in education in energy management.

This concept can easily be introduced at any level of the educational system, but an appropriate presentation must be devised for each case. The energy and mass balance, which underlies the basic principle of energy management and allows the solution of many practical problems, can be introduced in simple terms even in primary schools and can be a useful guideline for the comprehension of most phenomena.

Of course, the same principle will be better understood in the following years, when physics, chemistry, mathematics, thermodynamics and other specific courses will be encountered and the different aspects of energy transformation will be clarified also by practice at work.

General schooling must present the broad aspects of energy transformation; technical schools and universities must provide a detailed analysis of components and systems and finally a general overview to connect all these elements.

19.2 THE ROLE OF THE UNIVERSITY

According to the tradition of each country, the university is oriented to give a specialized or a more broadly based formation. In both cases, all the systems in which energy transformation occurs are the subject of courses, such as electrical and thermal sciences, thermodynamics including heat transfer and fluid mechanics, and the subsequent specialized courses on machines and plants. Other courses, not specifically related to energy, such as power electronics, information science and industrial automation are generally offered to complete the student's curriculum. The university has to correlate these courses to give students a synthesis of the various concepts and to prepare them for their future work.

The university is the right place to present the principles of energy transformation without any constraints due to production requirements and economic evaluation. The economic problems are not overlooked, but priority is given to technical aspects.

A clear and complete understanding of the technical facets of a problem is the base for a subsequent assessment of all the other factors, economic and human, which will be the decision criteria in industry, but which cannot be treated without regard for technical considerations.

Energy and mass balance can be assumed as the keystone for a wide comprehension of any phenomenon; if thoroughly understood, this concept becomes the basic means to solve any problem one may have to face in energy management.

19.3 PERSONNEL TRAINING IN THE FACTORY

Continuous training and motivation of both graduates and other technicians is the main key to the success of any energy management project.

Training means giving people of all ranks energy-related information to assist them in their day-to-day work and in planning.

Motivation means inducing people by economic and human incentives to feel very concerned about energy saving as a high priority.

Courses on different energy-management topics can be organized at different levels, depending on the background and staff position of each employee. But, as a primary goal, all workers must know what they are working for, how their contribution is used and what the main results of the project are.

Meetings for the presentation and discussion of energy-management projects and results of energy-accounting procedures can also constitute part of the personnel training program.

19.4 AWARENESS OF ENERGY MANAGEMENT AS AN INTERSECTORIAL DISCIPLINE

Fig. 19.1 shows a matrix presentation of the interaction between energy-management topics and the specialized courses traditionally offered in many universities and technical schools. The complexity of the interaction is evident.

Although energy-related topics have often been dealt with as parts of different cultural areas, energy management involves so many technical and economic aspect of life that it is now worth approaching it as an independent discipline.

The integration of several topics which must be known in detail, from basic concepts to plant design parameters, evidences the need for a systematically arranged program of studies which includes many traditional courses but which pays particular attention to energy-transformation concepts. This means changing the approach to the items of each individual course by emphasizing this common basis and by preparing students to focus attention on it. If necessary, individual courses can be integrated with one another by reorganizing the program so as to offer a specific energy-management curriculum.

Fig. 19.1 Integration between energy management topics and specialized courses (shaded areas evidence integration)

Specialized courses / Energy management topics	Measurement techniques	Power systems	Electrical drives	Computer science	Thermal machines	Heat & thermo-dynamics	Mechanics	Hydraulic systems	Economic analysis	Government policies
Utility supply contracts	■	■							■	■
Electrical substations	■	■	■	■		■			■	
Boiler plants	■	■		■	■	■			■	
Distribution lines	■	■	■	■		■			■	
Cogeneration plants	■	■	■	■	■	■	■	■	■	■
Lighting	■	■		■					■	
Water pumps, process pumps	■	■	■					■	■	
Air compressors	■	■	■			■	■	■	■	
Cold-production systems	■	■	■	■	■	■		■	■	
HVAC systems	■	■	■	■	■	■		■	■	
Heat recovery	■	■			■	■	■		■	
Pollution and industrial waste	■	■			■	■		■	■	■
Facilities and process-plant automation	■	■	■	■	■	■	■	■	■	
Feasibility studies	■	■	■	■	■	■	■	■	■	■
Energy accounting and balance	■	■	■	■	■	■	■	■	■	■

CHAPTER 20

PRINCIPLES OF ECONOMIC ANALYSIS OF ENERGY-SAVING INVESTMENTS

20.1 GENERAL CONSIDERATIONS ON ECONOMIC ANALYSIS

The economic analysis of investments is a critical step in an energy conservation program because monetary saving is generally the main factor leading to a decision. The same analysis can conveniently be used to choose among possibilities which may be equivalent from a technical point of view.

The main elements of an investment are: the capital cost or initial investment, the interest rate, the return on the investment and the life of the investment.

Several methods can be used, according to the company's internal evaluation criteria for investment, not only in the energy-saving field. Depending on the importance of the investment, more or less sophisticated methods can be used.

A short review follows of the main methods with special reference to energy-saving investments.

20.2 APPLICATION OF ECONOMIC ANALYSIS METHODS TO ENERGY-SAVING INVESTMENTS

Payback method

The payback period is the time required to recover the capital investment from net cash flow.

Payback = INVESTMENT/NET ANNUAL CASH FLOW

where

investment = the total capital cost

net annual cash flow = (annual energy cost saving + other cost saving – annual additional costs) referred to the year of the investment

The two terms of the payback ratio can be either before or after taxes depending on the requirements of the investor.

If the salvage cost, that is, the value of the equipment when it is retired or sold, is deducted from the initial capital cost, a payback with salvage is calculated. Of course, it is shorter than the payback without salvage as defined above.

The payback method does not consider the savings after the payback years; thus it penalizes projects that have a long potential life in comparison with those that offer high savings for a short period.

The payback method does not consider energy-pricing variation or the time value of the money.

Nevertheless, the payback method is very simple and it can serve as a yardstick to compare possible investments.

Investors' rate of return (ROR) method

ROR = NET ANNUAL CASH FLOW/INVESTMENT

This is the reciprocal of the payback and it is generally expressed as a percentage.

Methods using life-cycle costing

These methods are based on the conversion of investment and annual cash flow at various times to their equivalent present values and vice versa. The real interest or discount rate (r), that is the nominal discount rate less the inflation rate, and the number of years (n) of the evaluation period are introduced.

Several factors are used to accomplish these conversions:

— FUTURE WORTH FACTOR

This converts a single present amount (at year zero) to an amount at a future point in time

$FWF = (1 + r)^n$

— PRESENT WORTH FACTOR

This converts a future amount to an amount today (at year zero)

$$PWF = \frac{1}{(1+r)^n}$$

Present Worth method

The net present worth of a project is defined as the difference between the present worth of the total project revenues (energy cost saving + other cost savings – additional operating costs) and the present capital cost of the project.

$$PW = \sum_{1}^{n} {}_j \frac{1}{(1+r)^j}\left(C_{pel} \cdot Q_{el_j} + C_{p_{th}} \cdot Q_{th_j} \pm C_{pa} Q_{a_j}\right) - I_p$$

where

C_p = present monetary value (at year zero) of each unit of revenue (C_{pel}: value of unit of electrical energy saving; C_{pth}: value of unit of thermal energy saving; C_{pa}: value of unit of additional saving or cost of materials and working hours)

Q_j = annual revenues in physical units (Q_{el}: amount of electric energy saving; Q_{th}: amount of thermal energy saving; Q_a: amount of additional or saved materials and working hours, etc.). It is usually assumed that they become effective in year 1 (j = 1)

I_p = present investment at year zero. If the investment has been made in different years, the present investment can be calculated by using the PWF factor

$$= I_o + \sum_{1}^{n}{}_j \frac{I_j}{(1+r)^j}$$

In the case of energy-saving investments, all savings and additional operating costs can also be expressed as a percentage of the energy saving by increasing or decreasing the present monetary value of the energy (C_{pel}, C_{pth}).

The present investment I_p can be simply the effective capital cost if it is concentrated in the year zero or the present worth of the investment calculated by means of the Present Worth Factor if it is made in several years.

The expression given above can be adapted for wider use if a different inflation rate is introduced for each item of the revenues and for the capital investment. A widely accepted criterion takes into account two different values of the inflation rate for the energy and related revenues (f*) and for the investment (f).

A general expression for PW is:

$$PW = \sum_{1}^{n}{}_j \frac{\left(C_{pel} \cdot Q_{el_j} + C_{pth} Q_{th_j} \pm C_{pa} Q_{aj}\right) \cdot (1+f^*)^j}{(1+r)^j \cdot (1+f)^j} - I_p =$$

$$= \sum_{1}^{n}{}_j \frac{\left(C_{pel} \cdot Q_{el_j} + C_{pth} Q_{th_j} \pm C_{pa} Q_{aj}\right)}{(1+i)^j} - I_p$$

where $i = r + f - f^*$ for small values of r, f, f*.

In order to obtain a simpler expression, annual energy saving in physical units is assumed constant during the life of the investment, as are other savings and additional expenditures. In addition, the present investment I_p is assumed to be equal to the total investment I which is taken to be concentrated in the present year (year zero).

The simplified PW expression becomes as follows:

$$PW = \left(C_{pel} \cdot Q_{el} + C_{pth} Q_{th} \pm C_{pa} Q_a\right) \cdot \sum_1^n {}_j \frac{1}{(1+i)^j} - I =$$

$$= (C_{pel} \cdot Q_{el} + C_{pth} \cdot Q_{th} \pm C_{pa} Q_a) \cdot PAF - I$$

where:

$Q_{el}, Q_{th}, Q_a =$	annual amount of revenues in physical units assumed constant throughout the life of the investment. They start to become effective in year 1
$n =$	years of investment life
$i =$	$r + f - f^*$ $(i = r,$ if $f = f^*)$
$I =$	total investment concentrated in the year zero
$\sum_1^n {}_j \frac{1}{(1+i)^j} =$	Present Annuity Factor (PAF)

If PW is greater than zero, the project is valid since the revenues are enough to pay the interest and to recover the initial capital cost before the end of the life of the investment. If PW equals zero, the balance occurs at the end of the life, but the investment is scarcely attractive. PW less than zero means that the project is a bad one.
Projects can conveniently be compared by taking as a parameter the ratio between the present worth of the project and the related investment (PW/I).

Factors such as PWF and PFA can be evaluated either by using published tables (see Tables 20.1a and 20.1b) or by calculation on personal computers, with any software available.

Table 20.2 shows an example of PW calculation by using both the general and the simplified expressions.

Table 20.1a Present Worth Factor values

$$PWF = \frac{1}{(1+i)^j}$$

Year (j)	PWF values corresponding to coefficient i in %									
	1%	2%	3%	4%	5%	6%	7%	8%	9%	10%
1	0.990	0.980	0.971	0.962	0.952	0.943	0.935	0.926	0.917	0.909
2	0.980	0.961	0.943	0.925	0.907	0.890	0.873	0.857	0.842	0.826
3	0.971	0.942	0.915	0.889	0.864	0.840	0.816	0.794	0.772	0.751
4	0.961	0.924	0.888	0.855	0.823	0.792	0.763	0.735	0.708	0.683
5	0.951	0.906	0.863	0.822	0.784	0.747	0.713	0.681	0.650	0.621
6	0.942	0.888	0.837	0.790	0.746	0.705	0.666	0.630	0.596	0.564
7	0.933	0.871	0.813	0.760	0.711	0.665	0.623	0.583	0.547	0.513
8	0.923	0.853	0.789	0.731	0.677	0.627	0.582	0.540	0.502	0.467
9	0.914	0.837	0.766	0.703	0.645	0.592	0.544	0.500	0.460	0.424
10	0.905	0.820	0.744	0.676	0.614	0.558	0.508	0.463	0.422	0.386
11	0.896	0.804	0.722	0.650	0.585	0.527	0.475	0.429	0.388	0.350
12	0.887	0.788	0.701	0.625	0.557	0.497	0.444	0.397	0.356	0.319
13	0.879	0.773	0.681	0.601	0.530	0.469	0.415	0.368	0.326	0.290
14	0.870	0.758	0.661	0.577	0.505	0.442	0.388	0.340	0.299	0.263
15	0.861	0.743	0.642	0.555	0.481	0.417	0.362	0.315	0.275	0.239
16	0.853	0.728	0.623	0.534	0.458	0.394	0.339	0.292	0.252	0.218
17	0.844	0.714	0.605	0.513	0.436	0.371	0.317	0.270	0.231	0.198
18	0.836	0.700	0.587	0.494	0.416	0.350	0.296	0.250	0.212	0.180
19	0.828	0.686	0.570	0.475	0.396	0.331	0.277	0.232	0.194	0.164
20	0.820	0.673	0.554	0.456	0.377	0.312	0.258	0.215	0.178	0.149
	11%	12%	13%	14%	15%	16%	17%	18%	19%	20%
1	0.901	0.893	0.885	0.877	0.870	0.862	0.855	0.847	0.840	0.833
2	0.812	0.797	0.783	0.769	0.756	0.743	0.731	0.718	0.706	0.694
3	0.731	0.712	0.693	0.675	0.658	0.641	0.624	0.609	0.593	0.579
4	0.659	0.636	0.613	0.592	0.572	0.552	0.534	0.516	0.499	0.482
5	0.593	0.567	0.543	0.519	0.497	0.476	0.456	0.437	0.419	0.402
6	0.535	0.507	0.480	0.456	0.432	0.410	0.390	0.370	0.352	0.335
7	0.482	0.452	0.425	0.400	0.376	0.354	0.333	0.314	0.296	0.279
8	0.434	0.404	0.376	0.351	0.327	0.305	0.285	0.266	0.249	0.233
9	0.391	0.361	0.333	0.308	0.284	0.263	0.243	0.225	0.209	0.194
10	0.352	0.322	0.295	0.270	0.247	0.227	0.208	0.191	0.176	0.162
11	0.317	0.287	0.261	0.237	0.215	0.195	0.178	0.162	0.148	0.135
12	0.286	0.257	0.231	0.208	0.187	0.168	0.152	0.137	0.124	0.112
13	0.258	0.229	0.204	0.182	0.163	0.145	0.130	0.116	0.104	0.093
14	0.232	0.205	0.181	0.160	0.141	0.125	0.111	0.099	0.088	0.078
15	0.209	0.183	0.160	0.140	0.123	0.108	0.095	0.084	0.074	0.065
16	0.188	0.163	0.141	0.123	0.107	0.093	0.081	0.071	0.062	0.054
17	0.170	0.146	0.125	0.108	0.093	0.080	0.069	0.060	0.052	0.045
18	0.153	0.130	0.111	0.095	0.081	0.069	0.059	0.051	0.044	0.038
19	0.138	0.116	0.098	0.083	0.070	0.060	0.051	0.043	0.037	0.031
20	0.124	0.104	0.087	0.073	0.061	0.051	0.043	0.037	0.031	0.026

Table 20.1b Present Annuity Factor values

$$PAF = \sum_{1}^{n} {}_{j} \frac{1}{(1+i)^j}$$

Year (j)	PAF values corresponding to coefficient i in %									
	1%	2%	3%	4%	5%	6%	7%	8%	9%	10%
1	0.990	0.980	0.971	0.962	0.952	0.943	0.935	0.926	0.917	0.909
2	1.970	1.942	1.913	1.886	1.859	1.833	1.808	1.783	1.759	1.736
3	2.941	2.884	2.829	2.775	2.723	2.673	2.624	2.577	2.531	2.487
4	3.902	3.808	3.717	3.630	3.546	3.465	3.387	3.312	3.240	3.170
5	4.853	4.713	4.580	4.452	4.329	4.212	4.100	3.993	3.890	3.791
6	5.795	5.601	5.417	5.242	5.076	4.917	4.767	4.623	4.486	4.355
7	6.728	6.472	6.230	6.002	5.786	5.582	5.389	5.206	5.033	4.868
8	7.652	7.325	7.020	6.733	6.463	6.210	5.971	5.747	5.535	5.335
9	8.566	8.162	7.786	7.435	7.108	6.802	6.515	6.247	5.995	5.759
10	9.471	8.983	8.530	8.111	7.722	7.360	7.024	6.710	6.418	6.145
11	10.368	9.787	9.253	8.760	8.306	7.887	7.499	7.139	6.805	6.495
12	11.255	10.575	9.954	9.385	8.863	8.384	7.943	7.536	7.161	6.814
13	12.134	11.348	10.635	9.986	9.394	8.853	8.358	7.904	7.487	7.103
14	13.004	12.106	11.296	10.563	9.899	9.295	8.745	8.244	7.786	7.367
15	13.865	12.849	11.938	11.118	10.380	9.712	9.108	8.559	8.061	7.606
16	14.718	13.578	12.561	11.652	10.838	10.106	9.447	8.851	8.313	7.824
17	15.562	14.292	13.166	12.166	11.274	10.477	9.763	9.122	8.544	8.022
18	16.398	14.992	13.754	12.659	11.690	10.828	10.059	9.372	8.756	8.201
19	17.226	15.678	14.324	13.134	12.085	11.158	10.336	9.604	8.950	8.365
20	18.046	16.351	14.877	13.590	12.462	11.470	10.594	9.818	9.129	8.514
	11%	12%	13%	14%	15%	16%	17%	18%	19%	20%
1	0.901	0.893	0.885	0.877	0.870	0.862	0.855	0.847	0.840	0.833
2	1.713	1.690	1.668	1.647	1.626	1.605	1.585	1.566	1.547	1.528
3	2.444	2.402	2.361	2.322	2.283	2.246	2.210	2.174	2.140	2.106
4	3.102	3.037	2.974	2.914	2.855	2.798	2.743	2.690	2.639	2.589
5	3.696	3.605	3.517	3.433	3.352	3.274	3.199	3.127	3.058	2.991
6	4.231	4.111	3.998	3.889	3.784	3.685	3.589	3.498	3.410	3.326
7	4.712	4.564	4.423	4.288	4.160	4.039	3.922	3.812	3.706	3.605
8	5.146	4.968	4.799	4.639	4.487	4.344	4.207	4.078	3.954	3.837
9	5.537	5.328	5.132	4.946	4.772	4.607	4.451	4.303	4.163	4.031
10	5.889	5.650	5.426	5.216	5.019	4.833	4.659	4.494	4.339	4.192
11	6.207	5.938	5.687	5.453	5.234	5.029	4.836	4.656	4.486	4.327
12	6.492	6.194	5.918	5.660	5.421	5.197	4.988	4.793	4.611	4.439
13	6.750	6.424	6.122	5.842	5.583	5.342	5.118	4.910	4.715	4.533
14	6.982	6.628	6.302	6.002	5.724	5.468	5.229	5.008	4.802	4.611
15	7.191	6.811	6.462	6.142	5.847	5.575	5.324	5.092	4.876	4.675
16	7.379	6.974	6.604	6.265	5.954	5.668	5.405	5.162	4.938	4.730
17	7.549	7.120	6.729	6.373	6.047	5.749	5.475	5.222	4.990	4.775
18	7.702	7.250	6.840	6.467	6.128	5.818	5.534	5.273	5.033	4.812
19	7.839	7.366	6.938	6.550	6.198	5.877	5.584	5.316	5.070	4.843
20	7.963	7.469	7.025	6.623	6.259	5.929	5.628	5.353	5.101	4.870

Table 20.2 Comparison between general and simplified PW relationships

PW CALCULATION BY THE GENERAL RELATIONSHIP						
Year	Investment MU	Annual revenues MU	Yearly cash flow MU	PWF (i = 10%)	Present cash flow MU	PW of the project MU
0	1500	0	−1500		−1500.0	−1500.0
1	500	400	−100	0.909	−90.9	−1590.9
2	250	700	450	0.826	371.7	−1219.2
3	0	700	700	0.751	525.7	−693.5
4	0	700	700	0.683	478.1	−215.4
5	0	700	700	0.621	434.7	219.3
AT THE END OF THE LIFE					219.3	219.3
a	b	c	d = c − b	e	f = d · e	g = sum of yearly f

PW CALCULATION BY THE SIMPLIFIED RELATIONSHIP

Average annual revenues in 5 years equal to (400 + 700 + 700 + 700 + 700)/5 = 640 MU
Total investment concentrated in the year zero equal to 2250 MU
PAF (5 years, from year 1 to year 5; i = 10%) equal to 3.791

PW = 640 · 3.791 − 2250 = 176.2 MU

Notes
annual revenues include all kind of savings and additional operating costs
annual revenues become effective in the year 1

Internal Rate of Return (IRR)

IRR is the value of (i) such that PW = 0.

IRR is the internal interest rate that reduces to zero the Present Worth of the project.

This parameter, named IRR, can easily be compared with the company interest rate to evaluate the merit of the project. The calculation of IRR requires an iterative approach: selecting different values of the parameter (i) and calculating the PW, generally using the basic expression.

In conclusion, it should be pointed out that, to compare several projects, the value of the parameter (i) can be defined arbitrarily without regard to the formulations of the parameter (i) explained above.

20.3 CASE STUDIES

Several examples discussed from the technical point of view in the previous chapters are here examined to evaluate the economic validity of the correlated investment from the cost energy-saving angle.

The basic methods reported in Chapter 20.2 are used and the results for all the case studies examined throughout the book are shown in Table 20.3.

To facilitate the understanding of the Table, example values for the energy costs and for the investment are given. Notice that the ratio between electric and thermal energy costs, which is introduced as an example, has a general valididy because this ratio is roughly the same for many countries. The cost of unit of energy (MWh for electric energy, TOE for thermal energy) is expressed as MU/MWh or MU/TOE where U can be any currency; a reference cost of 100 U/kWh is introduced.

For each case, Table 20.3 shows the annual energy saving, electrical and thermal, and the corresponding cost saving. Expected investment life is also introduced; investment I is expressed in MU units.

The different kinds of investment are compared by using the payback parameter, the ratio PW/I (PW is calculated by the simplified expression) with different values of the real discount rate which changes according to the local situation and the parameter IRR.

Notice that only energy cost and saving have been considered; for a more detailed analysis, additional costs and savings (on working hours, maintenance, replacement of components during the investment life, etc.), which often markedly affect the economic evaluation, must be taken into account.

The same approach can be followed for any other energy-saving investment, in process and facility plants, and it is quite a good approach to correlate and to compare different investments inside a factory.

Table 20.3 Examples of case studies

Assumptions: $C_{pel} = 0.1$ MU/MWh; $C_{pth} = 0.16$ MU/TOE; additional saving and costs ignored; (*) only extra cost considered

Description of case studies	Chapter	Annual energy saving		Annual cost saving		Investment life	Investment	Economic evaluation parameters				
		Q_{el}	Q_{th}	$C_{pel} \cdot Q_{el}$	$C_{pth} \cdot Q_{th}$			PAYBACK	PW/I i% 3	PW/I i% 6	PW/I i% 9	IRR
		MWh/ year	TOE/ year	MU/year		year	MU	year				%
High–efficiency transformer	5.4	4.6	0.0	0.5	0.0	12	5 (*)	10.95	–0.09	–0.23	–0.35	1.40
High–efficiency boiler	6.12	0.0	187.0	0.0	29.9	10	200 (*)	6.69	0.28	0.10	–0.04	8.10
Power factor control	7.7											
– tab.7.2a CASE 4		35.2	0.0	3.5	0.0	10	15	4.26	1.00	0.73	0.51	42.70
– tab.7.2b CASE 4		8.8	0.0	0.9	0.0	10	5	5.69	0.50	0.29	0.13	26.50
Microwave instead of thermal dryer	7.7	–48	113	–4.8	18.1	10	125(*)	11.3	–0.09	–0.22	–0.32	1
High-efficiency electric motor	7.7	11.4	0.0	1.1	0.0	10	3	2.63	2.24	1.80	1.44	36.30
Wall insulation	8.5	0.0	23.0	0.0	3.7	10	4	1.09	6.85	5.77	4.90	92.00
Pipeline insulation	8.5	0.0	154.5	0.0	24.7	10	8	0.32	25.36	21.74	18.83	309.00
Cogeneration												
Steam turbine	9.7	6000.0	–610	600.0	–97.6	12	2000	3.98	1.50	1.11	0.80	23.10
Gas turbine	9.7	6000.0	–1062	600.0	–169.9	8	1500	3.49	1.01	0.78	0.59	23.1
Diesel gas	9.7	6000.0	–651.7	600.0	–104.3	8	1000	2.02	2.48	2.08	1.74	47.5
Non–dissipative pump regulation	10.6	20.5	0.0	2.0	0.0	10	5	2.44	2.49	2.01	1.63	39.50

Table 20.3 (continued)

Description of case studies	Chapter	Annual energy saving		Annual cost saving		Investment life	Investment	Economic evaluation parameters				
		Q_{el}	Q_{th}	$C_{pel} \cdot Q_{el}$	$C_{pth} \cdot Q_{th}$			PAYBACK	PW/I i% 3	PW/I i% 6	PW/I i% 9	IRR
		MWh/year	TOE/year	MU/year		year	MU	year				%
Air compressor plant	11.7											
– pressure reduction		95.5	0.0	9.6	0.0	10	25	2.62	2.26	1.81	1.45	36.50
– heat recovery		0.0	18.1	0.0	2.9	10	10	3.45	1.47	1.13	0.86	26.10
Refrigeration plant	12.10											
– replacement of absorption system		–1140	902.0	–114.0	144.3	10	305	10.06	–0.15	–0.27	–0.36	—
– heat recovery		0.0	80.7	0.0	12.9	10	20	1.55	4.51	3.75	3.14	64.00
HVAC	13.7											
– recovery from exhaust air		–21.7	422	–2.2	67.5	10	600	9.2	–0.07	–0.20	–0.3	1.5
– computer control		240.0	90.0	24.0	14.4	8	400	10.4	–0.33	–0.40	–0.47	—
Lighting	14.6											
– HF fluorescent lamps		249.3	0.0	24.9	0.0	8	160	6.4	0.09	–0.03	–0.14	5.1
– mercury and sodium		178.2	0.0	17.8	0.0	8	30 (*)	1.7	3.17	2.69	2.29	58
Heat recovery	15.5											
– water/water		0.0	474.3	0.0	75.9	10	75	1.0	7.63	6.45	5.49	101
– air/air		0.0	84.0	0.0	13.4	5	25	1.9	1.46	1.26	1.09	45.5
Plant control	17.10	278.1	101.3	27.8	16.2	8	500	11.4	–0.38	–0.45	–0.51	—

CHAPTER 21

CONCLUSIONS

Having reached the end of the book, readers should now be well acquainted with energy transformations in the factory and with the main possibilities of energy saving. Basic principles, of which often uncertain memories remain from wider studies in the past, should now be clear in their essential meaning and should be usable as guidelines for the comprehension of most phenomena. Energy and mass balance, presented throughout the book as the key to solve any problem, should be now a quite familiar tool and the flow of energy, from factory boundaries to end users, should be clearly understood in all its transformations.

Students will have had an opportunity to correlate various items which are scattered through the university curriculum.

Technicians, not specifically concerned in designing plants, will have had access to basic data for a quick evaluation of many problems.

Managers will have found a guide to the understanding of many investment proposals and maybe a spur to solicit new ones.

The author hopes that the goals of the book have been achieved and that readers will have obtained both answers to their problems and a global comprehension of the energy flow inside a factory, or at least a clear recapitulation of energy management principles and a stimulus to study particular topics in more detail.

REFERENCES

C.R. Westaway, A.W. Loomis, Cameron Hydraulic Data, Ingersoll-Rand Company, USA, 1977

Turner Malloy, Thermal Insulation Design Economics for Pipes and Equipment, McGraw-Hill Book Company, New York, 1980

Craig B. Smith, Energy Management Principles, Pergamon Press, USA, 1981

Wayne C. Turner, Energy Management Handbook, Wiley-Interscience Publication, John Wiley & Sons, USA, 1982

A.W. Loomis, Compressed Air and Gas Data, Ingersoll-Rand Company, USA, 1982

Various Authors, ENEA, Metodologie di risparmio energetico, Editore Ulrico Hoepli, Milano, Italy 1984

Various Authors, Manuale dell'ingegnere. Nuovo Colombo, Editore Ulrico Hoepli, Milano, Italy 1985

Various Authors, A User Guide on Process Integration for the Efficient Use of Energy, The Institution of Chemical Engineers, England, 1985

Tyler G. Hicks, Power Plant Evaluation and Design Reference Guide, McGraw-Hill Book Company, New York, 1986

Jean-François Guilmot, David McGlue, Pierre Valette, Christian Waeterloos, Energy 2000, Commission of the European Communities, Great Britain, 1986

Albert Thumann, P.E., C.E.M., Plant Engineers and Managers Guide to Energy Conservation, Third Edition, USA, 1987

John R. Howell, Richard O. Buckius, Fundamentals of Engineering Thermodynamics, McGraw-Hill Book Company, Singapore, 1987

Alessandro Brunelli, Misure di Portata, Associazione Italiana Strumentisti, Milano, Italy 1987

Various Authors, The Essentials of Heat Transfer 1, Staff of Research and Education Association, Dr. M. Fogiel, Director, USA, 1987

Various Authors, The Essentials of Heat Transfer 2, Staff of Research and Education Association, Dr. M. Fogiel, Director, USA, 1987

Michael J. Moran, Howard N. Shapiro, Fundamentals of Engineering Thermodynamics, John Wiley & Sons, Inc., Canada, 1988

F. William Payne, Integration of Efficient Design Technologies, The Fairmont Press, Inc. Prentice Hall, USA, 1988

Staff of Research and Education Association, Dr. M. Fogiel, Director, The Thermodynamics Problem Solver, REA, USA, 1989

Various Authors, Ashrae Handbook Series 1989 Fundamentals, SI Edition, USA, 1989

Commission of the European Communities Directorate-General for Energy, Energy in Europe Energy for a New Century: The European Perspective, Office for Official Publications of the European Communities, Luxembourg, 1990

Association Technique pour l'Efficacité Energétique, L'Eclairage Econome et Performant des Locaux Industriels et Tertiaires, Institut Français De L'Energie, Paris, 1990

H. Freeman Editor, Hazardous Waste Minimization, McGraw-Hill Publishing Company, USA, 1990

The Institute of Electrical and Electronics Engineers – IEEE Power Engineering Society, Publication Guide for Power Engineers, USA, 1990

M. Theodore Gresh, Compressor Performance Selection, Operation, and Testing of Axial and Centrifugal Compressors, Butterworth-Heinemann Edition, USA, 1991

INDEX

GPSR Compliance

The European Union's (EU) General Product Safety Regulation (GPSR) is a set of rules that requires consumer products to be safe and our obligations to ensure this.

If you have any concerns about our products, you can contact us on ProductSafety@springernature.com

In case Publisher is established outside the EU, the EU authorized representative is:

Springer Nature Customer Service Center GmbH
Europaplatz 3
69115 Heidelberg, Germany

Batch number: 10755284

Printed by Printforce, the Netherlands